Field and Laboratory Methods in Primatology

SECOND EDITION

Building on the success of the first edition and bringing together contributions from a range of experts in the field, the second edition of this guide to research on wild primates covers the latest advances in the field, including new information on field experiments and measuring behaviour. It provides essential information and advice on the technical and practical aspects of both field and laboratory methods, covering topics such as ethnoprimatology; remote sensing; GPS and radio-tracking; trapping and handling; dietary ecology; non-invasive genetics and endocrinology. This integrated approach opens up new opportunities to study the behavioural ecology of some of the most endangered primates and to collect information on previously studied populations.

Chapters include methodological techniques; instructions on collecting, processing and preserving samples/data for later analysis; ethical considerations; comparative costs and further reading, making this an invaluable tool for postgraduate students and researchers in primatology, behavioural ecology and zoology.

JOANNA M. SETCHELL is Senior Lecturer in Evolutionary Anthropology at Durham University, and Director of the MSc in Evolutionary Anthropology. Her research integrates behaviour, morphology and demographic studies with genetics and endocrinology to address questions relating to reproductive strategies, life history and sexual selection in primates.

DEBORAH J. CURTIS is Honorary Research Associate in the Department of Anthropology and Geography at Oxford Brookes University. Her interests focus on lemur biology and she has conducted research on functional anatomy, behaviour and ecology, reproductive endocrinology, food chemistry, activity patterns and genetics. She also works as a scientific and medical translator.

Field and Laboratory Methods in Primatology

A Practical Guide

SECOND EDITION

Edited by

JOANNA M. SETCHELL
Durham University

AND

DEBORAH J. CURTIS
Oxford Brookes University

CAMBRIDGE
UNIVERSITY PRESS

University Printing House, Cambridge CB2 8BS, United Kingdom

Cambridge University Press is part of the University of Cambridge.

It furthers the University's mission by disseminating knowledge in the pursuit of education, learning and research at the highest international levels of excellence.

www.cambridge.org
Information on this title: www.cambridge.org/9780521142137

© Cambridge University Press 2011

First published 2011

A catalogue record for this publication is available from the British Library

ISBN 978-0-521-19409-9 Hardback
ISBN 978-0-521-14213-7 Paperback

To Ymke Warren, dedicated primatologist and conservation biologist, 1970–2010

Contents

Contributors

Email addresses are included for corresponding authors only.

Marc Ancrenaz
Kinabatangan Orang-utan Conservation Project, PO Box 3109, Sandakan, 90734 Sabah, Malaysia

Nicola Anthony
Department of Biological Sciences, University of New Orleans, 2000 Lakeshore Dr., New Orleans, LA 70148, USA

Wendi Bailey
Liverpool School of Tropical Medicine, Pembroke Place, Liverpool L3 5QA, UK

Simon K. Bearder
Nocturnal Primate Research Group, Department of Anthropology, School of Social Sciences and Law, Oxford Brookes University, Oxford OX3 0BP, UK
skbearder@brookes.ac.uk

James R. Bell
Department of Plant and Invertebrate Ecology, Rothamsted Research, West Common, Harpenden, Hertfordshire AL5 2JQ, UK

Michael W. Bruford
School of Biosciences, Cardiff University, Cardiff CF10 3AX, UK
brufordmw@cardiff.ac.uk

Richard T. Corlett
Department of Biological Sciences, National University of Singapore, 14 Science Drive 4, Singapore 117543, Republic of Singapore

Deborah J. Curtis
Department of Anthropology and Geography, School of Social Sciences and Law, Oxford Brookes University, Gipsy Lane, Headington, Oxford OX3 0BP, UK
curtis@scimedtrans.co.uk

Brian Darvell
Faculty of Dentistry, Health Sciences Centre, PO Box 24923, Safat 13110, Kuwait

J. Lawrence Dew
Department of Biological Sciences, University of New Orleans, 2000 Lakeshore Dr., New Orleans, LA 70148, USA
jldew@uno.edu

Nathaniel J. Dominy
Department of Anthropology, 6047 Silsby Hall, Dartmouth College, Hanover, NH 03755, USA

Gregory A. Engel
Washington National Primate Research Center, University of Washington, Seattle, WA 98195, USA. and Swedish/Cherry Hill Family Medicine Residency, Seattle, WA 98122, USA

Hans G. Erkert
Institute for Zoology/Animal Physiology, University of Tübingen, Auf der Morgenstelle 28, D-72076 Tübingen, Germany
hans.erkert@t-online.de

Hafejee C. Essackjee
Department of Physiology, S.S.R. Medical College, Belle Rive, Mauritius

Anna T. C. Feistner
WWF Dzanga Sangha Protected Areas, Bayanga, Central African Republic
anna@feistner.com

Agustin Fuentes
Department of Anthropology, University of Notre Dame, Notre Dame, IN 46556, USA

Jörg U. Ganzhorn
Animal Ecology and Conservation, Institute of Zoology and Zoological Museum, Martin-Luther-King Platz 3, 20146 Hamburg, Germany
ganzhorn@zoologie.uni-hamburg.de

Thomas Geissmann
Anthropological Institute, University Zürich-Irchel, Winterthurerstrasse 190, CH-8057, Zürich, Switzerland
thomas.geissman@aim.uzh.ch

Louise Glew
Centre for Environmental Sciences, School of Civil Engineering and the Environment, University of Southampton, Highfield, Southampton SO17 1BJ, UK

Benoît Goossens
Danau Girang Field Centre, Kinabatangan Wildlife Sanctuary, Sabah, Malaysia and School of Biosciences, Cardiff University, Cardiff CF10 3AX, UK

Colin Groves
School of Archaeology and Anthropology, Australian National University, Canberra, ACT 0200, Australia
colin.groves@anv.edu.au

Joanne Harding
School of Archaeology and Anthropology, Australian National University, Canberra, ACT 0200, Australia

Michael Heistermann
Department of Reproductive Biology, German Primate Center, Kellnerweg 4, 37077 Göttingen, Germany
mheiste@gwdg.de

J. Keith Hodges
Department of Reproductive Biology, German Primate Center, Kellnerweg 4, 37077 Göttingen, Germany

Paul E. Honess
Department of Veterinary Services, University of Oxford, Oxford OX1 3PT, UK
paul.honess@vet.ox.ac.uk

Kathryn Jeffery
Agence National des Parcs Nationaux, Libreville, Gabon

Mireille Johnson-Bawe
School of Biosciences, Cardiff University, Cardiff CF10 3AX, UK

Clifford J. Jolly
Department of Anthropology, New York University, Washington
Square, New York, NY 10003, USA
clifford.jolly@nyu.edu

Lisa Jones-Engel
Washington National Primate Research Center, University of
Washington, Seattle, WA 98195, USA
jonesengel@wanprc.org

Peter W. Lucas
Faculty of Dentistry, Health Sciences centre, PO Box 24923,
Safat 13110, Kuwait
peterwlucas@gmail.com

David W. Macdonald
Wildlife Conservation Research Unit, Department of Zoology,
University of Oxford, Oxford OX1 3PS, UK

Robert D. Martin
Department of Anthropology, The Field Museum, 1400 S Lake
Shore Drive, Chicago, IL 60605, USA
rdmartin@fieldmuseum.org

Julian C. Mayes
Head of Training, MeteoGroup UK, 292 Vauxhall Bridge Road,
London SW1V 1AE, UK
julian.mayes@meteogroup.com

Alexandra E. Müller
Anthropological Institute & Museum, University of Zürich,
Winterthurerstrasse 190, 8057 Zürich, Switzerland

Marc Myers
Primate Conservation Inc, 1411 Shannock Rd, Charlestown, RI
02813, USA
mmyers@primate.org

K. Anne I. Nekaris
Nocturnal Primate Research Group, Department of Anthropology, School of Social Sciences and Law, Oxford Brookes University, Oxford OX3 0BP, UK
anekaris@brookes.ac.uk

Patrick E. Osborne
Centre for Environmental Sciences, School of Civil Engineering and the Environment, University of Southampton, Highfield, Southampton SO17 1BJ, UK
p.e.osborne@soton.ac.uk

Daniel Osorio
School of Life Sciences, University of Sussex, Falmer, Brighton BN1 9QG, UK

Claire M. P. Ozanne
Centre for Research in Ecology, School of Human & Life Sciences, Roehampton University, Holybourne Avenue, London SW15 4JD, UK
c.ozanne@roehampton.ac.uk

Stuart Parsons
School of Biological Sciences, University of Auckland, Private Bag 92019, Auckland 1142, New Zealand
s.parsons@auckland.ac.nz

Nicholas Pepin
Department of Geography, Portsmouth University, Lion Terrace, Portsmouth, Hampshire PO1 3HE, UK
nicholas.pepin@port.ac.uk

Jane E. Phillips-Conroy
Anatomy and Neurobiology, Washington University School of Medicine, Box 8108, 660 S. Euclid Avenue, St. Louis, MO 63110, USA

Jonathan F. Prinz
TNO Nutrition and Food Research, Franseweg 55b, 3921 DE, Elst, The Netherlands

S. Jacques Rakotondranary
Biozentrum Grindel, Martin-Luther-King Platz 3, 20146 Hamburg, Germany

Lawrence Ramsden
School of Biological Sciences, University of Hong Kong, Pokfulam Road, Hong Kong

Yedidya R. Ratovonamana
Département de Biologie et Ecologie Végétales, Faculté de Science, BP 906, Université d'Antananarivo, Madagascar

Nigel Reeve
The Royal Parks, Holly Lodge, Richmond Park, Richmond, Surrey TW10 5HS, UK and Centre for Research in Ecology, Roehampton University, Holybourne Avenue, London SW15 4JD, UK
nreeve@royalparks.gsi.gov.uk

Pablo Riba-Hernandez
Proyecto Carey, Bahia Drake, Peninsula de Osa, Puntarenas, Costa Rica

Caroline Ross
Centre for Research in Evolutionary Anthropology, Department of Sciences, Roehampton University, Holybourne Avenue, London SW15 4JD, UK
c.ross@roehampton.ac.uk

Noel Rowe
Primate Conservation Inc, 1411 Shannock Rd, Charlestown, RI 02813, USA
nrowe@primate.org

Jutta Schmid
Department of Experimental Ecology, University of Ulm, Albert Einstein Allee 11, D-89069 Ulm, Germany
jutta.schmid@uni-ulm.de

Joanna M. Setchell
Anthropology Department, Durham University, South Road, Durham DH1 3LE, UK
joanna.setchell@durham.ac.uk

Kathryn E. Stoner
Centro de Investigaciones en Ecosistemas, Universidad Nacional Autónoma de México, Apartado Postal 27-3 (Xangari), Morelia, Michoacán, Mexico

Mauricio Talebi
Biological Sciences Department, Federal University of Sao Paulo, Diadema Campus, SP, Brazil
talebi@unifesp.br

Steven Unwin
Conservation Medicine Division, Chester Zoo (North of England Zoological Society), Caughall Road, Upton-by-Chester, Chester CH2 1LH, UK
s.unwin@chesterzoo.org

Daniel G. Weaver
Centre for Research in Ecology, School of Human & Life Sciences, Roehampton University, Holybourne Avenue, London SW15 4JD, UK

Elizabeth A. Williamson
Behaviour and Evolution Research Group, Department of Psychology, University of Stirling, Stirling FK9 4LA, UK
e.a.williamson@stir.ac.uk

Roman M. Wittig
School of Psychology, University of St Andrews, St Mary's College, South Street, St Andrews, Fife KY16 9JP, UK and Budongo Conservation Field Station, Masindi, Uganda

Nayuta Yamashita
Department of Anthropology, University of Southern California, Grace Ford Salvatori Bldg, 3601 Watt Way, Suite 120, Los Angeles, CA 90089–1692, USA

Klaus Zuberbühler
School of Psychology, University of St Andrews, St Mary's College, South Street, St Andrews, Fife KY16 9JP, UK and Budongo Conservation Field Station, Masindi, Uganda
klaus.zuberbuehler@wiko-berlin.de

Foreword

It is a pleasure and a privilege to contribute to this second edition of *Field and Laboratory Methods in Primatology*. The first edition was published only eight years ago, in 2003, so it is a clear sign of success that a second edition should follow so rapidly in its wake. I know from personal experience that many primate field workers, particularly those embarking on their maiden study, found the first edition very useful. So this new, updated version will surely be very welcome. Those who appreciated the well-coordinated team effort in the first edition will be pleased to see that the line-up of chapters and authors has remained very much the same. The editors, Joanna Setchell and Deborah Curtis, have once again marshalled the contributions with a sure hand. The original twenty-one chapters have been retained, with minor shifts in authorship here and there. The only substantial change is the addition of a new chapter (Chapter 11) by Zuberbühler & Wittig on field experiments, in recognition of the growing importance of this approach. The editors note that a primary aim of this volume is 'to encourage fields of research that are currently under-exploited . . . and to address the (often neglected) broader cultural and legal implications of fieldwork'. It is important here that the editors therefore hope not only to provide a basic practical guide to methods but also to influence the future course of research. Whereas the first edition was undeniably very successful as a guidebook for researchers, it is less clear whether it has had a significant impact on overall strategy. In my view, after only eight years it is too early to know; but I believe that in the future we will look back and see this book as a crucial milestone in the development of primate field research.

Findings from field studies of primate behaviour and ecology have not only made many significant contributions to primatology as such but have also fed into discussions of human evolution. In an entirely different direction, they have also become increasingly important for conservation biology. Primates typically inhabit tropical or subtropical forests, and their relatively high profile makes them prominent indicator species for the frighteningly rapid process of deforestation that is a mounting problem for conservation generally. Reliable data on primate behaviour and ecology raise awareness and constitute one of the tools used in modern biology as increasing resources are devoted to the battle against extinction. Encouragingly, over the eight years that have elapsed since the first edition was published, primate field workers have devoted increasing time and energy to conservation issues.

Field studies have progressively expanded to cover the entire order Primates at varying depths, yielding an impressive array of data. Yet, despite this veritable explosion in primate field studies, relatively little attention has been devoted to synthetic treatments of methods. An early guide produced by the Subcommittee on Conservation of Natural Populations in Washington (1981), which focussed on ecological aspects, became unavailable long ago. Apart from that, prior to publication of the first edition of *Field and Laboratory Methods in Primatology*, there was no single published source for even a basic set of methods required for primate field studies. That practical guide to field methods and certain allied laboratory procedures, including contributions from many key players, was therefore sorely needed and warmly welcomed. Primate field workers have since had a single source providing invaluable practical guidance. It is of, course, becoming increasingly easy to track down basic documentation through the internet. This was already obvious in 2003 from the numerous websites cited in individual chapters of the first edition, and dramatic improvements have taken place over the past eight years. But carefully distilled advice for primate fieldworkers remains an extremely valuable commodity found only in *Field and Laboratory Methods in Primatology*.

Primate field studies have made remarkable progress over the past 75 years since anecdotal accounts first gave way to systematic data collection. The initial transition to objective reporting was marked by Carpenter's classic field study of the behaviour and social relations of howler monkeys (1934). This was followed by a series of detailed field studies of the natural behaviour of individual primate species, focussing on Old World monkeys living in relatively open habitats (e.g., baboons, patas monkeys, vervets) and our closest relatives, the apes (notably chimpanzees). For many years, a binoculars-and-notebook

approach prevailed, but the emphasis was very much on careful com-
pilation of detailed behavioural observations over extended periods.

Some 30 years passed before another major transition, spurred
by innovative field studies of birds, was marked by the seminal publication of Crook and Gartlan (1966). These authors were the first to
generate a basic but comprehensive classification of primate social
systems and to examine possible evolutionary developments in relation to ecological factors. This quickly led on to the key concept of
primate socioecology and instigated a new wave of primate field
studies in which quantitative data were systematically collected for
both behaviour and selected ecological variables (notably those connected with diet). Eisenberg *et al.* (1972) further consolidated the basic
aim of generating an overall comparative framework for primate field
studies.

At this stage, it was recognized that reliable methods for collecting unbiased quantitative data are crucially important. Careful attention to objective, well-defined methods to guarantee collection of
reliable quantitative data was undoubtedly a key development in the
history of primate field studies. Altmann's fundamental (1974) publication on behavioural sampling surely played a pivotal rôle in transforming the approach taken to data collection in the field. Martin and
Bateson (1986) later provided an influential guide to methods of behavioural observation for use both in the field and in captivity. Although
now taken almost for granted, the shift towards focal animal sampling
as the method of choice – because it is explicitly designed to exclude
observational bias – was one of the most prominent features of primate
field studies after 1974. In parallel, it was also realized that careful
quantification of ecological variables is essential. This, in turn, led to
the clear realization that a field study of any chosen primate species
should cover at least one annual cycle in order to encompass seasonal
variation. Indeed, because year-to-year differences can be marked, primate field studies should ideally cover a period of several years for
proper assessment of the rôle of ecological factors. Long-term field
studies of individual primate species are in any case especially valuable, as evolutionary arguments depend on reproductive success,
which is optimally measured over an individual's entire lifetime.
Because primates are relatively long-lived mammals, the ideal requirements for long-term study are very difficult to meet. But an increasing
number of primate field studies are beginning to meet that very exact-
ing standard, and it is encouraging to see that further progress has been
made in this respect over the past eight years.

It did not take long before quantitative data from several detailed, long-term studies of primate behaviour and ecology were brought together in an influential book edited by Clutton-Brock (1977). This book, in addition to presenting overviews from individual field studies, provided results from overall comparative analyses of data on group size and composition, ranging behaviour and diet. These comparisons revealed several general quantitative principles of primate ecology and social organization (see also Clutton-Brock & Harvey, 1977; Martin, 1981). A decade later, the process of synthesizing quantitative data from a wide range of primate field studies was further advanced by another edited volume (Smuts *et al.*, 1987). More recently, a synthetic work produced by Sussman (1999; 2000) systematically presented information from a very wide spectrum of field studies. This was followed in 2007 by a 44-chapter book described as the first edited volume to offer a comprehensive overview of primatology since 1987 (Campbell *et al.*, 2007). The value of that volume as a reference work is underlined by the fact that a second, revised and expanded, edition has just been published (Campbell *et al.*, 2011).

In an accompanying development, primate field studies were progressively transformed from basic fact-finding exercises to problem-oriented investigations. This was directly reflected in the title of another edited volume (Sussman, 1979) that was published soon after Clutton-Brock's 1977 compendium. Sussman's volume marked a significant shift towards primate field studies specifically designed to investigate particular issues that were seen to be of theoretical importance, such as intraspecific variation in behaviour between study sites or distinctions between sympatric species. A number of single-author synthetic treatises have also been produced over the years, notably by Jolly (1985; the second edition of a book originally published in 1972), Fedigan (1982), Richard (1985) and Dunbar (1988).

For various reasons, coverage of primate species both in field studies and in synthetic works has generally been conspicuously uneven. For many years, there was a heavy emphasis on higher primates, predominantly on Old World monkeys and apes. At first, studies concentrated on species that either inhabit open habitats (particularly savannah-living Old World cercopithecine monkeys) or are relatively closely related to humans (i.e. apes). Thereafter, increasing attempts were made to include a number of New World monkeys, which are all exclusively forest-living. Carpenter's classic study of howler monkeys in Panama (1934) was, in fact, notable not only for its pioneering rôle but also because it exceptionally involved a forest-living New World monkey species. Despite the fact that they account for a quarter of extant primate species and markedly extend the adaptive and temporal range of primates, prosimians (lemurs, lorises and tarsiers) were largely neglected for some time. The few prosimian species selected for initial detailed studies were monkey-like in being relatively large-bodied and diurnal in habits. One practical reason for this is that most prosimians are small-bodied, nocturnal and forest-living and therefore relatively difficult to study. However, there was also a measure of deliberate neglect because of the myopic view that prosimians are of marginal importance. In fact, studies of prosimians yield a far richer foundation for comparative studies.

In the 1970s, an important insight emerged from the first detailed studies of nocturnal prosimian species: these 'solitary' primates actually have well-developed social networks involving occasional encounters between individuals at night and nest sharing by day (Charles-Dominique, 1977; Bearder & Martin, 1980; Bearder, 1987). Prosimians show certain special features that are completely lacking among higher primates, particularly with respect to activity pattern. As noted by Erkert in his chapter on chronobiology (Chapter 18), most primates are either clearly diurnal (active from dawn to dusk) or clearly nocturnal (active from dusk to dawn). Most prosimians are nocturnal. Simian primates (monkeys and apes) are typically diurnal; owl monkeys (*Aotus*) provide the only exception. Thus, study of adaptations for nocturnal life among primates is necessarily largely confined to prosimians. Moreover, some prosimians (*Eulemur* and *Hapalemur*) show a highly unusual pattern of activity, including both diurnal and nocturnal phases. In a relatively late development, this distinct activity pattern was clearly recognized and defined as cathemerality (Tattersall, 1988). Field studies have only recently begun to reveal the significance of cathemeral behaviour (Curtis & Rasmussen, 2006). Studies of nocturnal prosimians also led to certain methodological innovations, most notably radio-tracking, which is still far more commonly used for nocturnal than for diurnal primates. One welcome development has been partial eradication of the previous general neglect of prosimians in primate field studies. It is noteworthy for instance, that the first volume of Sussman's synthetic survey (1999) was devoted almost entirely to prosimian primates. This reflects both increased awareness and marked growth in studies devoted to prosimians in recent years. It is particularly pleasing to see a dramatic improvement in our understanding of nocturnal tarsiers, most notably through the studies conducted by Gursky (2007).

A further extension to problem-oriented primate field studies arose through explicit attention to sociobiological hypotheses. Various core concepts of sociobiology, most notably kin selection, inclusive fitness, reciprocal altruism and sexual selection, are clearly of direct relevance to primate behaviour. It is therefore understandable that field studies have increasingly been designed to test sociobiological hypotheses. However, it should be noted that it is exceedingly difficult to collect adequate data for convincing tests of sociobiological interpretations from long-lived, slow-breeding animals such as primates. Only a few primate studies have been continued long enough to yield data on lifetime reproductive success, which are ideally required to test many arguments about selective advantage of individual behaviour patterns.

Lest it be thought that the development of primate field studies has followed a simple linear trajectory of unmitigated progress, it should also be emphasized that some important aspects have to some extent fallen by the wayside. One prominent example is provided by studies designed to explore proximate causation, which typically require some form of field experimentation. Although this is a very promising approach, it has been relatively little used (Kummer, 2002). A number of excellent studies have illustrated the particular value of field playback experiments employing natural vocalizations and manipulating them in various ways (see, for example, Cheney & Seyfarth, 2007; Semple, 1998; Semple et al., 2002). But the experimental approach remains relative rare in primate field studies. So the addition of a new chapter specifically dealing with field experiments in this second edition of Field and Laboratory Methods of Primatology is a well-chosen improvement. Effective experiments require very careful design, so the guidelines provided are extremely useful. As various studies have shown, experiments with primates under natural conditions provide unique opportunities for probing the primate mind.

The most recent methodological milestone in primate field studies, still ongoing, has arisen through the availability of various relatively new techniques. Some of these new techniques are technological in nature (e.g. coupling of radio-tracking with automatic activity recording, Chapters 10, 18; GPS, GIS and remote sensing, Chapter 4); others are attributable to a combination of new methods with non-invasive collection of samples (e.g. hormone assays, Chapter 20; genetic typing, Chapter 21). In particular, a tripartite combination of field observations with hormone assays and genetic testing (essential for reliable inference of paternity) has opened up new possibilities for testing long-established notions regarding the relationship between social behaviour and reproduction. Excellent examples of combined approaches are provided by studies of female cycles, mating behaviour and paternity in free-living Hanuman langurs and Barbary macaques (Heistermann *et al.*, 2001; Brauch *et al.*, 2008).

Despite the wide array of available methods reviewed in the first edition of *Field and Laboratory Methods in Primatology*, helpfully updated in this revised version, there is still considerable scope for greater integration into new studies. It is, for instance, regrettable that detailed investigation of food items has not moved ahead faster and on a broader front. As Lucas and colleagues note in Chapter 13, threats to wild primate populations mean that many opportunities to collect valuable data may be missed for ever. Numerous possibilities for studying the physical and chemical properties of food items are effectively reviewed in Chapters 13 and 14 and deserve to be applied more widely. In my view, the best way to develop a feel for these topics is to read the excellent book that Lucas published in 2004, after the first edition of the present volume appeared. (For some reason, Lucas was too modest to cite his own book in either of the two chapters on which he is lead author.) There are also many opportunities for new research in the realm of primate vocalizations, notably in the ultrasonic range. As Geissmann and Parsons note in Chapter 16, there have been rapid advances with respect to suitable equipment. Indeed, things have moved so fast that some state-of-the-art equipment recommended in 2003 is no longer manufactured. Last but not least, it would surely be rewarding for field primatologists to make greater use of techniques such as GPS, GIS and remote sensing. My own dream is that one day it might be possible to sit in my office in Chicago and enjoy real-time monitoring of primate behaviour and ecology, with data streaming back through satellite links.

Primate field studies have become increasingly complex because of the high standards now set for objective, bias-free data collection and the steadily growing spectrum of available ancillary techniques. The need for thorough advance preparation has hence become particularly acute. It is increasingly difficult for a single investigator to cope with the demands of a modern field study, so there is an increasing trend towards field studies involving teamwork. Extensive advance preparation has now become obligatory across the board, and the chapters in this book provide many useful indications to help in this. It cannot be emphasized enough that timely application for research approval and permits is a key part of any preparation. The basic design of any study naturally requires very careful thought. As it is now also customary for a primate field study to address particular issues or hypotheses that have been identified in advance, there are fairly strict constraints on study design from the outset. Ways of achieving statistical independence in any data collected must be carefully considered. As several authors in this volume stress, advance preparation should also include training in relevant skills, expert advice and prior familiarization with particular techniques and equipment where appropriate. If animals are to be trapped, appropriate training before departure for the field is essential.

Increasing needs for careful advance planning of primate field studies are challenging because it is also important to maintain some degree of flexibility when dealing with unpredictable field conditions. The best-laid plans can be upset by unexpected circumstances and developments in the field. It is all very well to invest time and energy in advance preparation and training, but field studies also demand adaptability and the ability to respond with inspired improvization. For this and several other reasons, as Curtis *et al.* note in the introductory chapter, it is best to conduct a pilot study at the proposed field study site. A dry run prior to the main study is particularly necessary when there is no prior history of comparable fieldwork at that site.

This volume highlights an issue of central importance for any modern field study in clearly emphasizing ethical responsibilities and appropriate early decision-making by fieldworkers. Awareness of ethical aspects has progressively increased over the years, partly because of rapidly growing concern about conservation issues but also because of mounting sensitivity to animal welfare in a very general sense. In the first place, proper compliance with regulations for permissions and permits is mandatory for ethical, not just practical, reasons. Fieldworkers are 'ambassadors'. They have a duty to promote understanding for conservation and animal welfare by their actions both at home and abroad. Applications for CITES permits, in particular, may be time-consuming and occasionally vexatious (notably when the thorny topic of export/import of urine and faecal samples is involved), but it is imperative to show respect for international legislation that was introduced to control international trade in endangered species. Suitable prior training in certain techniques (e.g. handling of animals and taking blood samples) is also important in this context, as it can significantly enhance animal welfare.

Another key area that involves several ethical aspects is the decision whether or not to capture animals (see, for example,

Chapters 7, 8, 10 and 19). In this respect, too, fieldworkers serve as ambassadors. So it is essential that any capture programme should be properly authorized and conducted with evident concern for animal welfare. If the decision is made to capture animals as part of a field study, every effort should be made to maximize information that can be collected without prejudice to the subjects. While a captured animal is appropriately immobilized, it is possible to take standard measurements, check on health and reproductive condition and take various samples (e.g. dental casts, ectoparasites, plucked hairs, blood samples, X-rays; Chapters 7, 8 and 9). Indeed, the fieldworker is arguably bound by obligation to the scientific community to maximize collection of biological information from captured animals, provided that physical intervention and risk are minimized. One productive possibility that usually requires retention of animals in captivity is measurement of basal metabolic rate (BMR). As explained by Jutta Schmid (Chapter 19), BMR can be measured under field conditions using portable equipment. Here, a modified approach offers advantages regarding the level of disturbance imposed on study subjects (Genoud et al., 1997). A nestbox with a built-in thermal jacket can be used as a metabolic chamber, and air can be sampled and assayed by drawing rather than pumping it through the box. This avoids the need for hermetic sealing of the metabolic chamber, with all the attendant complications. Ideally, any measurements should be conducted over a single rest period. Retention of an animal in captivity should always be kept to a strict minimum because it may disrupt social interactions.

One of the greatest benefits of capturing animals is undoubtedly the opportunity to apply various kinds of marking to facilitate recognition of individuals, provided that such marking is conducted with a proper concern for animal welfare (Chapter 10). In particular, radio-tracking can be used, dramatically increasing the reliability of individual identification and the ease with which study animals can be located. This may facilitate the process of habituation if stress is minimized during capture. However, prominent questions of animal welfare arise in this context too: careful consideration must be given to the size and method of attachment of the transmitter package. Moreover, timely recapture must be reliably feasible in order to remove the package in due course. Removal of transmitter packages from all animals by the end of a field study is absolutely essential, but regrettably this is a topic that is still inadequately reported or discussed (Chapter 10).

Another crucial area is that of taxonomy (Chapter 9). All primate fieldworkers should give due attention to taxonomy in designing and conducting their studies. Findings may contribute to the general data-base that feeds into continued refinement of primate taxonomy. In addition, correct identification of the taxon under study may directly affect the outcome. An example is provided by Richard's study of two sifaka populations in Madagascar (1974). At the time, those two populations were identified as subspecies of *Propithecus verreauxi* (*P. v. verreauxi* and *P. v. coquereli*). The field study was specifically designed to examine variability in behaviour and ecology within a single primate species. However, twenty-five years later analyses of mitochondrial DNA sequences revealed that these two populations in fact belong to two distinct clades of sifakas. The degree of genetic divergence between *verreauxi* and *coquereli* suggests a species-level divergence (Pastorini *et al.*, 2001). Two separate species are now recognized in a widely used primate classification (Groves, 2005). Hence, a study explicitly designed to examine intraspecific variability has turned out to be an investigation of interspecific differences instead.

Ethical issues also arise with taxonomy. As Groves and Harding (Chapter 9) discuss, taxonomic aspects are fundamental to any field-work on primates. This is particularly true if any contribution is made to conservation measures, so investigators may be said to have an obligation to collect data and specimens that will help refine taxonomy. But this is precisely where a major ethical issue arises. Current museum-based taxonomic practice is explicitly based on a strategy of collecting reference specimens, and type specimens play a fundamental rôle in formal recognition of species and subspecies. The question that must now be posed is whether it is ethically justifiable to continue to collect primate specimens (other than incidental finds) from the field specifically to contribute to reference collections. This question is, of course, particularly acute where severely threatened populations are involved.

In fact, many investigators would simply rule out further collection of primate specimens for museum reference purposes, even for species that are still relatively common. This matter still needs proper debate among primatologists, indeed among zoologists generally, to establish generally acceptable guidelines. The fact of the matter is that formal recognition (including publication) of a new primate species or subspecies currently requires at least one reference specimen. In the absence of such a specimen, newly discovered primate taxa may remain in limbo. In principle, it should now be possible to design an acceptable reference base (e.g. combining photographs, measurements taken on the live animal, dental casts, field X-rays, plucked hairs and genetic typing) that no longer requires collection and preparation of a cadaver. However, there is a pressing need for formal international agreement on some alternative system if we are to accept that collection of museum specimens is no longer mandatory for taxonomic recognition. As Groves and Harding state, the fourth edition of the *International Code of Zoological Nomenclature*, which took effect on 1 January 2000, specifies that a type specimen must be 'an animal, or any part of an animal'. Some progress has been made since the publication of the first edition of this book. DNA samples have been used as type specimens for a few primate species and even photographs have been accepted for this purpose. If it should prove possible to design an acceptable system to replace collection of classical reference specimens, the need for effective broad-based sampling from live animals in primate field studies would become even greater.

In closing, I want once again to emphasize the need to consider what happens when the investigator returns home after completing a long-term primate field study in a distant foreign country. Quite apart from its scientific benefits, a long period spent abroad for this purpose has a particular value as a stimulating (if often challenging) life experience that broadens the investigator's horizons and permits an external perspective on that investigator's own culture. However, it is not often realized that the investigator may suffer a fairly acute form of culture shock on returning home. This 're-entry syndrome' (a term coined by one of the respondents) was specifically reported by Hinde (1979) in a hard-to-obtain publication based on replies to a questionnaire that he distributed to gauge the overseas experiences of recent graduates and others. Completed questionnaires concerning the first field experience lasting 5–24 months were received from 65 people in four groups (30 primatologists, 7 social anthropologists, 5 modern linguists and 21 participants in Voluntary Service Overseas), most of whom replied to the questionnaire within two years after returning from the field.

Hinde's survey incidentally revealed some important basic points about primate field studies. Significantly, respondents generally emphasized the value of prior training in methods of data collection and of previous field experience, thus emphatically underlining the value of appropriate preparation and pilot work. In Hinde's words: 'Proper preparation is crucial. Problems must be formulated, data collection methods outlined and practised, and the fieldworker trained in the skills necessary for his project.' Prior language training was also identified as a valuable asset. Most of the primatologists were very positive about their field experiences, although they reported that the first three months or so in the field tended to be quite problematic. It is noteworthy that half of the primatologists thought that they worked too hard because of self-imposed work schedules (which may account for 'Fieldworker's Procrastination Syndrome', recognized by Bearder and Nekaris in Chapter 22). Many of the primatologists and anthropologists thought that a break after three to six months would be a sensible measure. Academic isolation was a serious problem for many respondents, and a key factor was presence or absence of effective local support at the field study site. Hinde noted: 'It is desirable, but not usually possible, for research supervisors to visit research students in the field – if possible in the early phases of their study. If that is not possible, it is an asset to have a supervisor who knows the study area, and who keeps in as close touch as possible by correspondence.'

The most striking point to emerge from Hinde's survey (1979), however, was the high frequency with which respondents reported feeling some kind of disorientation or alienation after returning home. This was specifically mentioned by 23 primatologists (77%), by all of the anthropologists, by 3 modern linguists (60%) and by 13 VSO workers (62%). The 're-entry syndrome' thus identified involved culture shock in relation to 'civilized society' and the perception of its fast pace, difficulties with social interaction and difficulties in settling down to work (reported by half the primatologists). Clearly, re-adapting to the home culture after completion of a field study is also a serious issue that deserves attention as part of the experience of fieldwork. At the very least, recognition of the fact that this is a widespread phenomenon and hence a shared experience may make it easier for returning primatologists to cope with this additional challenge.

As a final note, I would like once again to make a personal tribute to all of the dedicated primate fieldworkers who have braved tough and demanding conditions – including 're-entry syndrome' – to build the truly impressive hoard of information that is now available. They have performed a major service to the academic community and are increasingly contributing in a major way to conservation biology. I pass on my very best wishes to the next generation of fieldworkers, who will surely be involved in studies of ever-increasing complexity, but who at least will have the revised version of this comprehensive methodological survey to guide them.

Robert D. Martin

REFERENCES

Altmann, J. (1974). Observational study of behaviour: sampling methods. *Behaviour* 49, 227–65.

Bearder, S. K. (1987). Lorises, bushbabies, and tarsiers: diverse societies in solitary foragers. In *Primate Societies*, ed. B. B. Smuts, D. Cheney, R. M. Seyfarth, R. Wrangham & T. Struhsaker, pp. 11–24. Chicago, IL: Chicago University Press.

Bearder, S. K. & Martin R. D. (1980). The social organization of a nocturnal primate revealed by radio-tracking. In *A Handbook on Biotelemetry and Radio Tracking*, ed. C. J. Amlaner & D. W. Macdonald, pp. 633–48. Oxford: Pergamon Press.

Brauch, K., Hodges, K., Engelhardt, A. *et al.* (2008). Sex-specific reproductive behaviours and paternity in free-ranging Barbary macaques (*Macaca sylvanus*). *Behav. Ecol. Sociobiol.* 62, 1453–66.

Campbell, C. J., Fuentes, A., Mackinnon, K. C., Bearder, S. K., & Stumpf, R. M. (2011). *Primates in Perspective*. (2nd edn.) Oxford: Oxford University Press.

Campbell, C. J., Fuentes, A., MacKinnon, K. C., Panger, M. & Bearder, S. K. (eds.) (2007). *Primates in Perspective*. Oxford: Oxford University Press.

Carpenter, C. R. (1934). A field study of the behaviour and social relations of howling monkeys. *Comp. Psychol. Monogr.* 10, 1–168.

Charles-Dominique, P. (1977). *Ecology and Behaviour of Nocturnal Primates*. (Translated by R. D. Martin.) London: Duckworth.

Cheney, D. L. & Seyfarth R. M. (2007). *Baboon Metaphysics: The Evolution of a Social Mind*. Chicago, IL: University of Chicago Press.

Clutton-Brock, T. H. (ed.) (1977). *Primate Ecology*. London: Academic Press.

Clutton-Brock, T. H. & Harvey P. H. (1977). Primate ecology and social organization. *J. Zool., Lond.* 183, 1–39.

Crook, J. H. & Gartlan J. S. (1966). On the evolution of primate societies. *Nature, Lond.* 210, 1200–3.

Curtis, D. J. & Rasmussen M. A. (2006). The evolution of cathemerality in primates and other mammals: a comparative and chronoecological approach. *Folia Primatol.* 77, 178–93.

Dunbar, R. I. M. (1988). *Primate Social Systems*. London: Croom Helm.

Eisenberg, J. F., Muckenhirn, N. A. & Rudran, R. (1972). The relation between ecology and social structure in primates. *Science* 176, 863–74.

Fedigan, L. M. (1982). *Primate Paradigms: Sex Roles and Social Bonds*. Montréal: Eden Press.

Genoud, M., Martin, R. D. & Glaser, D. (1997). Rate of metabolism in the smallest simian primate, the pygmy marmoset (*Cebuella pygmaea*). *Am. J. Primatol.* 41, 229–45.

Groves, C. P. (2005). Order Primates. In *Mammal Species of the World: A Taxonomic and Geographic Reference*, vol. 1, ed. D. E. Wilson & D. M. Reeder, pp. 111–84. Baltimore, MD: Johns Hopkins University Press.

Gursky, S. L. (2007). *The Spectral Tarsier*. Upper Saddle River, NJ: Pearson Prentice Hall.

Heistermann, M., Ziegler, T., van Schaik, C. P. *et al.* (2001). Loss of oestrus, concealed ovulation and paternity confusion in free-ranging Hanuman langurs. *Proc. R. Soc. Lond.* B 268, 2445–51.

Hinde, R. A. (1979). Report on replies to a questionnaire concerning the experiences of recent graduates and others abroad. *Primate Eye* 11 (Suppl.), 1–18.

Jolly, A. (1985). *The Evolution of Primate Behavior*. (2nd edn.) New York: Macmillan.

Kummer, H. (2002). Topics gained and lost in primate social behaviour. *Evol. Anthropol.* 11 (Suppl. 1), 73–4.

Lucas, P. W. (2004). *Dental Functional Morphology: How Teeth Work.* Cambridge: Cambridge University Press.

Martin, P. & Bateson P. (1986). *Measuring Behaviour.* Cambridge: Cambridge University Press.

Martin, R. D. (1981). Field studies of primate behaviour. *Symp. Zool. Soc. Lond.* 46, 287–336.

Pastorini, J., Forstner, M. R. J. & Martin, R. D. (2001). Phylogenetic history of sifakas (*Propithecus*: Lemuriformes) derived from mtDNA sequences. *Am. J. Primatol.* 53, 1–17.

Richard, A. F. (1974). Intra-specific variation in the social organization and ecology of *Propithecus verreauxi. Folia Primatol.* 22, 178–207.

Richard, A. F. (1985). *Primates in Nature.* New York: W.H. Freeman.

Semple, S. (1998). The function of Barbary macaque copulation calls. *Proc. R. Soc. Lond.* B 265, 287–91.

Semple, S., McComb, K., Alberts, S. & Altmann, J. (2002). Information content of female copulation calls in yellow baboons. *Am. J. Primatol.* 56, 43–56.

Smuts, B. B., Cheney, D., Seyfarth, R. M., Wrangham, R. & Struhsaker, T. (eds.) (1987). *Primate Societies.* Chicago, IL: Chicago University Press.

Subcommittee on Conservation of Natural Populations (1981). *Techniques for the Study of Primate Population Ecology.* Washington, D.C.: National Academy Press.

Sussman, R. W. (ed.) (1979). *Primate Ecology: Problem Oriented Field Studies.* New York: John Wiley.

Sussman, R. W. (1999). *Primate Ecology and Social Structure,* vol. 1, *Lorises, Lemurs and Tarsiers.* Needham Heights, MA: Pearson Custom Publishing.

Sussman, R. W. (2000). *Primate Ecology and Social Structure,* vol. 2, *New World Monkeys.* Needham Heights, MA: Pearson Custom Publishing.

Tattersall, I. (1988). Cathemeral activity in primates: a definition. *Folia Primatol.* 49, 200–2.

DEBORAH J. CURTIS, JOANNA M. SETCHELL AND
MAURICIO TALEBI

Introduction

The ideal program for the future study of animal behaviour would involve the coordination of field and laboratory investigation.

Schaller (1965, p. 624)

In the early days of expeditions, a naturalist's field equipment consisted of little more than a gun and the means to preserve specimens. Later on, with the shift towards collecting information on primate behaviour and ecology, rather than collecting the animal itself, field primatologists relied on pencil and paper, binoculars, a compass and, if studying nocturnal species, a torch. In 1974, Jeanne Altmann's paper on observational sampling methods, intended 'as a guide to thinking, planning and design' (Altmann, 2010, p. 48), led to systematic, quantitative studies of behaviour. More recently, a shift in emphasis towards integration of methods has led to collaboration between laboratory and field researchers working on wild primates. Technological advances have presented fieldworkers with the opportunity to collect more sophisticated data, replace check-sheets with hand-held computers, store samples for later laboratory analysis, analyse samples in the field and collect information remotely. This has led to an increase in data collected, using, for example, non-invasive techniques for DNA analyses and hormonal assays. Ecological methods and techniques available for monitoring primate habitats have also improved, with the application of remote sensing, mapping (Global Positioning Systems), and data integration (Geographic Information Systems). These methods open up possibilities to collect new information on previously studied populations, and a means to collect data on species that cannot be habituated for behavioural observations.

The technological advances described above come at a time when 48% of primate species are under threat and insufficient data are

Field and Laboratory Methods in Primatology: A Practical Guide, ed. Joanna M. Setchell and Deborah J. Curtis. Published by Cambridge University Press. © Cambridge University Press 2011.

available to assess the status of a further 15% of species (IUCN, 2008). Knowledge of the basic behavioural ecology of a species is essential for conservation, and this makes an integrated approach combining laboratory and field techniques all the more important, particularly as some of the most endangered species are those about which least is known, such as the drill (*Mandrillus leucophaeus*) on the front cover of this second edition. Habitat destruction is rising exponentially and the effective and practical implementation of modern, state-of-the-art environmental laws appears a long way off – a paradoxical situation. The legal apparatus has evolved quickly but human culture and practices are still adapting to these modern laws. For example, in Brazil, the hunting of wild animals (including primates) is a deeply rooted practice stemming from 10 000 years of tradition in local indigenous peoples, as well as the habits of European invaders, slaves and settlers over the past 500 years. Hunting has only been illegal since 1991 and, unfortunately, continues, even for threatened primate species, at the end of the first decade of the twenty-first century.

Although fieldworkers are generally well prepared in terms of basic behavioural and ecological methodology, prior to the publication of the first edition of this volume there was no easily accessible source of information concerning the wide variety of additional data and samples that can be collected for subsequent laboratory analysis; methods usually had to be compiled from the primary literature. This was an almost impossible task for researchers and students in many habitat countries, with often very limited access to journals, and brings us to another change that has taken place relatively recently in primatological research.

We find ourselves in the midst of a shift from the traditional 'power bases' of primate research to research that is increasingly initiated and conducted by nationals of the countries where non-human primates occur. This is nicely demonstrated by research in the Neotropics, where some field studies have now been running for over 25 years. These studies were generally initiated, conducted, coordinated and funded by non-Latin-American researchers and institutions. Over the past two decades, a growing number of Latin-American researchers have completed their academic training either in their country of origin or abroad. Most of these latter scientists have returned home, bringing with them the knowledge and techniques they have learned and, most importantly, contributing significantly towards developing primatology in habitat countries. As a result of this development, a new trend has arisen in Neotropical primate research: studies are now led by local researchers and researchers from abroad are looking to collaborate with them. However, regular capacity-building programmes and field training courses are required for the continuation of this trend.

The second edition of this volume attempts to build on what has been achieved since the publication of the first edition in 2003. Although many of the chapters have simply been brought up to date, we (DJC and JMS) have done our best to address criticisms levelled at the first edition, for example, by including a new chapter on behavioural methods and field experiments, trying to ensure that complex terminology and jargon is explained within the individual chapters and that all chapters provide references appropriate to primatological applications, and ensuring that the 'tone' of the chapters is inclusive and addresses all primatologists. That said, this second edition adheres to the guiding principles of the first edition, as the feedback has been very positive. We have also retained the individual approach taken by the authors to each topic as we believe that the imposition of too much structure would be counter-productive, given the diverse nature of the research methods covered in this book. It will become clear to the reader that primatology has moved on, but we wonder whether primatologists have moved with the changes. There remains so much to be done, with little time left to do it in the case of many species. The first edition clearly served to broaden horizons with reference to some methods, but other topics seem to have remained side-lined, as is outlined in the preface to this second edition. For example, we know little more about food mechanics than we did prior to the first edition, and the role of ultrasound in primate communication has yet to be investigated. We hope that this will change over the coming years and that this second edition will serve not only as a guide for students and researchers wishing to enter new fields, but also encourage people to 'think outside the box'. The world is becoming more and more competitive – you might just get your hands on the funds that have eluded you for so long by trying something novel or trying a new combination in your pursuit of further knowledge of primates.

In this introductory chapter, we briefly illustrate the methodological paradigm shifts that have occurred in the history of primate field studies with a case study (the aye aye). We then give our own advice to those preparing for the field, with a general overview of how to plan and conduct a field study, before detailing the aims of this volume and presenting an overview of the chapters.

METHODOLOGICAL PARADIGM SHIFTS IN FIELD PRIMATOLOGY: THE AYE AYE

Several methodological paradigm shifts have occurred in the history of primate field studies (Strum & Fedigan, 2000; see also the preface to this volume). Until the early twentieth century, the driving force behind field studies was classification, through the collection and dissection of specimens, and comparative anatomy. Knowledge of behaviour in the wild increased much more slowly and, until the 1920s, comprised incidental observations by adventurers, seafarers, hunters, explorers and missionaries, some of whom were naturalists. This interest in behaviour, albeit anecdotal, constituted the first paradigm shift leading towards modern day behavioural and ecological studies. This and subsequent changes in methodological approach are well illustrated by studies of the aye aye (*Daubentonia madagascariensis*; family Daubentoniidae).

The aye aye was discovered by Sonnerat (1782, cited in Owen, 1863) in Madagascar during a voyage to the East Indies and China. Sonnerat described it as bearing resemblance to a squirrel, but also having characters that allied it to lemurs and monkeys. Cuvier (1798, cited in Owen, 1863) classified the aye aye as a rodent, and although Schreber (cited in Peters, 1865) placed it within the lemurs as early as 1803, it was not definitively recognized as such until 60 years later, with the publication of Owen's monograph on aye aye anatomy in 1863.

Expeditions aimed almost uniquely at the collection of aye aye specimens for anatomical study continued into the early twentieth Century (reviewed by Lavauden, 1933). Pollen (1863) neatly summed up the reason for the lack of information on the natural history of the Malagasy vertebrates, and his explanation remains generally applicable today:

> *It remains incomprehensible for those who do not know the obstacles faced by scientific explorations, that one knows so little of the natural history of this large country, even though a number of travellers have visited different parts of it. But, if one notes the numerous difficulties that these enterprising travellers have taken in their stride,* ... *one is no longer surprised. On the contrary, one is obliged to pay homage to these zealous travellers who have risked their lives for science and sacrificed their health for the collection of the limited number of specimens they managed to obtain* (translated from the original French).

Early descriptions of aye aye behaviour stem from animals kept in captivity in La Réunion (Vinson, 1855) and Mauritius (Sandwith, 1859 in Owen, 1863). Sandwith (1859 in Owen, 1863) provides an excellent description, stating:

> I ... was much struck with the marvellous adaptation of the creature to its habits, shown by his acute hearing, which enables him to aptly distinguish different tones emitted from the wood by his gentle tapping; his evidently acute sense of smell, aiding him in his search; his secure footsteps on the slender branches, to which he firmly clung by his quadrumanous members; his strong rodent teeth, enabling him to tear through wood; and lastly, by the curious slender finger, unlike that of any other animal, and which he used alternately as a pleximeter, a probe, and a scoop.

Lamberton (1911) carried out the longest such study, observing a captive individual in Madagascar for over a year. Given its nocturnal habits, Lamberton concluded that the only way to study the aye aye's natural behaviour was in captivity, and was ahead of his time in remarking that:

> ... individuals taken to Europe are certainly in bad condition. Made feeble by the voyage, embittered by suffering, sick due to the change in climate and food, how to distinguish between what is natural and what is a consequence of the conditions in which the animal finds itself? Observations made on aye ayes held in captivity in their country of origin certainly approach reality far more ... (translated from the original French).

The earliest descriptions of aye aye behaviour in the wild come from missionaries in Madagascar, who compiled information by interviewing the Malagasy (see, for example, Baron, 1882). Lavauden (1933) made the earliest qualitative behavioural observations of wild aye ayes, including photographs of wild individuals, a nest and a larval tunnel exposed by an aye aye. The first comprehensive description of aye aye behaviour and ecology was published by Petter in 1962, but no quantitative studies were conducted in the wild until the late 1980s (Ancrenaz et al., 1994; Sterling, 1993). In terms of the integration of laboratory and field techniques, only diet and nutrition have been investigated for the aye aye, by combining field data with chemical analyses of food items (Sterling et al., 1994). Our knowledge of aye aye biology could be greatly expanded by the application of more of the methods detailed in this volume.

PLANNING AND CONDUCTING A FIELD STUDY

Field trips are expensive, and may well be once-in-a-lifetime opportunities for both the researcher and, unfortunately, the study population.

For example, DJC's study population of mongoose lemurs (*Eulemur mongoz*) at Anjamena, Madagascar, has declined by 85% over a 13-year period from 1995 (16 groups/km^2) to 2008 (3 groups/km^2) (Shrum, 2008). Even more worrying is that of all the areas surveyed by Shrum (2008), Anjamena exhibited the highest population density. It is therefore imperative to prepare well and to maximize the information collected. In this section we provide some guidance on how to deal with some of the non-scientific aspects of fieldwork. However, this is by no means to be regarded as covering all eventualities, as it is mainly based on our personal experience and on conversations with colleagues. Many of the following chapters give specific advice on various problems.

Preparation

Before you leave for the field, ensure that you know what your objectives are. The stated aims of your project must be fulfilled by your planned data collection. It is likely that these aims will change over time, usually because you have been too ambitious at the outset, but you need to start somewhere. Think about how you will analyse your data before you go into the field. Make sure that you are familiar with your study species by referring to previous work, contacting others who have worked on the same animals, and, if possible, spending time at a zoo observing them to give you a feel for their behaviour and general characteristics. You can also try out behavioural sampling methods at the zoo, although you might need to change things once you are in the field if, for example, your contact time with the animals is low or visibility is poor. If possible, test your ecological methods before you go; if your study requires particular techniques (e.g. collection and preparation of botanical samples, blood sampling), take expert advice, get the necessary training and practise the techniques you will be using in a laboratory. Make sure that you are familiar with all your equipment before you leave. If your study will involve capturing your study animals, training and preparation are essential and you should take the opportunity to collect as much information as you can about each captured individual (e.g. morphometrics, health status and biological samples). Even if these data, or other samples collected, are not directly related to your current research questions, they may constitute a future research project, or provide opportunities for collaboration.

Gather as much information on potential study sites as you can, read up on previous studies carried out at the same or similar study sites and contact colleagues who have worked in these areas. Bear in mind that those running a study site have invested a great deal of time in energy in it, and respect this. Inform yourself of particular problems that may pertain to the study area(s) (e.g. difficult terrain; climate; health hazards; antagonism from local people; political stability) and don't hesitate to ask for advice from those with more experience. Inoculations, applying for ethical permission, permits, dealing with travel arrangements and setting up collaborations all take time. You may also need to sort out travel arrangements some time in advance if you will be travelling during busy periods. Insurance, at the very least medical cover, is expensive but essential in case something goes wrong.

Terrain and climate

Ensure that you are fit enough to cope with difficult terrain, and that you can be evacuated within a reasonable time scale, should you injure yourself and need hospital treatment. This will apply generally, but particularly in remote areas, and being sensible will greatly reduce the probability of an accident. For example, employ local assistants to climb trees, should this be necessary, as they will usually be much better at it than you are. Make sure that you have adequate protection from the elements, such as warm clothing, clothing to protect you from the sun, and rainwear.

Health hazards

Assess the health hazards that you may face and take the necessary precautions. Get a medical check-up before you leave for the field, have all the recommended vaccinations, take the necessary medication with you, and remember to take it. If you are travelling from abroad, it is also advisable to seek the advice of a medic once you reach your destination (especially if you will be carrying out a long-term study), as a local doctor will know more about local diseases and the precautions or prophylaxes that may be necessary. For example, consider using locally produced malarial prophylaxes, which are perfectly adequate in most cases, but always consult a doctor first. Furthermore, you are less likely to contribute to the growing resistance of the parasite to drugs that are too expensive for local people to afford. As is the case the world over, see a medic before you take antibiotics (if possible), and finish the course – again, to avoid contributing to the ever-increasing problem of antibiotic-resistant bacteria.

If you will be handling animals and/or working at some distance from medical services, then rabies vaccinations are a good idea. However, if you are bitten by an animal that may be rabid, then you must seek medical attention immediately, as the vaccination does not give you full protection. It is also a good idea to have a selection of syringes and needles with you, as well as the necessary equipment for perfusions, to be used if you are admitted to an under-equipped hospital. You will have been inoculated against Hepatitis A and B as a standard procedure, but HIV is rife the world over and if you don't have your own equipment with you, then you may run the risk of infection from recycled syringes.

Take large amounts of standard first aid kit items with you (e.g. plasters, bandages, painkillers). This is not just for your own use; people may come to you for help before they go to the nearest medic, particularly if you are working in a remote area. In Madagascar, DJC was asked to help in cases of serious ear and eye infections, septic wounds, sexually transmitted diseases, malaria, minor cuts and bruises and headaches, as well as sick zebu cattle. In most cases, all you can do is send your 'patients' to the local doctor, in others, a painkiller or plaster will do the world of good and you will have made a friend. Think about taking a first aid course before you leave, and finally, if you are employing local assistants, think about acting as their health insurance (we think this is only reasonable) and pay for any medical and dental treatment. In addition to moral considerations, it is to your own advantage to do so, as healthy employees are more inclined to work.

Precautions against dangerous animals require common sense, vigilance and, for example, thick leather garters as a protection against venomous snakes, and long trousers and long-sleeved shirts as protection against insects. In remote places we recommend that you keep snake anti-venin at your field camp (and know how to use it if required), because this needs to be taken within a few hours after a snake bite occurs, and you won't have time to get to a medical centre.

Look after your mental health, ensure you take regular breaks from the field and don't push yourself beyond your limits. Remember, you will be working hard, but you are not supposed to return from the field a mental and physical wreck – again, use your common sense. Take advantage of being in a new place: meet people and learn about the culture and traditions of the country and region you are working in. Little things can help you get through bad times – DJC found the shortwave radio useful, as well as the luxury of an inexpensive reading lamp in the tent, powered by a car battery charged using solar panels; JMS kept a diary; MT made good friends in the local village for weekly meals and monthly Sunday barbecues.

Political stability

If you will be working abroad, check your government's website for detailed information on the political stability of the country and area you intend to work in, and follow their advice. If you will be working in a highly volatile area, then ensure you have your escape route planned in advance, should you need to leave in a hurry. Even if a country is stable, it is always a good idea to register your presence with your local embassy or consulate of your country of origin (if you are a foreigner), and let them know your itinerary. Whether you are a foreigner or a national of the country you are working in, always inform the local authorities of your presence and what you are doing. This is common courtesy and will also help guarantee your safety.

The pilot phase

If you can afford it, then a pilot phase is invaluable for finding a study site, setting up collaborations, arranging permits and testing your methodology. If you are travelling around to find an appropriate study site, then make contacts with people working and living in the area to get to know the advantages and disadvantages of working there. Spend some time at the site to assess the logistics of provisioning. If you will be working in the forest, assess the visibility, and density of the forest and, above all, try to find and follow your study animals to test your methods. DJC visited three potential study sites during her pilot phase to study the cathemeral mongoose lemur. One location was out of the question as the density of the study animals was low and the forest highly degraded. The second had the advantage that it was at a field station, and transport and provisioning would have been relatively easy to deal with. However, the density of mongoose lemurs was again low. In one area of forest near the field station where visibility was good, the flora was mostly introduced. In a second area the forest was dense, visibility low and the canopy very high. A third area where visibility was good contained no study animals. The study was eventually carried out in the third location, because the density of study animals was extraordinarily high, the forest was open enough that no transects needed to be cut, the canopy was not too high and visibility was good. The downside was that the site was fairly remote, making provisioning complicated, and transport and access difficult, particularly during the rainy season.

Use your pilot phase to test your sample collection and storage methods. For example, the practicalities of collecting faeces, or food specimens, from small canopy primates may only become apparent when sitting below subject animals in a 30-m canopy (keep your mouth shut). A few representative samples taken back to the lab at this stage will allow you to test preservation methods and will save a lot of trouble at a later stage. Check out the possibilities for getting access to power, and adapt your methods accordingly, including a 'plan B'. Bear in mind that logistical arrangements, agreements, etc., can all change between your pilot study and your next major field season. Finally, look into how you will communicate with home, colleagues, advisors and/or collaborators. Establish this before returning to the field, so that everyone knows what's happening.

Working with other people

Studying primates inevitably involves interactions with other people, be they other researchers (in the field and the lab), field assistants, local populations or government officials. When working with other researchers and with field assistants, make sure that everyone (including you!) has a clear picture of what to expect from the project and the roles different people play. If you are in a situation where you need to employ assistants personally (e.g. you are not working on a pre-existing field station), then we recommend drawing up some kind of contract. This can prevent a lot of problems and also makes it clear who is the boss. You and your assistants and employees are not a team (this observation stems from bitter experience!), and any working situation usually needs some kind of hierarchy in order to function efficiently. Field biologists are often passionate and prepared to live on very little, in challenging conditions, but you can't assume that your employees will want to do the same.

Respect the culture of the area you are in. Good relations with people from top officials down to workers and villagers are very important, both for your study and for researchers coming after you. Bureaucracy can be a pain, but rules are there for a reason, and you wouldn't break the law at home. Whether you are a foreigner or from a different part of the country you are working in, working with local collaborators will help with language and cultural understanding, as well as bureaucracy. At some point you may encounter antagonism from local people. Much of this can be prevented by respecting local customs (e.g. dress and food prohibitions). Ensure that people living in your study area are aware of what you are doing and why you are doing it. You may need to visit the heads of surrounding villages, and perhaps give a talk to the villagers – don't rely on officials to do the job for you. Your paperwork might be in order, but that will not be enough. Make some kind of contribution to the local economy, i.e. don't try to set up a self-contained study site, but employ some of the local population (not just strangers to the area) and buy local produce. This will also help spread the word of what you are doing in the area. Be aware of payment scales, and avoid over- or under-paying for supplies and assistance.

Longer-term studies

Primate field studies often last the one or two years necessary to acquire data for a PhD thesis. Less often, studies last longer, either to address questions arising from the initial investigation or because the initial research period did not allow the researcher(s) to successfully accomplish their goals. Long-term studies of wild primates are even rarer, although primatology is characterized by some of the longest-running studies of wild animals. These did not necessarily begin with the aim of long-term data collection, but have developed over time. Several factors may contribute to a study becoming long-term. For example, institutional networks of researchers may work at the same site, investigating different questions (e.g. plant ecology, behavioural ecology) but collaborating to attract funding for critical aspects of the field site, such as infrastructure, equipment, logistics and human resources. In other cases, primate species are faced with such a high local anthropogenic threat that researchers avoid terminating a field study because the study is the only thing preventing the species' local extinction.

Running a long-term site

Running a long-term field site obviously requires the ability to maintain good records of the study species and publish papers, but also requires an ability to maintain good relationships with local people and the authorities. This isn't always simple, but you may even be able to overcome a lack of stable funding, and bridge gaps in financial support, if you can achieve good local support for a project. You need to gain the confidence and trust of staff members, field assistants and others involved in your project, as well as local people and the authorities. For example, MT heard many reasons why local people at his study site disliked other researchers, including perceived arrogance, or even worse in their eyes: 'our invisibility to a researcher that views him/herself as too important to spend time or to talk with us'. These comments were not related to nationality and therefore had nothing to do with language abilities. Good relations with and publicity in the local media, such as newspapers and radios, talks in local schools, and so forth are useful. Finally, it is critically important to keep the environmental authorities happy with meetings and frequent, informative research reports, submitted on time.

When you get back

Once you are back from the field, laboratory work or data analysis will quickly take over your life, but make sure that you get a medical check-up. Some parasites can remain undetected in your body for years, but cause serious harm to your health. Watch out for strange symptoms and always inform your doctor that you have been working in the field. In addition to your physical health, be aware of the 're-entry syndrome' described in the preface to this volume.

AIMS OF THIS VOLUME

The increasingly competitive nature of scientific research, in combination with space limitations in scientific journals, means that it is often difficult to come by information on methodological problems, troubleshooting and, in particular, negative results. Journal articles tend to present a clear, logical development of ideas, hiding the methodological problems encountered along the way. Furthermore, as many governments increasingly insist on returns from research in terms of either technological or biomedical advances that are beneficial to economic growth, it is becoming more and more difficult to obtain funding for studies rooted in behaviour and ecology. Interdisciplinary is a buzzword, and the combination of field and laboratory techniques may well make a study more interesting to funding bodies.

This book brings together expert authors to cover methods at the field–laboratory interface and techniques used in the study of primates and non-primates, guide in the selection of appropriate methods and produce a volume that can act as a reference book in the field. One of the central aims of the first edition was to encourage fields of research that were under-exploited in primatological and zoological fieldwork, and to address the (often neglected) broader cultural and legal implications of fieldwork. Although the book is primarily aimed at primatologists, many of the methods are also applicable to other groups of animals. The volume does not include an introduction to the primates or statistics, as these subjects are amply and ably covered by other authors (e.g. Field, 2009; Fleagle, 1999; Siegel & Castellan, 1988; Strier, 2007; Campbell *et al.*, 2007). No topic is reviewed comprehensively, as this would fill several volumes. As noted by a number of the contributors, you will still need to refer to the specialist literature for additional information, but the chapters provide an excellent starting point. The focus is on the technical and practical aspects of each method or area, rather than on research results, and chapters include the following information, where appropriate:

- Point by point instructions for sample collection, processing and preservation for later analysis.
- The time and equipment required for work in the field and laboratory.
- Methodology and techniques, including the difficulties of various methods and how to overcome them.
- The ideal situation/equipment and what will do if necessary.
- Exactly what data the samples will provide the field worker with.
- Ethical considerations.
- Comparative costs.
- Selected useful references.
- Useful Internet sites.

Where authors mention the use of commercially available equipment and supplies, this does not constitute endorsement either by the editors or by the publisher. Similarly, we have endeavoured to ensure that the URLs for external websites referred to in this book are correct and active at the time of going to press. However, neither we nor the publisher has any responsibility for the websites and we can make no guarantee that a site will remain live or that the content is or will remain appropriate.

OVERVIEW OF CHAPTERS

It is now almost impossible to study non-human primates without becoming aware of contact (and conflict) between human and non-human primates, whether via hunting, tourism, pet ownership, food competition, crop-raiding or research. Jones-Engel *et al.* (Chapter 1) discuss the practicalities of studying such interactions between human and non-human primates (ethnoprimatology). They cover the design of surveys, conducting interviews and sampling both human and non-human primate subjects, with a particular emphasis on cultural fluency and the ethics of working with people in habitat countries. This is followed by Williamson and Feistner (Chapter 2), who review the advantages and disadvantages of methods for habituating primates. Chapters 3–5 concern the methods used to describe a habitat. Ganzhorn *et al.* (Chapter 3) discuss vegetation as the biotic matrix for the evolution of life history traits, detailing methods for forest classification, habitat structure and phenology. Osborne and Glew (Chapter 4) then address the potential applications of two modern technologies in primate research: geographical information systems and remote sensing. These have, as yet, rarely been applied in field studies of primates, and Osborne and Glew consider how they may be applied to problems in primatology, cover methods for collecting primate location data, discuss sources of environmental data layers and give examples of analyses that integrate the two. Weather and climate have important influences on ecosystems; Mayes and Pepin (Chapter 5) describe the types of local weather information that can be collected in field studies. They demonstrate how weather can vary spatially and temporally and give methods for measuring local atmospheric conditions. Ross and Reeve (Chapter 6) then introduce some of the key issues to consider in primate population studies and discuss survey and census methods for estimating the population and population density of a species.

Many field studies involve capturing primates, for example to mark or radio-collar study animals, or to collect morphological or physiological data. In Chapter 7, Jolly *et al.* cover the practicalities of trapping primates of all shapes and sizes, while Unwin *et al.* (Chapter 8) present the veterinary aspects of handling primates, including anaesthesia, health evaluation and methods for collecting biological samples. Both chapters emphasize the necessity of collecting as much information as possible when animals are captured, owing to the risks involved in the procedure. Following on from capture methods, Groves and

Harding (Chapter 9) introduce morphology, morphometrics and taxonomy, with details of how to preserve dead specimens and measure both
dead primates and anaesthetized live animals. Honess and MacDonald
(Chapter 10) then discuss the advantages and disadvantages of various
methods for marking study animals and the uses and practicalities of
radio-tracking primates, satellite tracking and biotelemetry.

New to the second edition, Zuberbühler and Wittig (Chapter 11)
provide a brief introduction into the methods used to collect data on
behaviour and provide detailed instructions on how to conduct behavioural experiments in the field. The next four chapters are concerned
with food habits and nutritional ecology. In Chapter 12, Dew introduces the study of feeding ecology and seed dispersal, detailing methods for the observation of feeding behaviours, the collection and
classification of feeding remains and faecal contents analysis. This is
followed by two contributions by Lucas et al. (Chapters 13 and 14)
covering dietary analysis. Chapter 13 concentrates on the physical
properties of food items as they might influence feeding behaviour,
including methods for measuring food geometry (size and shape),
colour, texture and mechanical properties. Chapter 14 then deals
with the chemical properties of food, including sample preservation
and extraction and chemical tests for food components. In both contributions Lucas et al. emphasize the value of in situ analysis, in a 'field
laboratory', but they also include information on laboratory analyses if
this is not possible. Food availability is an important factor in primate
diet and methods of estimating fruit and leaf availability to frugivorous
and folivorous primates are covered in Chapter 3, while Ozanne et al.
(Chapter 15) cover methods of assessing (and identifying) the insect
prey available to insectivorous species.

The next two chapters give practical advice on equipment and
methods available for recording study animals. Geissmann and Parsons
(Chapter 16) discuss how to record primate vocalizations. This is important in terms of archival data, for sound analysis to determine the
properties of vocalizations, and also for playback experiments. In
Chapter 17, Rowe and Myers discuss still and video photography of
primates, with advice on choosing, using, protecting, and travelling
with photographic equipment.

Chapters 18–20 concern primate physiology. Erkert (Chapter 18)
introduces the subject of chronobiology, the study of biological
rhythms, and cautions all primate fieldworkers to be aware of the
implications of circadian rhythms, zeitgebers and masking effects
and our own anthropocentric measurement of time for studies of

physiology and behaviour. He also describes methods for measuring luminance and recording activity patterns in the field. Schmid (Chapter 19) introduces methods for the study of energy expenditure and body temperature in wild animals, discussing various techniques and the advantages and disadvantages of each. Hodges and Heistermann (Chapter 20) then discuss methods for monitoring hormonal changes in free-ranging primates using non-invasive samples (faeces and urine). More non-invasive techniques are described in Chapter 21 (Goossens *et al.*), which covers the collection, storage and analysis of non-invasive genetic material and includes advice for those able to obtain more traditional blood or tissue samples. Finally, in a more light-hearted style, Bearder and Nekaris (Chapter 22) close the book with an A–Z of tips to make fieldwork safer and more comfortable.

CONCLUDING COMMENTS

Although most research on primates relies almost entirely on observation, the methods that can be useful in the field are clearly limited only by the investigator's ingenuity. Cheney et al. (1987, p. 8)

This is a book about methods, which field and laboratory workers are continually developing and improving through practice, trial and error. Both primatology and primate conservation are dependent upon people who are prepared to go into the field, put in hard work and get dirty. Fieldwork can be demanding and exhausting, but it is also highly rewarding (see any fieldworker's diary). Although this book advocates the use of technology, it should be integrated with standard field techniques, and certainly not replace traditional behavioural and ecological data collection. Field conditions have not changed dramatically since the early days (Fig. 1) although our tools have: handheld computers, pencil and paper have replaced guns (Greif & Schmutz, 1995). It is still only by travelling on foot (like the early naturalists) that new populations, and even new species, are discovered. If this book proves useful to field and laboratory workers, acts to stimulate research and understanding of primates in their natural state, and through that increased knowledge can make some small contribution to primate conservation, then it will have achieved its aim.

ACKNOWLEDGEMENTS

The editors are indebted to the contributors to this book for their patience, enthusiasm and encouragement during the editorial process,

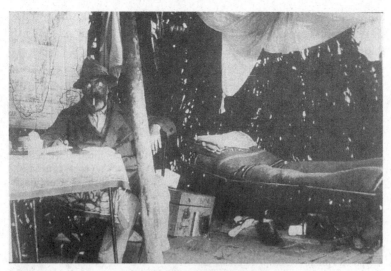

Fig. 1. Above: Hans Bluntschli in his field camp, Madagascar 1931–1932 (from the archives of the Anthropological Institute and Museum, University of Zürich). Below: Field camp at Anjamena, Madagascar 1994–1995 (Curtis & Zaramody, 1998).

Martin Griffiths at Cambridge University Press for suggesting a second edition, Rachel Eley, Lynette Talbot and Megan Waddington for guiding us through the process of publication, other primatologists and reviewers for their encouragement and feedback, Richard Thomas for redrawing illustrations, and Hans Schmutz for information on Hans

Bluntschli. DJC is grateful to Urs Thalmann, with whom she discussed a similar volume many years ago. JMS and MT thank their parents for their support.

REFERENCES

Altmann, J. (1974). Observational study of behavior: sampling methods. *Behaviour* 49: 227-67.

Altmann, J. (2010). Motherhood, methods and monkeys: an intertwined professional and personal life. In *Leaders in Animal Behavior: The Second Generation*, ed. L. Drickamer & D. Dewsbury, pp. 39-58. Cambridge: Cambridge University Press.

Ancrenaz, M., Lackman-Ancrenaz, I. & Mundy, N. (1994). Field observations of aye-aye (*Daubentonia madagascariensis*) in Madagascar. *Folia Primatol.* 62, 22-36.

Baron, L. (1882). Notes on the habits of the aye-aye of Madagascar in its native state. *Proc. R. Soc. Lond.* XLIII, 639-40.

Campbell, C. J., Fuentes, A., MacKinnon, K. C., Panger, M. & Bearder, S. K. (2007). *Primates in Perspective*. Oxford: Oxford University Press.

Cheney, D. L., Seyfarth, R. M., Smuts, B. B. & Wrangham, R. W. (1987). The study of primate societies. In *Primate Societies*, ed. B. B. Smuts, D. L. Cheney, R. M. Seyfarth, R. W. Wrangham & T. T. Struhsaker, pp. 1-8. Chicago and London: University of Chicago Press.

Curtis, D. J. & Zaramody, A. (1998). Group size, home range use and seasonal variation in the ecology of *Eulemur mongoz*. *Int. J. Primatol.* 19, 811-35.

Field, A. P. (2009). *Discovering Statistics using SPSS*. (3rd edn.) London: Sage Publications Ltd.

Fleagle, J. G. (1999). *Primate Adaptation and Evolution*. San Diego, CA: Academic Press.

Greif, R. & Schmutz, H.-K. (1995). Hans Bluntschli als Morphologe. *Gesnerus* 52, 133-157.

IUCN (International Union for Conservation of Nature) (2008). *IUCN Red List of Threatened Species*. www.redlist.org. Accessed 31 Jan 2010.

Lamberton, C. (1911). Contribution à l'étude des moeurs du Aye-aye. *Bull. Acad. Malg.* VIII, 129-40.

Lavauden, L. (1933). Le Aye-aye. *La Terre et la Vie Paris* 3, 77-9.

Owen, R. (1863). On the aye-aye. *Chiromys*, Cuvier. *Trans. Zool. Soc. Lond.* 68, 33-101.

Peters, W. (1865). Ueber die Säugethiergattung Chiromys. *Abh. Akad. Wiss. Berlin*, 79-100.

Petter, J. J. (1962). Recherches sur l'écologie et l'éthologie des Lémuriens malgaches. *Mém. Mus. Nat. Hist. Nat. Sér. A, Zoologie* 27, 115-20.

Pollen, F. (1863). Enumération des animaux vertébrés de l'île de Madagascar. *Nederl. Tijdschr. Dierkunde* 1, 277-345.

Schaller, G. B. (1965). Field procedures. In *Primate Behavior: Field Studies of Monkeys and Apes*, ed. I. De Vore, pp. 623-9. New York: Holt, Rinehart and Winston.

Shrum, M. (2008). Preliminary study of the lemur diversity in the Mahavavy-Kinkony Wetland Complex and Ankarafantsika National Park in northwest Madagascar: with emphasis on the reassessment of the status of the mongoose lemur (*Eulemur mongoz*) and its habitat. Unpublished MSc thesis, Oxford Brookes University.

Siegel, S. & Castellan, N. J. (1988). *Nonparametric Statistics for the Behavioural Sciences.* (2nd edn.) New York: McGraw-Hill.

Sterling, E. J. (1993). Behavioral ecology of the aye-aye (*Daubentonia madagascariensis*) on Nosy Mangabe, Madagascar. Doctoral thesis, University of Yale.

Sterling, E. J., Dierenfeld, E. S., Ashbourne, C. J. & Feistner, A. T. C. (1994). Dietary intake, food composition and nutrient intake in wild and captive populations of *Daubentonia madagascariensis. Folia Primatol.* 62, 115–24.

Strier, K. 2007. *Primate Behavioural Ecology.* (3rd edn.) Boston, MA: Allyn & Bacon.

Strum, S. C. & Fedigan, L. M. (2000). Changing views of primate society: a situated North American perspective. In *Primate Encounters: Models of Science, Gender and Society*, ed. S. C. Strum & L. M. Fedigan, pp. 3–49. Chicago, IL: University of Chicago Press.

Vinson, A. (1855). Observations sur un Aye-aye gardé deux mois à la Réunion. *C. R. Acad. Sci.* XLI, 640.

FURTHER READING AND USEFUL INTERNET SITES

www.rgs.org – for the Royal Geographical Society (U.K.)'s expedition planning handbook and other publications.

LISA JONES-ENGEL, GREGORY A. ENGEL
AND AGUSTIN FUENTES

1

An ethnoprimatological approach to interactions between human and non-human primates

AN ETHNOPRIMATOLOGICAL DIMENSION

Twenty years ago it was generally considered unthinkable (as well as unfundable and unpublishable) to carry out research on non-human primates (hereafter 'primates') that were in contact with human populations. Such contexts were seen as abnormal/aberrant situations, which distorted 'natural' primate behaviour and ecology. In the last decade this bias has gradually yielded to an appreciation that the human–primate interface is not only a legitimate area of research, it is an area of critical importance to both primate conservation and human health. In fact, as human encroachment on primate habitat squeezes the remaining populations of wild primates into ever shrinking areas, the human–primate interface increasingly represents the rule, rather than the exception, of interspecies interaction.

Ethnoprimatology refers to the study of the ecological and cultural interconnections between humans and primates. In a sense, all research on primates, from behavioural studies on free-ranging animals to research on laboratory-based primates, has an ethnoprimatological dimension. This is at least in part because our own cultures so completely colour every aspect of how and why we carry out our scientific inquiry. Additionally, culture, as it relates to the economic, social and political structures of society, profoundly influences how people living in habitat countries interact with and respond to primates.

Increasingly, ethnoprimatological approaches are informing research on cross-species infectious agent transmission (Jones-Engel *et al.*, 2001, 2005, 2006a, b; Schillaci *et al.*, 2008) and its impact on primate conservation (Travis *et al.*, 2006; Jones-Engel *et al.*, 2006b; Wallis & Lee, 1999; Beck *et al.*, 2001; Gillespie *et al.*, 2008; Köndgen

Field and Laboratory Methods in Primatology: A Practical Guide, ed. Joanna M. Setchell and Deborah J. Curtis. Published by Cambridge University Press. © Cambridge University Press 2011.

et al., 2008; Goldberg *et al.*, 2008; Chapman *et al.*, 2005), human health (Jones-Engel *et al.*, 2008; Engel *et al.*, 2002, 2006), the interplay between primate and human behaviour (Fuentes & Wolfe, 2002; Fuentes, 2007; Fuentes *et al.*, 2007; Cormier, 2003; Riley, 2006), and the conservation and management of urban primates (Sha *et al.*, 2009a; Southwick *et al.*, 2005; Zhao, 2005). In this chapter we discuss the conduct of primatological research from an ethnoprimatological perspective. Our recommendations are based largely on our experiences researching the transmission of infectious agents between humans and NHPs in multiple countries and contexts in Asia and in Gibraltar. We recognize that each research project presents unique challenges and obstacles. Here we have tried to set out some general considerations for those contemplating research on primates and the human communities with which they come into contact.

Humans and primates interact in complex ways. As a result, the study of interspecies interactions is often facilitated by a multidisciplinary approach. Our research on bi-directional disease transmission has benefited from the contributions of team members who possess expertise in a variety of fields, including primatology, anthropology, human medicine, veterinary medicine, epidemiology and virology. These diverse perspectives have proved helpful at all stages of the work, from planning to data collection and analysis. The multidisciplinary nature of our research means that at times we have found ourselves surrounded by specialists in other fields. In a world of increasing specialization, the field primatologist occasionally finds her or himself in a critical, though sometimes awkward, rôle as a generalist at the cross-roads of different fields of study. While cutting-edge research frequently requires the expertise of scientists with unique and intensive knowledge of small areas of study, it is often the primatologist, employing ethnoprimatological techniques, who is uniquely able to integrate these multiple sources of data and give them meaning and relevance within the context of real populations of humans and primates where they come into contact.

RESEARCH COLLABORATIONS

Successful research involving human and primate interactions in habitat countries will almost certainly rely heavily on a strong collaborative relationship between you, as the visiting researcher, and your local collaborator(s). This holds for researchers who are citizens of the habitat country or members of a different ethnic group within that country, as well as foreign researchers. Often these relationships are established through professional or academic contacts. However the initial contact is made, you should consider discussing all issues regarding the research with your local collaborators prior to departure for the field. These issues can include timing of the field research, acquisition of necessary permits, local transport, recruitment of field assistants, local arrangements and arranging access to research subjects. We strongly recommend that you agree on compensation and acknowledgement of local collaborators prior to the start of collaboration. Academic recognition (i.e. co-authorship on manuscripts) and the need for financial compensation often loom large in any working collaboration. Ideally, a visiting researcher going into the field should know beforehand what constitutes appropriate compensation. For foreign researchers, there are potential unforeseen consequences in compensating local collaborators on the same monetary scale that you would use at home: over-paying (by in-country standards) a collaborator may create animosities at his or her institution, or make it difficult for future visiting researchers to work in that area.

Often one of the greatest challenges for the researcher is understanding the motives and agendas of their collaborators or field assistants. Sometimes these agendas run contrary to the stated objectives of the research. Perhaps this is one of the most critical areas for ethnoprimatological understanding. There is no substitute for good communication and cultural fluency.

CULTURAL FLUENCY

You must be adequately familiar with local cultures and customs to conduct research effectively. Cultural fluency must extend beyond linguistic competence to include a basic understanding of the subtext of the culture in which you are working. You must not only be able to work with local collaborators and research subjects but also to interact effectively with local and governmental officials. This can be difficult to achieve, especially in multi-ethnic societies where the primary collaborator's ethnicity may differ from that of human research subjects. In such cases it is advisable to hire local field assistants or go-betweens who are of the ethnicity of the area where the research is being conducted and who are familiar with the local traditions, customs and power structure. Your primary local collaborator is often the best resource for identifying and hiring experienced field assistants.

DESIGNING SURVEYS

The literature on designing ethnographic surveys is extensive (Schensul *et al.*, 1999). Your area of interest will dictate the survey's content. It is important to design your questionnaire with specific hypotheses in mind. For our research on cross-species transmission of infectious agents (Jones-Engel *et al.*, 2008), we gather data on: the contexts in which people and primates come into contact (i.e. pet ownership, temple monkeys, urban primates, performing monkeys, bushmeat hunting); primate species; the types of exposure (i.e. bites, scratches, mucosal splashes) and sequelae of exposures as well as demographic and health information from our human subjects relating to prior NHP bites and scratches. In general, take care to make survey questions simple, direct and specific. Avoid leading questions. A frequently encountered problem is misinterpretation of questions by subjects owing to cultural attitudes or poorly designed survey questions. Some subjects try to guess what the 'right' answer is in order to please the interviewer. In other situations the subject's responses may be constrained by cultural or legal considerations. For example, subjects who are questioned about their bushmeat hunting practices may be less than forthcoming if bushmeat hunting is illegal. One way to circumvent this tendency is to design the survey so that the most sensitive questions are asked in several different ways. It is also important to consider the context of the interview. Responses may differ if, for example, subjects are interviewed in the presence of family members, peers, elders, etc. Have your surveys checked for cultural relevance by local collaborators before using them in the field. In many countries, survey instruments will need to be approved by the visiting researchers' human subjects review boards (see below).

Even when going into the field with a survey instrument that has been meticulously researched and prepared, at times you will be tempted to modify the survey. However, in many cases it will have been reviewed and approved by a committee at your home institution and adding questions that have not been reviewed may not be permitted. Occasionally, it becomes apparent that a question is not culturally relevant, or even causes embarrassment to subjects. In such a situation it is usually permissible to omit the question. The survey should be as short as possible yet still allow you to access the information necessary for your study.

The construction and implementation of an ethnographic survey is another area where cultural fluency is critical. It took three field seasons before our standard ethnographic survey reached its final streamlined form. Optimally, the primary researcher should be responsible for administering the survey, assuming that he or she is fluent in the subject's language. This is especially important in early versions of a survey as shades of meaning, inconsistencies, misunderstandings and concepts that simply do not cross cultural boundaries are worked out. Strongly consider a pilot survey to clarify meanings and work out 'bugs'. One of our early surveys required respondents to indicate whether they 'agreed strongly, agreed, didn't know, disagreed or disagreed strongly' with the statement: 'monkeys can make you sick'. This very Western notion of stratified answers to a question was met with bewilderment by most of the respondents.

If you intend to do a quantitative analysis of the survey data, we suggest using a program such as Epi Info or Sudaan, which are designed for epidemiology and public health survey research (see 'Useful Internet sites', below).

CONDUCTING INTERVIEWS

It is not uncommon for primate hunting, primate pet ownership or feeding primates (see Sha *et al.*, 2009b) to be illegal in habitat countries. Because of this, we often noticed that interviewees were hesitant to provide information about some of their practices. We found that by consulting community leaders we were able to improve the efficiency of our data gathering. Upon arriving in a community we sought out the leaders in the communities where we were working and explained our research to them. Their knowledge of the communities both saved us valuable time in locating research subjects and broke down barriers to communication.

When interviewing people about their practices/beliefs towards NHPs it is also important to remember that the way that you present yourself is important and must be culturally sensitive. Wear conservative, respectful and more formal clothes when you interview subjects.

Consider a small honorarium as thanks for the time that subjects spend answering survey questions. Gauge the amount on the typical daily wage of the subject for that given local economy. Monetary gifts are potentially problematic. Our experience has been that in some communities receiving money is awkward for recipients, who, although they desire such gifts, are at pains to avoid the external appearance of being paid by the researcher whom they regard in one sense as a guest. Our solution to this has been to offer honoraria and to comment that the money is being provided for the owner's children or for monkey food. This practice has been well received. In urban and/or more developed areas we found that having on hand a generous supply of pens, pencils, and stickers with the project logo were well received and appropriate compensation.

ETHICAL CONSIDERATIONS OF GIFT GIVING

The practice of giving gifts is rooted in the practical realities of doing field research in many habitat countries. Without such gifts it can be extremely difficult to gain access to people, primates and information. There is also the impulse on the part of the researcher to engender goodwill towards themselves and towards the research. This is especially true when you plan to return to the community. However, the practice of giving gifts is not without its ethical dilemmas.

The researcher who believes that he or she can enter a community and do research without in some way changing that community is deluded. As visitors, it is impossible for us to know all the myriad ways in which our research and very presence impacts these communities. Our approach is to maximize our positive impacts and attempt to mitigate the negatives. Ultimately, if we were not convinced that our research benefited the animals, environments and communities where we work we would not continue.

BIOLOGICAL DATA COLLECTION

Ethnoprimatological research may involve collecting biological samples from primates, humans, or both. Our sampling protocol has evolved with our hypotheses as well as with technology. Collecting, labelling, storing, cataloguing and organizing samples is not a trivial task. We strongly recommend developing a detailed, step-by-step algorithm for the entire process – from sample acquisition to data analysis. A database such as Access (Microsoft) has helped us to store, organize and quickly retrieve the myriad data that can be collected in a project that looks at the human–primate interface.

Detailed protocols for handling, anaesthetizing, sampling, and giving NHPs health examinations are presented elsewhere in this volume (Chapter 8). When sampling free-ranging primates we use trapping protocols similar to those outlined in Chapter 7. Over the past several years we have successfully designed, refined and used a large 2 m × 2 m × 2 m portable trap that allows us to capture multiple individuals at a time. This approach has certain advantages over darting or individual trapping because it allows us to collect samples from a larger proportion of the population we are studying.

Collecting samples from humans has several pitfalls, which you need to consider before submitting your research to your institution for approval (see below). In addition it is also important to consider the specific cultural beliefs of subjects regarding biological specimens (e.g. blood, faeces, hair). Not everyone is accustomed to and comfortable with the notion of having body fluids analysed for diagnostic and therapeutic purposes, particularly individuals living in remote communities. Once again, this is an area where cultural fluency can prevent the researcher from going astray.

Access to cold storage (−20 °C and −80 °C) is a challenge in many field settings, yet can be critical for thorough sample analysis. Recent advances in sample storage include: lightweight cold boxes (SCA Thermosafe® Deep Chill™ Insulated Shipper) that can hold samples at −20 °C using dry ice or even blue ice for several days. RNAlater solution (Ambion, Inc.) can preserve RNA from blood and faecal samples at room temperature for long periods. Protein saver cards (Whatman 903® Protein Saver Cards) are convenient, lightweight and compact means of storing whole blood and faeces for later analysis for a variety of assays (Chapter 8; Mlambo et al., 2008).

PERMITS

Research that includes both human and primate subjects typically means that you need to negotiate at least twice the number of governmental agencies at both federal and local levels. Never assume that permits granted at one level will automatically be valid at another level of bureaucracy. In countries where the provincial governments are geographically removed from the central governments, you should be prepared to pursue additional permits for your research. Much of the negotiation for these permits is often better handled by your local collaborators, as they are typically more familiar with their governmental agencies. We can not stress enough that the permitting process be initiated well in advance of the intended start date for the field research. This applies to both the process in the fieldwork country and any procedures that are required from your own institution or government. Many universities require that any research on animals, whether it be in the laboratory or in the field, be reviewed and approved by a committee established by the university. Additionally, research that involves humans typically must be reviewed and evaluated by the university's Human Research and Review Committee or equivalent body (e.g. Ethics Committee). The review processes for research involving humans and primates can be time-consuming and arduous: plan accordingly.

Research that involves the collection and transportation of biological samples from humans and/or primates may be subjected to international regulations regarding the import and export of such samples (in particular the Convention on International Trade in Endangered Species, www.cites.org). Familiarize yourself with local, national and international regulations before undertaking this type of data collection.

TIMING

When considering field studies on human–primate interactions, several factors must go into the planning including the season, climate, holidays, and the political and economic stability of the field site country. For example, in Bali, Indonesia, we discovered that although the months of May through August worked very well for our schedules, this time period corresponded with the height of the tourist season. We encountered local resistance to our research, with guides, shopkeepers and guards expressing concern that our sampling during the tourist season might negatively impact tourist revenues. Consultation with local administrative and governmental officials, in conjunction with informing local villagers about our research and the potential positive and negative consequences of our project, was essential to allay some of these concerns.

The timing of local holidays can have a significant impact on field research. In Muslim countries, Ramadan, the Muslim fasting month, can profoundly affect the ability of a researcher to move around the country and interact with people. Similarly, access to governmental agencies and the permits they provide can be severely restricted during national or religious holidays.

When considering scheduling of interviews with farmers, you need to consider the planting season. Harvest times typically require intense and extended efforts by the farmers. During these periods they may be less inclined to take the time to sit and talk or answer survey questions. Similarly, if you are interested in documenting crop-raiding behaviours by primates it is imperative that you identify certain seasons that are associated with crop-raiding. In Northern Sulawesi, hunting pressure on primates is most severe in December, because Christmas monkey is a preferred feast dish, among the Minahasa. When planning a field season you should consult with your local collaborators about expected local weather conditions. Interviewing and collecting samples during the rainy season can be extremely challenging, if not impossible.

TRANSPORTATION AND EQUIPMENT

The importation and transportation of research equipment to your research site(s) can be a logistical nightmare. If your research is equipment-intensive and involves international transport, you will need to make provisions for this. For example make arrangements with airlines well in advance of departure and make sure that any valuable capital items that are to be brought into the host country have the appropriate permits and declarations. Make arrangements with your local collaborators to meet them immediately upon your arrival in the country. No matter how thorough your permits and documents may be, it is not wise to spend a lot of time standing around in the airport or customs surrounded by a mound of expensive equipment.

If your work requires that you move about the country, you can either rely on local transportation, which is generally inexpensive, or hire personal transport. The latter option is always more expensive but often more convenient, especially when transporting a great deal of equipment or going to remote and difficult to access areas. Keep in mind, though, that arriving in a village in an expensive automobile may immediately set up expectations from the villagers.

CAVEATS: CROSS-SPECIES TRANSMISSION OF INFECTION

Researchers who come into direct or even indirect contact with NHPs should be cognizant of the potential for pathogen transmission both from humans to primates and from primates to humans. Recognizing the risk of interspecific disease transmission is even more crucial when collecting biological samples. It is the responsibility of researchers to avoid or contain the risk of exposure, both to themselves and to their human and primate subjects.

Here we would like to emphasize that meticulous attention to technique is essential for safe sample collection, and good technique is attained through training. You should be well versed in all sample collection techniques before venturing out into the field. Field conditions are often suboptimal and can compound the difficulty of collecting samples. Weather conditions and poor lighting are common obstacles. Crowds pose a particular difficulty. In our experience, the very presence of scientists attracts curiosity, and sedating and sampling animals often causes a sensation. It was not uncommon to have a group of 15–100 people crowding around us as we worked. Of course, spectators are more than just inconvenient; they pose an infectious disease threat to the animals. In circumstances such as these it is important to maintain your composure, and request some member of the community in a position of authority to control the crowd.

CONCLUSION

Ethnoprimatology's focus on the ecological, cultural, economic and social interface between humans and primates is a complex field of research. You need to factor flexibility into your schedule, accounting for unpredictability in locating and sampling primates, and for the cultural, ritual and occupational commitments of your human subjects. Additionally, attention to permitting and data acquisition and storage requires significant pre-planning and focus. The results can be a wealth of integrated data that reflects the complex way that humans and primates interact in a world where that interface is increasingly dense.

REFERENCES

Beck, B.B., Stoinski, T.S., Hutchins, M. *et al.* (2001). *Great Apes and Humans. The Ethics of Coexistence.* Washington, D.C.: Smithsonian Institution Press.

Chapman, C., Gillespie, T. & Goldberg, T. (2005). Primates and the ecology of their infectious diseases: How will anthropogenic change affect host-parasite interactions? *Evol. Anthropol.* 14, 134–44.

Cormier, L.A. (2003). *Kinship with Monkeys: The Guaja Foragers of Eastern Amazonia.* New York: Columbia University Press.

Engel, G.A., Jones-Engel, L., Schillaci, M.A. *et al.* (2002). Human exposure to herpes virus B-seropositive macaques, Bali, Indonesia. *Emerg. Infect. Dis.* 8, 789–95.

Engel G., Hungerford, L., Jones-Engel, L. *et al.* (2006). Risk assessment: a model for predicting cross-species transmission of SFV from macaques (*M. fascicularis*) to humans at a monkey temple in Bali, Indonesia. *Am. J. Primatol.* 68, 934–48.

Fuentes A. (2007). Monkey and human interconnections: the wild, the captive, and the in-between. In *Where the Wild Things Are Now: Domestication Reconsidered*, ed. R. Cassidy & M. Mullin, pp. 123–45. New York: Berg.

Fuentes, A. & Wolfe, L. D. (2002). *Primates: Face to Face*. Cambridge: Cambridge University Press.

Fuentes, A., Shaw, E. & Cortes, J. (2007). Qualitative assessment of macaque tourist sites in Padangtegal, Bali, Indonesia, and the Upper Rock Nature Reserve, Gibraltar. *Int. J. Primatol.* 28, 1143–58.

Gillespie, T. R., Nunn, C. L. & Leendertz, F. H. (2008). Integrative approaches to the study of primate infectious disease: implications for biodiversity conservation and global health. *Am. J. Phys. Anthropol. Suppl.* 47, 53–69.

Goldberg, T. L., Gillespie, T. R., Rwego, I. B., Estoff, E. L. & Chapman, C. A. (2008). Forest fragmentation as cause of bacterial transmission among nonhuman primates, humans, and livestock, Uganda. *Emerg. Infect. Dis.* 14, 1375–82.

Jones-Engel, L., Engel, G. A., Schillaci, M. A., Babo, R. & Froehlich, J. (2001). Detection of antibodies to selected human pathogens among wild and pet macaques (*Macaca tonkeana*) in Sulawesi, Indonesia. *Am. J. Primatol.* 54, 171–8.

Jones-Engel, L., Schillaci, M. A., Engel, G. A., Paputungan, U. & Froehlich, J. W. (2005). Characterizing primate pet ownership in Sulawesi: implications for disease transmission. In *Commensalism and Conflict: The Primate-Human Interface*, ed. J. Patterson & J. Wallis, pp. 197–221. Norman, OK: American Society of Primatology.

Jones-Engel, L., Engel, G. A., Heidrich, J. *et al.* (2006a). Temple monkeys and health implications of commensalism, Kathmandu, Nepal. *Emerg. Infect. Dis.* 12, 900–6.

Jones-Engel, L. E., Engel, G. A., Schillaci, M. A. *et al.* (2006b). Considering human to primate transmission of measles virus through the prism of risk analysis. *Am. J. Primatol.* 68, 868–79.

Jones-Engel, L., May, C., Engel, G. *et al.* (2008). Diverse contexts of zoonotic transmission of simian foamy viruses in Asia. *Emerg. Infec. Dis.* 14, 1200–8.

Köndgen, S., Kühl, H., N'Goran, P. K. *et al.* (2008). Pandemic human viruses cause decline of endangered great apes. *Curr. Biol.* 18, 260–4.

Mlambo, G., Vasquez, Y., LeBlanc, R., Sullivan, D., Kumar, N. (2008). A filter paper method for the detection of *Plasmodium falciparum* gametocytes by reverse transcription-polymerase chain reaction. *Am. J. Trop. Med. Hyg.* 78, 114–16.

Riley, E. P. (2006). Ethnoprimatology: towards reconciliation of biological and cultural anthropology. *Ecol. Environ. Anthropol.* 2, 1–10.

Schensul J. J., Le Compte, M. D. &, Schensul, S. (1999). *The Ethnographer's Toolkit*. 7 volumes. Walnut Creek, CA: AltaMira Press.

Schillaci, M. A., Jones-Engel, L., Engel, G. & Fuentes, A. (2008). Characterizing the threat to the blood supply associated with nonoccupational exposure to emerging simian retroviruses. *Transfusion* 48, 398–401.

Sha, J. C., Gumert, M., Lee, B. *et al.* (2009a). Status of the long-tailed macaque (*Macaca fascicularis*) in Singapore and implications for management. *Biodivers. Conserv.* 18, 2909–26.

Sha, J. C., Fuentes, A., Gumert, M. *et al.* (2009b). Macaque–human interactions and the societal perceptions of macaques. *Am. J. Primatol.* 71, 1–19.

Southwick, C. H., Malik, I. & Siddiqi, M. F. (2005). Rhesus commensalisms in India: problems and prospects. In *Commensalism and Conflict: The Human-Primate Interface*, ed. J. D. Paterson & J. Wallis, pp. 240–57. Norman, OK: American Society of Primatologists.

Travis, D. A., Hungerford, L., Engel, G. A. & Jones-Engel, L. (2006). Disease risk analysis: a tool for primate conservation planning and decision making. *Am. J. Primatol.* 68, 855–67.

Wallis, J. & Lee, D. R. (1999). Primate conservation: the prevention of disease transmission. *Int. J. Primatol.* 20, 803–26.

Zhao, Q. K. (2005). Tibetan macaques, visitors and local people at Mt. Emei: problems and countermeasures. In *Commensalism and Conflict: The Human-Primate Interface*, ed. J. D. Paterson & J. Wallis, pp. 376–99. Norman, OK: American Society of Primatologists.

USEFUL INTERNET SITES.

http://www.cdc.gov/epiinfo/ – for Epi Info (shareware).
http://www.rti.org/SUDAAN/ – to order Sudaan software.

ELIZABETH A. WILLIAMSON AND
ANNA T. C. FEISTNER

2

Habituating primates: processes, techniques, variables and ethics

INTRODUCTION

Field biologists adopted the term habituation from physiology, as the relatively persistent waning of a response as a result of repeated stimulation that is not followed by any kind of reinforcement (Thorpe, 1963). Repeated neutral contacts between primates and humans can lead to a reduction in fear, and ultimately to the ignoring of an observer. Historically, the techniques and processes involved were rarely described, as habituation was generally viewed as a means to an end (Tutin & Fernandez, 1991). As we become increasingly aware of the potential effects of observer presence on primate behaviour, and especially the potential risks of close proximity with humans, it behoves us to measure as much about the habituation process as possible. However, most recent studies that have quantified primate behaviour in relation to habituators have focussed on great apes (see, for example, Ando et al., 2008; Bertolani & Boesch, 2008; Blom et al., 2004; Cipolletta, 2003; Doran-Sheehy et al., 2007; Sommer et al., 2004; Werdenich et al., 2003), with little information available for other primate taxa (but see Jack et al., 2008).

There are limits to what studies of unhabituated primates can achieve: it is difficult to observe at close range, so subtle or cryptic behaviour such as facial expressions and soft vocalizations may be missed, and even individual identification may be difficult, resulting in analyses based only on age–sex classes. Primates disturbed by the presence of observers will show altered patterns of behaviour and it may not be possible to follow groups on the move. Habituation enables you to approach closely, because subjects no longer flee and do not respond overtly to observers, allowing you to become familiar with individuals, and to observe fine-level behaviours such as subtle social

Field and Laboratory Methods in Primatology: A Practical Guide, ed. Joanna M. Setchell and Deborah J. Curtis. Published by Cambridge University Press. © Cambridge University Press 2011.

interactions or food processing. You can then sample behaviour consistently (Chapter 11). However, there are costs to habituation, borne largely by the animals themselves, including increased risks of disease (Köndgen *et al.*, 2008; Chapters 1 and 8, this volume), changes in behavioural ecology and increased vulnerability to poaching (see 'Ethical issues' below).

In this chapter we examine habituation methods, factors affecting the success of habituation, and associated ethical issues. Many behavioural responses are taxon-specific, and these should be taken into account when habituating human-naïve wild primates. We have first-hand experience with a range of primates, from marmosets (*Callithrix* spp.) to gorillas (*Gorilla* spp.), which we have used to make this chapter as broadly applicable as possible.

METHODS

Preparation

Knowing your primate

To prepare for a study involving primates, first read as much about your target species and its environment as possible to see what others have achieved. Familiarize yourself with your species' basic behavioural repertoire. Second, try to get exposure to the species itself, perhaps by visiting a zoo to observe the animals directly. The more naturalistic the physical and social environment of the captive primate, the better its potential as a model for wild conspecifics. Spending time observing captive primates will increase your familiarity with their basic locomotor patterns and postures, facial expressions and other behaviours, including foraging and food processing, grooming, play, reproductive and agonistic interactions. Pay particular attention to vocalizations, since these may help you to locate wild primates and to interpret behaviour in the absence of clear observation. Concentrate on alarm and display behaviours too, as these are likely to be the initial responses to human presence. Observing captive primates will also provide an opportunity to learn to distinguish individuals and will give you a head start in recognizing primates in the wild.

Equipment

In the early stages, you need little equipment: binoculars, a notebook, a compass, maps and a GPS (Chapter 4). Use binoculars and cameras with care, as pointing a big 'eye' at primates can make them uneasy.

Using a trail system

Trails can facilitate movement within the study area and make it easier to follow primate groups, but this depends greatly on vegetation type and even more on the area to be covered, which will be determined by home-range size. For example, researchers studying chimpanzees (*Pan troglodytes verus*) at Taï, Ivory Coast, worked alone, and did not cut transects or trails but learnt to orient themselves in the forest with a compass. We took the same approach with lowland gorillas (*Gorilla gorilla gorilla*) at Lopé, Gabon, where an additional factor was fast-growing vines in the family Marantaceae that would have required labour-intensive trail maintenance. Learning the layout of the forest, using features such as streams and vegetation types, without trail cutting keeps disturbance to a minimum. In African forests with high elephant (*Loxodonta africana*) density, you can also use established networks of elephant trails.

Some researchers studying monkeys with home ranges smaller than those of great apes have found a trail system to be invaluable. Kaplin judged a trail system crucial to keeping up with blue monkeys (*Cercopithecus mitis*) and therefore to habituating them, and that 'building it as we went was key. It was created to facilitate animal follows, and improved as we came to know the animals and began to predict where we could find them. The trail system allowed us to move more quickly through the forest, and to open access to places where the monkeys had travelled previously' (B. Kaplin, personal communication, 2002). Other researchers have established trails on a grid system (e.g. Bezamahafaly, Madagascar) or worked in previously logged areas where an old trail or grid system existed (e.g. Kirindy, Madagascar) or at sites where oil or mineral prospection has left transects (e.g. Menabe, Madagascar).

Finding your primate

Frequent contacts with the same individuals are necessary to achieve habituation, so locating a known group or individual daily is important. Some researchers have trapped and radio-tagged primates as an aid to locating them (Chapters 7 and 10, this volume). This option must be carefully considered for several reasons, including ethics. In addition, depending on the species and procedures used, catching and marking may actually set back habituation (Sterling, 1993). Assuming that you don't use catching and marking, you need to locate primates by other means. Searching at random in the hope of encountering your primate is unlikely to be a successful strategy, especially when animals are in small and cryptic groups or occur at low densities (for example, lowland gorillas usually occur at about one group per 10 km^2). Prior knowledge about the behavioural ecology of your species is extremely helpful, but when this is not possible, some practical decisions and a number of clues may help in locating primates and ensuring regular contact.

Auditory clues

Often the first indication of primate presence will be vocalizations such as long-calls (e.g. chimpanzees, *Pan troglodytes*; gibbons, *Hylobates* spp.; howler monkeys, *Alouatta* spp.; indri, *Indri indri*), alarm calls or warning barks. Taking a compass bearing and heading in that direction should facilitate your approach. Alternatively, sounds of movement in the vegetation as animals travel may alert you to their presence. Groups of arboreal guenons and colobines can be quite noisy as branches and leaves move with their passage; species such as *Colobus* and *Propithecus* make a thump as they land on larger branches and trunks; or you may hear sounds of food processing (including nut cracking) or plant parts dropping to the ground. You can locate chimpanzees from their drumming. Indirect evidence of primate presence includes nests (great apes), feeding signs, dung and scent- or gouge-marks.

Waiting at key sites

An alternative to searching for your primates is to wait for them to come to you by identifying key sites where it may be possible to encounter them regularly. Being at an animal's sleeping site before it wakes means that you can locate your animal early in the day and become part of the surroundings it sees on waking up. Most primates sleep above ground, so aim to be close to nests, sleeping cliffs, sleeping trees or tree holes before dawn (before dusk for nocturnal primates). Predictable food resources, such as figs, salt licks, water sources and seasonally available foods with limited distribution, are also good places to locate primates.

It is possible to establish key sites through provisioning, which is the use of artificial feeding as a positive incentive to tolerate human presence. If you provide food at limited sites, animals will be attracted to those sites and may be easier to observe and habituate than trying to locate them in the forest. However, provisioning has several disadvantages (Chapter 11), including the modification of natural behaviour. For example, changes in chimpanzee activity budgets, aggression and territoriality were recorded at Gombe and Mahale, Tanzania (Wrangham, 1974). Provisioning can also affect non-target species: baboons (*Papio cynocephalus*) have been habituated as a side-effect of provisioning chimpanzees (A. Collins, personal communication, 2002).

Approaching your animals

How contact is established is one of the most important elements of successful habituation, and various factors should be taken into account. Above all, avoid surprise; sudden contacts that frighten the animals will always have a negative impact. Be seen as often as possible and always in a calm, relaxed posture. Initially, it is desirable to be clearly visible at a distance greater than that which invokes alarm and flight. Obviously the practicalities will depend on habitat variables (discussed below).

Habituation can only be achieved when the primate sees you, but try to choose the moment. Once you make your presence known, the animals are likely to leave; you will need to sacrifice observation to achieve progress. Many primates can identify human faces, so the same observers should contact the animals, at least during the initial stages of habituation. It is also helpful to be consistent in your appearance (wear the same clothing, hat, rucksack, etc.). Behave calmly and attempt to 'reassure' animals by remaining still and/or mimicking natural behaviours, such as grooming or feeding. A crouching or sitting position was less intimidating to bonobos (*Pan paniscus*) than a walking observer (Krunkelsven *et al.*, 1999).

Signalling your presence

It is often useful to adopt a signal that communicates your presence and which the primates learn to associate with your approach. This signal should be a specific noise, which becomes identified with a non-threatening presence. 'Belch' or 'tongue-clacking' vocalizations are used with gorillas.

Keeping your distance

It is difficult to make generalizations, but err on the side of caution, while trying to maintain good visibility between primate and observer. Appropriate distances will depend on the environment, especially vegetation density, visibility, species concerned and risks of disease transmission. For example, the regulation minimum distance for observing mountain gorillas is 7 m (Homsy, 1999).

What not to do

Avoid making loud noises, sudden gestures, or surreptitious movements, although tolerance will depend on species and habitat. Primates do not like to be crept up on or followed, so avoid hiding and help them to keep track of your location by vocalizing. Many species respond better to arriving upon an observer. Moreover, 'pushing' primates ahead of you by following them may cause them to incur unnecessary energetic costs. Avoid pointing with telephoto lenses, sticks or guns.

Primates are also sensitive to the number of people present, so it is better to habituate with only one or two observers, preferably the same people. Even habituated primates may alter their behaviour with more observers: stumptail macaques (*Macaca arctoides*) reacted differently to a team of observers than to the presence of one or two (Rasmussen, 1991), Sulawesi crested black macaques (*Macaca nigra*) were more likely to flee or climb trees when visited by larger groups of people (Kinnaird & O'Brien, 1996) and mongoose lemurs (*Eulemur mongoz*) exhibited signs of agitation if unknown observers were present or more than one or two people followed them (D. Curtis, personal communication, 2002).

Recording data

Systematic records are vital for assessing progress towards habituation, so we recommend recording the following: (1) time at which search started, (2) time animals were located, (3) how animals were located, (4) observer–primate distance (often useful to measure perpendicular distance and height, e.g. guenons at 40 m and at 15–20 m height), (5) animals' activity when contacted, (6) animals' behaviour in response to observer, (7) time at end of contact, and (8) how contact ended. As a minimum, collect cumulative contact time (after your presence has been detected), observation time and location of subjects.

Table 2.1 *Typical reactions of primates to observer presence during the habituation process*

Behaviour	Definition
Flight	Rapid, often noisy, panicked departure coupled with alarm or fear vocalizations, but no display
Avoidance	Groups are relatively calm, silent, and disappear quickly without displaying; they 'melt' into the forest
Curiosity	Responses range from brief monitoring (surveillance), to moving to acquire a clearer view of an observer, to approaching the observer
Display	Vocalizations and species-typical displays (e.g. chest-beats, branch bouncing, yawns) are directed at the observer
Ignore	Animal is aware of the observer's presence but shows no obvious response

Habituation can only occur when the same individuals are contacted on a regular basis, so it will be important to know whether you are seeing the same individuals. In addition to sex and age class, try to record shape of face, ears and nose pattern, pelage pattern and any scars, notches in ears, missing fingers or toes, broken canines, bent/broken tails.

Changes in behaviour during habituation provide valuable information for quantifying the process and measuring the changing impact of observer presence. Choose simple categories of behaviour that can be easily quantified (Table 2.1).

Knowing when your primate is habituated

Ideally your subjects will be aware of your presence yet learn to ignore you. During habituation, both primate and observer behaviour is likely to change. As your familiarity with the species and its habitat increases, your ability to move around the study area, pick up clues to primate location, and identify individuals should improve, reducing search time. Moreover, as the primates become habituated to your presence, there should be an increase in the duration of average contact time, and decreases in flight, avoidance and display behaviours. Be aware that daily path length or nightly travel distance may increase during initial stages of habituation as animals flee, but travel decreases as habituation progresses (Geoffroy's tamarins, *Saguinus geoffroyi*,

Rasmussen, 1998; lowland gorillas, Cipolletta, 2003; spectral tarsiers, *Tarsius spectrum*, S. Gursky-Doyen, personal communication, 2009). The ratio of habituation time to observation time should change, but be patient and do not have high expectations: mean contact time with lowland gorillas increased from about 7 min to only 20 min over three years (Blom *et al.*, 2004). 'Hand-over' periods may be helpful if different people will be involved in habituation, although these should be avoided during the early phases. Rasmussen (1991) reported that observers who were at ease and not intimidated by stumptail macaques evoked a different response than neophyte observers. It's worth sounding a note of caution here: can we ever assume that a study animal behaves as it would if no observer was present? Rasmussen (1991) also found that stumptail macaques adjusted their travel in the presence of observers even after 14 years of almost daily contact, while Jack *et al.* (2008) pointed out that white-faced capuchin monkeys (*Cebus capucinus*) still reacted to observers after 20 years of research and suggested that 'we can never become a truly neutral presence to our study animals' (p. 494).

Habituation is an ongoing process, especially as group composition changes with immigration, emigration, births and deaths. The time taken to habituate depends on a range of factors, which may be taxon-specific, influenced by the environment, or linked to previous experience of humans. A rough guide to habituation time is presented in Table 2.2. Most nocturnal primates (bushbabies, *Galago* spp.; sportive lemurs, *Lepilemur* spp.; woolly lemurs, *Avahi* spp.; and dwarf lemurs, *Cheirogaleus* spp.; but not aye ayes or tarsiers) appear to need no habituating at all, they simply continue their activity after noticing an observer. Diurnal and cathemeral lemurs that have not been hunted and are accustomed to humans passing through the forest can be habituated in less than a month (Andrews & Birkinshaw, 1998; Curtis, 1998). It took about three months to be able to follow arboreal mangabeys (*Cercocebus albigena*) without disturbing them or causing them to flee, but six months before there was no noticeable effect of observer presence on their behaviour (R. Kormos, personal communication, 2002). Similarly, it took three months before a group of patas monkeys (*Erythrocebus patas*) could be followed all day, and six months before they could be reliably observed from 50 m (Chism & Rowell, 1988). Hamadryas baboons (*Papio hamadryas*) could be approached to 60 m within a year, but to be able to walk among them took about two years (Kummer, 1995). Only after a year of repeated contacts did lowland gorillas' reactions change from

Table 2.2 *Approximate times to habituate non-provisioned primates that have not been hunted*

Time	Primate taxa	Source
<1 hour –	Bushbabies (*Galago* spp.)	S. Bearder, pers. comm., 2002
<1 week	Most nocturnal lemurs	A. Mueller, pers. comm., 2002; U. Thalmann, pers. comm., 2002
~1 month	Most diurnal and cathemeral lemurs	A. Feistner, pers. obs.
	Orangutan (*Pongo pygmaeus*)	Rodman, 1979
	Spectral tarsiers (*Tarsius spectrum*)	S. Gursky-Doyen, pers. comm., 2009
2–5 months	Aye aye (*Daubentonia madagascariensis*)	Sterling, 1993
	Potto (*Perodicticus potto*)	E. Pimley, pers. comm., 2002
	Marmosets (*Callithrix* spp.) and tamarins (*Saguinus* spp.)	Passamani, 1998; Rasmussen, 1998; Rylands, 1986
	Chacma and olive baboons (*Papio ursinus, P. anubis*)	Barton & Whiten, 1993; Cowlishaw, 1997; Y. Warren, pers. comm., 2002
3–8 months	Colobus (*Colobus satanus*)	M. Harrison, pers. comm., 2002
	Guenons (*Cercopithecus l'hoesti, C. mitis*)	B. Kaplin, pers. comm., 2002
	Mangabey (*Cercocebus albigena*)	R. Kormos, pers. comm., 2002
	Muriqui (*Brachyteles arachnoides*)	Strier, 1999
6+ months	Patas (*Erythrocebus patas*)	Chism & Rowell, 1988
	Siamang (*Hylobates syndactylus*)	Chivers, 1974
	Yellow baboon (*Papio cynocephalus*)	Rasmussen, 1979
1–2 years	Hamadryas baboon (*Papio hamadryas*)	Kummer, 1995
	Mountain gorilla (*Gorilla beringei*)	Butynski, 2001; Schaller, 1963
2–5 years	Bonobo (*Pan paniscus*)	Susman, 1984
5–15 years	Chimpanzee (*Pan troglodytes*)	Bertolani & Boesch, 2008
	Lowland gorilla (*Gorilla gorilla*)	Blom *et al.*, 2004; Doran-Sheehy *et al.*, 2007

aggression to ignore (Blom *et al.*, 2004), yet titi monkeys (*Callicebus personatus personatus*) were habituated without systematic effort in about 12 weeks (E. Price, personal communication, 2002).

FACTORS AFFECTING PRIMATE HABITUATION

Many factors influence the success of habituation: some can be determined by the observer, such as contact distance and observer behaviour, while there are others which you cannot control, such as species-specific reactions, habitat variables and previous experience with humans. Many of these factors interact; here we attempt to evaluate some of them.

Species-specific factors

Species differences

In general, opportunistic species, such as macaques (*Macaca* spp.) and baboons (*Papio* spp.), are relatively easy to habituate because they are extrovert and adapt readily to changing circumstances. Moreover, if one macaque or baboon approaches with no negative consequences, others may learn from watching that 'non'-interaction. Primates that live in stable groups also tend to be easy because when you encounter one individual you have usually found the whole group. In contrast, chimpanzees travel in small groups and their fission–fusion society means that if you follow two or three one day you may encounter a different subgroup next day.

Diet is an important factor. Arboreal folivores such as colobus tend to be easier to habituate than arboreal frugivores such as guenons. This is likely to be related to the former having smaller home ranges, and their food being available in larger patches so that more individuals can feed together and increased time is spent resting.

Home-range size is also important: primates with large home ranges are more difficult to encounter consistently, partly explaining differences between mountain and lowland gorillas, and perhaps the difficulty of habituating mandrills (*Mandrillus sphinx*) and drills (*M. leucophaeus*).

Even when sympatric, species vary in their ease of habituation. For example, white-throated capuchins (*Cebus capuchinus*) have habituated to people, whereas squirrel monkeys (*Saimiri oerstedi*) in the same area of Costa Rica have not (Boinski & Sirot, 1997). In addition, monkeys that sometimes occur in polyspecific groups may react differently depending on the circumstances: habituated mangabeys fled without looking when unhabituated guenons alarm called; but were much bolder in polyspecific groups (R. Kormos, personal communication, 2002).

Finally, evolutionary history can influence ease of habituation. It may be that Madagascan strepsirrhines are remarkably easy to habituate as a result of the reduced predator assemblage in Madagascar, combined with the very recent arrival of humans, some 2000 years ago.

Sex and age differences

Responses also differ according to primate sex and age. In multi-male multi-female or one-male multi-female groups, adult males tend to be larger, more aggressive, and play patrolling or sentinel rôles in groups. Thus they are more overt and often more exposed to observer contact than females carrying or protecting young, while juveniles generally respond with more curiosity than adults. Bertolani and Boesch (2008) found that male chimpanzees habituate faster than females, and cycling females faster than non-cycling females; however, there were strong individual differences with some females not fully habituated after 15 years.

Habitat factors: visibility

A clear view of the observer is one of the most important factors in habituation. Primates living in more open habitats (such as baboons) are easier to habituate than those living in dense forest (such as mandrills), both within and between taxa. In open habitats, such as savanna, you can gradually move closer as animals see you from a distance. However, some open habitats still have poor visibility; patas monkeys, for example, are notoriously difficult to habituate (Chism & Rowell, 1988).

In the Virunga volcanoes, low vegetation and uneven topography provide ideal conditions to observe mountain gorillas (*Gorilla beringei beringei*) from the opposite side of a ravine. In contrast, visibility in lowland forest is poor, and lowland gorillas are usually obscured even within 10 m of an observer, unless they climb trees. Sudden contacts are difficult to avoid in dense forest, and probably hinder habituation by frightening the animals.

Issues of visibility may explain differences between terrestrial and arboreal forest primates. Arboreal guenons are generally easier to habituate than terrestrial ones. Terrestrial sun-tailed guenons (*Cercopithecus solatus*) seem almost impossible to habituate owing to their large home ranges and cryptic habits. L'Hoest's guenons (*Cercopithecus l'hoesti*) are wary when followed on the ground, in contrast to blue monkeys, which seem less concerned by observer presence when high in the trees (B. Kaplin, personal communication, 2002). The semi-terrestrial pigtail macaques (*Macaca nemestrina*) are more difficult to habituate than arboreal long-tailed macaques (*Macaca fascicularis*).

ETHICAL ISSUES

Risks to humans

During habituation, aggression initially increases, peaks and then diminishes (Blom *et al.*, 2004); thus, in the early stages, habituators may be subject to intimidating displays. These are more likely to be unpleasant than dangerous; however, people have been bitten while habituating gorillas.

Over-habituation also poses risks. A primary aim of habituation is for the observer to become a neutral element in the environment. However, primates are highly intelligent, complex, socially manipulative animals, and over-familiarity with the observer may change the observer from being a 'piece of the furniture' to a social tool, available for inclusion in their social relations. Being 'used' by your study animals is rarely mentioned in the literature, although it probably happens regularly. Mountain gorillas sometimes redirect aggression towards a more vulnerable observer rather than a conspecific, and baboons and macaques use observers in displays or to avoid aggression from others. Very habituated lemurs also occasionally redirect aggression at observers (J. Razafimahaimodison, personal communication, 2009). Even seemingly harmless interactions such as a juvenile approaching an observer can quickly lead to incite-screaming, leaving the observer in trouble.

Finally, loss of fear of humans can lead to negative interactions with local people: baboons and chimpanzees attack and steal from villagers; habituated mountain gorillas feed in fields outside the national park and have attacked people in fields. Over-habituation and/or poorly managed contacts pose severe problems for local human populations, ecotourism programmes and primate conservation, to the extent that Human–Wildlife Conflict has become a science (see, for example, Hockings & Humle, 2009).

Risks to primates

You should seriously evaluate whether or not to attempt habituation in terms of the long-term effects on normal activity, behaviour and vulnerability.

Disease transmission between humans and non-human primates

Close proximity with researchers increases the risks of disease transmission (Chapters 1 and 8), particularly for terrestrial primates. Certain human pathogens, both respiratory (measles, herpes, pneumonia) and enteric (polio, salmonella), can infect apes; deaths have occurred in wild populations (Butynski, 2001). Studies of captive apes show that they have a definite susceptibility to human diseases, but not the same resistance as humans, and illnesses to which animals have never been exposed are potentially the most dangerous (Homsy, 1999). Commonsense observer behaviours can reduce the risks: avoiding physical contact, careful management of waste products, avoiding excretion in the forest and not working if feeling unwell. Quarantine periods for field staff and minimum distance regulations should be in place for all habituated primates.

Stress

Stress provoked during habituation could potentially reduce reproductive success or result in immunosuppression and increased susceptibility to disease (Woodford et al., 2002). Jack et al. (2008) documented decreasing cortisol levels in capuchin monkeys during habituation, and measuring cortisol non-invasively from faecal samples could be useful for monitoring habituation over time (Nizeyi, 2005).

Generalization to other humans

Since habituation is basically the loss of fear of humans, we make animals more easily approachable by hunters and poachers and render them extremely vulnerable in dangerous situations, as evidenced by the deliberate killing of mountain gorillas in the Democratic Republic of Congo (British Broadcasting Corporation, 2007). Thus, you should seriously consider the future of your study population, and your ability to sustain protection, before undertaking habituation.

Other impacts on primate behaviour

The presence of an observer can have direct or indirect effects on primate behaviour by changing the target species' interactions with its conspecifics, predators, prey and other species in its environment. Activity budgets may change, as feeding and other activities are disrupted while the animals move away from habituators, and groups fleeing to avoid human contact will expend more energy. For example, 'chimpanzees sit down and wait for us to catch up if we're slow or disappear into deep thickets to lose us' (J. Setchell, personal communication, 2001). Regular contact with humans may also hinder reproduction, alter inter-group dynamics and impede transfer between groups. Habituated and unhabituated primates respond differently to people, so a human presence may well alter the nature of interactions between neighbouring groups, unintentionally benefiting a study group. For example, a naïve group may flee, enabling the habituated group to take over a food resource (see, for example, Rasmussen, 1991). Finally, carnivores (hyenas, *Hyaena hyaena*; lions, *Panthera leo*; leopards, *Panthera pardus*; ocelots, *Leopardas pardalis*; and fossa, *Cryptoprocta ferox*) are unlikely to approach when a human is present, so habituated groups may inadvertently be protected from predation (see, for example, Isbell & Young, 1993; Rasmussen, 1979).

CONCLUSIONS

In this chapter we have explored factors affecting the habituation of primates to human observers. The key to success is persistent, regular and frequent neutral contact with the same individuals. Habituation usually requires considerable time investment, from months to years, depending on species and environment (with Madagascan strepsirrhines the exception). In addition, habituated animals are more vulnerable to human predators, so when we habituate primates, we must accept responsibility for their protection for the remainder of their lives (Goldsmith, 2005). We suggest that anyone intending to habituate primates should (1) think carefully about the pros and cons, especially the potential risks and need for long-term commitment, (2) familiarize themselves with their study species prior to going to the field,

(3) engage and use their common sense, and (4) record, measure and evaluate the habituation process. Habituation is usually hard work but the rewards, both personal and academic, are great: being in the company of wild primates that have accepted you into their environment and allow you an insight into their daily lives is a privilege.

ACKNOWLEDGEMENTS

We thank Jim Anderson, Anthony Collins, Sharon Gursky-Doyen, Rebecca Kormos, Mike Harrison, Beth Kaplin, Eluned Price, Urs Thalmann and Ymke Warren for sharing their experiences with us.

REFERENCES

Ando, C., Iwata, Y. & Yamagiwa, J. (2008). Progress of habituation of western lowland gorillas and their reaction to observers in Moukalaba-Doudou National park, Gabon. *Afr. Study Monographs* Suppl. 39, 55–69.

Andrews, J. R. & Birkinshaw, C. R. (1998). A comparison between daytime and night time diet, activity and feeding height of the black lemur, *Eulemur macaco* (Primates: Lemuridae), in Lokobe Forest, Madagascar. *Folia Primatol.* 69, 175–82.

Barton, R. A. & Whiten, A. (1993). Feeding competition among female olive baboons, *Papio anubis. Anim. Behav.* 46, 777–89.

Bertolani, P. & Boesch, C. (2008). Habituation of wild chimpanzees (*Pan troglodytes*) of the south group at Taï Forest Côte d'Ivoire: empirical measure of progress. *Folia Primatol.* 79, 162–71.

Blom, A., Cipolletta, C., Brunsting, A. R. H. & Prins, H. H. T. (2004). Behavioural responses of gorillas to habituation in the Dzanga-Ndoki National Park, Central African Republic. *Int. J. Primatol.* 25, 179–96.

Boinski, S. & Sirot, L. (1997). Uncertain conservation status of squirrel monkeys in Costa Rica, *Saimiri oerstedi oerstedi* and *Saimiri oerstedi citrinellus. Folia Primatol.* 68, 181–93.

British Broadcasting Corporation (2007). Concern over mountain gorilla 'executions'. http://news.bbc.co.uk/2/hi/science/nature/6918012.stm

Butynski, T. M. (2001). Africa's great apes. In *Great Apes & Humans: The Ethics of Coexistence*, ed. B. B. Beck, T. S. Stoinski, M. Hutchins *et al.*, pp. 3–56. Washington, D.C.: Smithsonian Institution Press.

Chism, J. & Rowell, T. H. (1988) The natural history of patas monkeys. In *A Primate Radiation: Evolutionary Biology of the African Guenons*, ed. A. Gautier-Hion, F. Bourlière, J.-P. Gautier & J. Kingdon, pp. 412–38. Cambridge: Cambridge University Press.

Chivers, D. J. (1974). The Siamang in Malaya: a field study of a primate in tropical rain forest. *Contr. Primatol.* 4, 1–335.

Cipolletta, C. (2003). Ranging patterns of a western gorilla group during habituation to humans in the Dzanga-Ndoki National park, Central African Republic. *Int. J. Primatol.* 24, 1207–26.

Cowlishaw, G. (1997). Trade-offs between foraging and predation risk determine habitat use in a desert baboon population. *Anim. Behav.* 53, 667–86.

Curtis, D. J. (1998). Group size, home range use, and seasonal variation in the ecology of *Eulemur mongoz*. *Int. J. Primatol.* 19, 811–35.

Doran-Sheehy, D. M., Derby, A. M., Greer, D. & Mongo, P. (2007). Habituation of western gorillas: the process and factors that influence it. *Am. J. Primatol.* 69, 1–16.

Goldsmith, M. (2005). Habituating primates for field study: ethical considerations for African great apes. In *Biological Anthropology and Ethics: From Repatriation to Genetic Identity*, ed. T. R. Turner, pp. 49–64. New York: SUNY Press.

Hockings, K. & Humle, T. (2009). *Best Practice Guidelines for Avoidance and Mitigation of Conflict between Humans and Great Apes*. Gland, Switzerland: IUCN/SSC Primate Specialist Group. 48 pp. www.primate-sg.org/BP.conflict.htm

Homsy, J. (1999). *Ape Tourism and Human Diseases: How Close Should We Get?* Report to the International Gorilla Conservation Programme, Nairobi. www.igcp.org/library/

Isbell, L. A. & Young, T. P. (1993). Human presence reduces predation in a free-ranging vervet monkey population in Kenya. *Anim. Behav.* 45, 1233–5.

Jack, K. M., Lenz, B. B., Healna, E. *et al.* (2008). The effects of observer presence on the behavior of *Cebus capucinus* in Costa Rica. *Am. J. Primatol.* 70, 490–4.

Kinnaird, M. F. & O'Brien, T. G. (1996). Ecotourism in the Tangkoko DuaSudara Nature Reserve: opening Pandora's box? *Oryx* 30, 65–73.

Köndgen, S., Kühl, H., N'Goran, P. K. *et al.* (2008). Pandemic human viruses cause decline of endangered great apes. *Curr. Biol.* 18, 260–4.

Krunkelsven, E. van, Dupain, J., Elsacker, L. van & Verheyen, R. (1999). Habituation of bonobos (*Pan paniscus*): first reactions to the presence of observers and the evolution of response over time. *Folia Primatol.* 70, 365–8.

Kummer, H. (1995). *In Quest of the Sacred Baboon*. Princeton, NJ: Princeton University Press.

Nizeyi, J.-B. (2005). Non-invasive monitoring of adreno-cortical activity in free-ranging mountain gorillas of Bwindi Impenetrable National Park in southwestern Uganda. PhD thesis, Makerere University, Uganda.

Passamani, M. (1998). Activity budget of Geoffroyi's marmoset (*Callithrix geoffroyi*) in an Atlantic forest in southeastern Brazil. *Am. J. Primatol.* 46, 333–40.

Rasmussen, D. R. (1979). Correlates of patterns of range use of a troop of yellow baboons (*Papio cynocephalus*). I. Sleeping sites, impregnable females, births, and male emigrations and immigrations. *Anim. Behav.* 27, 1098–112.

Rasmussen, D. R. (1991). Observer influence on range use of *Macaca arctoides* after 14 years of observation? *Lab. Prim. Newsl.* 30, 6–11.

Rasmussen, D. R. (1998). Changes in range use of Geoffroy's tamarins (*Saguinus geoffroyi*) associated with habituation to observers. *Folia Primatol.* 69, 153–9.

Rodman, P. S. (1979). Individual activity patterns and the solitary nature of orangutans. In *Perspectives On Human Evolution*, Vol. 5, *The Great Apes*, ed. D. A. Hamburg & E. R. McCown, pp. 235–55. Menlo Park, CA: Benjamin/Cummings Publishing Co.

Rylands, A. B. (1986). Ranging behaviour and habitat preference of a wild marmoset group, *Callithrix humeralifer* (Callitrichidae, Primates). *J. Zool. Lond.* 210, 489–514.

Schaller, G. B. (1963). *The Mountain Gorilla: Ecology and Behaviour*. Chicago, IL: University of Chicago Press.

Sommer, V., Adanu, J., Fauscher, I. & Fowler, A. (2004). Nigerian chimpanzees (*Pan troglodytes vellerosus*) at Gashaka: two years of habituation efforts. *Folia Primatol.* 75, 295–316.

Sterling, E. J. (1993). Behavioral Ecology of the Aye-Aye (*Daubentonia madagascariensis*) on Nosy Mangabe, Madagascar. Doctoral thesis, University of Yale.

Strier, K. B. (1999). *Faces in the Forest: The Endangered Muriqui Monkeys of Brazil.* Cambridge, MA: Harvard University Press.

Susman, R. L. (1984). The locomotor behavior of *Pan paniscus* in the Lomako Forest. In *The Pygmy Chimpanzee. Evolutionary Biology and Behavior*, ed. R. L. Susman, pp. 369–93. New York: Plenum Press.

Thorpe, W. H. (1963). *Learning and Instinct in Animals.* London: Methuen.

Tutin, C. E. G., & Fernandez, M. (1991). Responses of wild chimpanzees and gorillas to the arrival of primatologists: behaviour observed during habituation. In *Primate Responses to Environmental Change*, ed. H. O. Box, pp. 187–97. London: Chapman & Hall.

Werdenich, D., Dupain, J., Arnheim, E. *et al.* (2003). Reactions of chimpanzees and gorillas to human observers in a non-protected area in south-eastern Cameroon. *Folia Primatol.* 74, 97–100.

Woodford, M., Butynski, T. M. & Karesh, W. (2002). Habituating the great apes: the disease risks. *Oryx* 36, 153–60

Wrangham, R. W. (1974). Artificial feeding of chimpanzees and baboons in their natural habit. *Anim. Behav.* 22, 83–93.

JÖRG U. GANZHORN, S. JACQUES RAKOTONDRANARY
AND YEDIDYA R. RATOVONAMANA

3

Habitat description and phenology

INTRODUCTION

Habitat characteristics represent the matrix for the evolution of morphological, physiological and behavioural adaptations and life history traits of animals. These include bottom-up factors (such as distribution and abundance of food) and top-down constraints (such as protection from predators). The role of different habitat components in the survival and reproduction of individuals is of prime interest and is relevant for evolutionary understanding as well as for primate conservation. So far, this has been studied by linking habitat characteristics with phenotypic traits of animals, assuming that these traits represent adaptation to a given habitat constraint. However, we should bear in mind that the characteristics we observe today may represent adaptations to constraints that acted in the past (the Epaminondas effect, named after the little boy who always did the right thing for the previous situation; or, in evolutionary terminology: animals are tracking fitness optima by trying to climb the sides of shifting adaptive peaks, Cody, 1974). However, testing hypotheses based on these 'ghosts of past constraints' has so far been impossible.

The present chapter focuses on the description of the vegetation of forest habitats as the biotic matrix for the evolution of life history traits. References are to textbooks or publications that illustrate applications of the methods and do not always refer to the original work.

BEFORE YOU START

Any statistical analysis requires quantitative measurements of habitat and animal characteristics. Most habitats are extremely complex and there are many features that can be measured, spending endless time

Field and Laboratory Methods in Primatology: A Practical Guide, ed. Joanna M. Setchell and Deborah
J. Curtis. Published by Cambridge University Press. © Cambridge University Press 2011.

and money, without measuring the right variable. Although it is no guarantee that you will measure the relevant variables, consider the following issues before getting started (this applies to all research projects but seems to be neglected particularly in studies of relationships between animals and their habitat).

- What is the question? Don't just measure habitat features simply because somebody else did it (e.g. do you need to sample herbaceous vegetation and bushes for primate species that rarely leave the canopy? How relevant is a $10 \times 100 \, m^2$ plot of forest to a group of chimpanzees that range over several km^2?).
- What is the scale of the study?
- What is the unit of analysis?
- Which habitat variable do you need to measure to answer your question?
- How accurately do you need to measure variables?
- Which statistical test will be applied?
- Are your samples independent?
- What sample size is required for statistical analysis?

Primate body mass ranges from 0.03 to 100 kg, home ranges from 0.01 to 400 km^2, and habitats from snow-covered mountains to deserts and evergreen tropical rain forests, and it is obvious that there is no universal sampling method appropriate to answering all questions. All methods can be adjusted to satisfy the needs and special circumstances of the study in question. Don't develop new methods just to do something new, as studies gain from the number of other studies they can be compared with.

Regardless of the method to be applied, it is important to relate samples to unit areas, otherwise the study will remain an isolated case of limited value. Georeference all samples using a Global Positioning System (GPS) (Chapter 4). Although this may not be of immediate relevance to your study, it will add tremendous potential for future studies to monitor environmental or other long-term changes. As a general rule, statistical comparisons are facilitated by sampling more smaller units rather than fewer large ones.

FOREST CLASSIFICATION

Forests can be classified based on their overall structure, phenology, leaf shape, climate, geology/soil characteristics and history of human use. Structural classifications are, for example, closed- or open-canopy forests, mixed forest with open or closed under-storey, woodland or wooded grassland. Phenological, climatic, soil–water relations and leaf characterizations are included in descriptions such as wet, moist or dry forest, gallery or mountain forest, seasonally inundated forest, peat or swamp forest, evergreen or deciduous forest, broad-leaved or spiny forest (Whitmore, 1990). The history of human use is included in terms such as primary or secondary forest. In most studies these classifications are rather subjective and based on the fieldworker's overall impression, yet they give a reasonable impression of the physiognomy of different forests.

Floristic characterization of forests is often beyond the skills of zoologists. Field guides are available for woody plants (APG, 1998; Gentry, 1993; Keller, 1996; Schatz, 2001), but plant species identification usually requires field collaboration with botanists or at least collection of herbarium specimens. In most cases fertile material (flowers and fruits) are required for identification and you should make arrangements for help from experts prior to sample collection (see also Chapter 12).

You can base a rough but useful forest classification on families of plants or just trees. Within communities, you can assign families (or species) different degrees of dominance. Based on the percentage of individuals in the community, a family (or species) is called 'eudominant' if it accounts for over 10% of the individuals. Other categories include 'dominant' (5%–10%) and 'subordinate' (2%–5%). Another standard classification for vegetation cover has been developed by Braun-Blanquet (in Kent & Coker, 1992) as:

+:   < 1%
1:   1%–5%
2:   6%–25%
3:   26%–50%
4:   51%–75%
5:   76%–100%.

Other measures of diversity are summarized in most textbooks (e.g. Brower *et al.*, 1990; Krebs, 1998). The most comprehensible form of documentation of floristic characteristics is a simple table or graph that shows relative abundance of families or species.

HABITAT STRUCTURE

There are three basic methods used to describe the density, frequency, coverage and biomass of plants in forest habitats: plot methods, transects and point-quarter sampling (Brower *et al.*, 1990). All three methods have advantages and disadvantages.

Plot methods

In plot methods, you count, measure and identify all individual plants within a given area. Consider a tree to be in the plot if the centre of the trunk is within the limits of the plot. We recommend rectangular plots with sides in a 1:2 ratio. The size of the plot depends on the vegetation to be sampled. We recommend 0.71 m × 1.41 m (1 m²) for herbaceous vegetation; 2.24 m × 4.47 m (10 m²) for bushes, shrubs and trees up to 3–4 m tall; 7.07 m × 14.14 m (100 m²) for trees taller than 3–4 m. For tropical forests with large trees, you will need to extend the size of the sampling area. The standard measurement for trees is the Diameter at Breast Height (DBH), related to the circumference (*C*) as DBH = *C* / 3.14, and measured at a height of 1.3 m. Measure DBH of buttressed and stilt-rooted trees above the buttresses/stilts, or by estimating the diameter of the trunk without the buttresses. Forestry suppliers provide 'DBH-tapes' that give tree diameters when measuring circumference. Different studies use different lower size limits and size categories for measuring trees due to different forest structure and structural requirements of primates in different forests. However, a lower limit of 10 cm DBH and 5 or 10 cm increments will allow comparisons with most other descriptions of vegetation.

For tropical forests, we propose modified Whittaker plots for multi-scale vegetation sampling (Fig. 3.1). Nested subplots of different sizes within a larger plot allow the development of species–area curves and estimation of the number of species in a larger unsampled area. The Smithsonian's Program for Monitoring & Assessment of

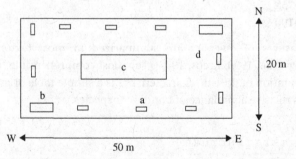

Fig. 3.1 Modified Whittaker plot consisting of nested subplots (after Stohlgren & Chong, 1997). a, b, c, and d indicate subplots of different size.

Biodiversity have applied this method in Peru (Stohlgren & Chong, 1997). For primate habitats, identify and measure all trees $\geq$ 10 cm DBH in the largest plot. Smaller sized trees (< 10 cm DBH) are rarely considered, and might make inclusion of the small sized nested subplots unnecessary. However, the principle of the sampling design remains valid and can be adjusted to include differently sized trees. For example, you could record trees $\geq$ 10 cm DBH in subplots of 100 m^2 nested within larger plots in which only trees $\geq$ 30 cm DBH are registered. As a rule of thumb, you need about 5–10 plots per stratum to characterize a vegetation formation if within-stratum variation is low. If within-stratum variation is high, you need more sites (Stohlgren & Chong, 1997).

Other plots consist of series of uniformly sized square or rectangular plots in which trees are identified and measured (see, for example, Schwarzkopf & Rylands, 1989; Ganzhorn, 1989). The only difference from the methods described above is that these plots are not nested and are all of the same size. This has been the most commonly used method in primatological studies.

Advantage: Plot methods are most efficient for all kinds of density-related features. You can adjust plot sizes to the size of microhabitats relevant for your study questions, such as analyses of used versus unused habitats (Southwood & Henderson, 2000).

Cautionary note: Plots of different shapes are likely to give different density estimates even if they cover the same surface (e.g. a 2 m × 50 m plot is likely to give a higher density estimate of trees than a 10 m × 10 m plot).

Transect samples

A belt transect is a long rectangular plot. In primatological studies, 0.1 ha (e.g. 10 m × 100 m) transects have been found to be useful as they are of manageable size and provide a good idea of the structure and floristic composition of the forest. You can subdivide this belt into intervals (e.g. 10 m × 10 m) to facilitate recordings and check for species–area relationships. However, you cannot use the single contiguous subplots as independent measures for statistical analyses because they are likely to be autocorrelated. Transects have also been used to analyse gibbon ranging patterns (Whitten, 1982).

Transects seem more appropriate than plots in heterogeneous habitats as they sample a greater variety of microhabitats. In principle, they can also describe environmental gradients, although these are difficult to analyse in conjunction with primate ranging patterns. Analyses are facilitated if you use clearly defined categories of vegetation.

The most complete transect descriptions that are available for comparison are 'Gentry transects'. These measure $2\,m \times 50\,m$ and were designed to sample floral diversity in many different countries (Gentry, 1995). The Smithsonian Institution are installing large-scale plots (50 ha) at various sites for comparative descriptions of different forests and long-term monitoring (www.ctfs.si.edu).

Advantage: Transects are less susceptible to patchy distributions of certain plant species than plot samples. In tropical forests, sample sites are rarely selected at random as required by statistical analyses, but rather by accessibility. You can reduce this bias by sampling along transects. While the choice of the starting point of the transect is unlikely to be random, subsequent sites are likely to provide a more representative picture of the habitat than plots.

Cautionary note: Subsequent subplots aligned along continuous transects suffer from autocorrelations. You can reduce this bias by spacing subplots along transects.

Point-quarter sampling

The point-quarter method (= point-centred quadrants) is a plot-less method. For this, you select a number of randomly determined points. These points can be distributed throughout the habitat, or along a transect line running through the habitat to be described. Each point represents the centre of four compass directions that divide the sampling point into four quarters. In each quarter, measure the distance from the centre of the nearest plant to the sampling point (d_1 - d_4; Fig. 3.2). You need to define which plants to measure (e.g. only trees $\geq 10\,cm$ DBH). Since it is difficult to measure from the centre of a large tree, add the radius of the tree trunk to the distance from the tree to the central sampling point.

Calculate densities as follows: Compute the mean of d_1 - d_4 per sampling plot (= d). You can use this as the unit for further analyses. The mean area in which a single plant occurs is equal to the mean distance squared. Thus, theoretically, each plant covers an area of d^2. The density of plants per unit area (A) is then: A / d^2 (example: mean distance of trees $\geq 10\,cm$ DBH to the sampling point $= 5.24\,m$. Density per hectare $(1\,ha = 100\,m \times 100\,m = 10\,000\,m^2) = 10\,000 / 5.24^2 = 364$ trees / ha). This method has been used to contrast primary forest with various kinds of secondary vegetation formations or to monitor small scale variation within primary forest (Ganzhorn, 1995), and can be modified to measure the density of single species or of any other structural habitat component.

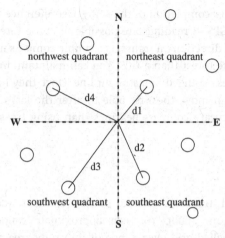

Fig. 3.2 Point centred quadrants or point-quarter sampling (after Brower *et al.*, 1990).

Advantage: The point-quarter method is rapid and yields large numbers of independent data points that you can use for statistical analyses. It avoids the tedious work of setting up plots and the decision of whether trees along the boundaries are in or out of the plot. Statistical results seem to indicate differences in the structure between habitats appropriately.

Cautionary note: Under field conditions, and given that primatologists normally have interests other than measuring plants, it is almost impossible to achieve a random set of sampling points. The tendencies are either to select a sapling that can be marked with flagging tape as the centre of the four quadrants, or to select points at some distance from the nearest trees. If the same person identifies the sampling point, these biases do not average out. Although the statistical results seem to indicate differences in the structure between habitats appropriately, the calculated values of tree densities can deviate substantially from the actual density, especially at small sample sizes (< 50 sampling points). Thus statistical differences indicate relative rather than absolute differences.

Equipment

You can measure plots and transects with a tape or hip chain and mark them temporarily with flagging tape or permanently with numbered aluminium tags. If you use flagging tape it should be biodegradable and you should remove it after completion of the study. Georeference the corners of plots with a GPS (if readings are possible in dense forest), although to keep a given direction on transects a simple compass may be easier to handle in the forest than a GPS. You can maintain directions by using three poles arranged in a straight line. Once they have been established, you can move the first pole to after the last one. This results in more accurate compass bearings than using trees as landmarks.

Plots or transects?

In principle, plots and transects should be set up at random (Southwood & Henderson, 2000). You can approximate random 'choice' by putting a labelled grid over a map of the study area and drawing a predefined number of grid cells. You can use the predefined corner of a given cell as the starting point for vegetation analyses. You can generate random numbers using most computers or by taking the last two digits of telephone numbers. Realistically, randomness is never achieved in studies of primate natural history.

If you know the boundaries for different vegetation communities and their extension, plot methods might provide more information, such as patchiness or species–area relationships in a homogeneous habitat. You can then extrapolate the results of the plots to the whole surface. Vegetation units could be swamp forest as compared with forest on solid ground, primary forest as compared with secondary forest, or areas used versus areas that are not used by primates. In this case, you should locate the plots at randomly selected sites within the vegetation formation (stratified random sampling; e.g. Oates *et al.*, 1990). However, primate home ranges are often not known precisely before the study begins, 'uniform' vegetation types are not known or delineated, the existing trail system cannot be extended (e.g. in National Parks), or new trails must be installed where possible logistically and not where random numbers fall. In these cases, transects seem to be the better compromise between data requirements, statistical rigour and time spent on vegetation issues.

Canopy cover

You can measure canopy cover by taking photos from ground level and measuring the percentage of open sky. Given solar panels, portable computers and digital cameras, you can now measure canopy cover at many sites almost online, even at stations with very basic infrastructure. You need to calibrate estimates of canopy cover based on photographic documentation to account for distortions. You can obtain estimates of the percentage of the sky covered with any Geographic Information System (GIS) program (Chapter 4). Aerial photographs also allow classification of canopy cover but it is absolutely essential that you ground-truth these classifications (Chapter 4). A low-budget alternative to photographic documentation involves recording whether or not the open sky is visible through a pipe held vertically at fixed intervals along a transect (e.g. every 1 m). The advantage of this method over the photographic measurements is that you can estimate canopy cover at different heights or for different strata.

Vertical and horizontal structure

The line-intercept technique is another transect method used to describe forest structure. Here, you record contacts of vegetation with a vertical line from 100 points spaced at regular intervals along a defined transect (Gautier *et al.*, 1994; Fig. 3.3a). The length of the baseline should be about 10 times the height of the canopy. Messmer *et al.* (2000; Fig. 3.3b) provide a fine example of this method. They measured the height of contact and species encountered above each of 100 census points per transect with an 8 m pole marked at 20 cm intervals and held vertically. At each sampling point, they recorded the height at which all plants came in contact with the pole. They estimated heights of contact above 8 m. For dense foliage, they recorded the lower and upper points of the foliage mass. Alternatively, you can take readings in the vertical direction using a telelens focussed at defined intervals. This method allows very detailed analyses of forest structure that you can illustrate in graphs. You can compare the relative densities of vegetation from different transects with respect to their horizontal or vertical orientation by using Kolmogorov–Smirnov tests. However, this method does not sample an area (it can be adapted to sample unit areas by sampling a surface rather than along a straight line; but there are no comparative databases in primatology that have used this method).

You can measure the lateral density of vegetation by positioning a square or longitudinal checkerboard at different heights in the forest. A reasonable size checkerboard is about 50 cm × 50 cm with 10 cm × 10 cm black and white squares to measure vegetation density on a small scale. For larger distances you can subdivide a board of 25 cm × 200 cm in quadrats of 25 cm. Record how many (what

(a)

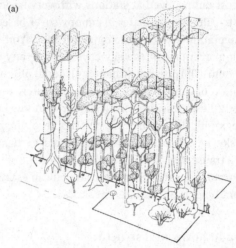

(b)

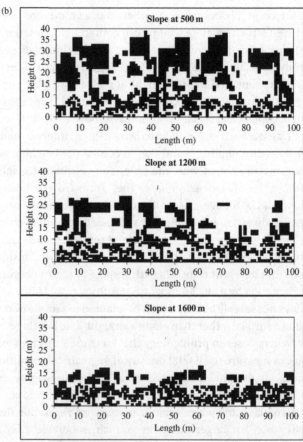

Fig. 3.3 Illustration of horizontal and vertical forest structure based on linear sampling. (Continued on p. 61.)

percentage) of the different squares can be seen from specified distances. This method is not applied routinely in primatological studies as it is difficult to interpret its biological relevance. It might be useful to evaluate 'thickness of vegetation' (e.g. as protection from predators or as potential hides for predators).

Population dispersion / patchiness

You can assess the dispersion patterns of plants by dividing the variance of tree densities by the mean density per plot (variance-to-mean ratio). Values <1 indicate uniform or overdispersed distribution, values =1 indicate random distributions and values >1 indicate clumped (underdispersed, clustered) distributions. This method is labour-intensive, as you need inventories of many different plots. The variance-to-mean ratio is sensitive to population size and plot size, requiring some thought on the scale of the study that is biologically relevant. Brower *et al.* (1990) provide an introduction to measurements and indices for population dispersions, and Ganzhorn (1993) provides primate examples. Ludwig and Reynolds (1988) describe plot-less (distance) methods for analysing spatial patterns.

Single-tree-related structural data

We list some useful data that might be collected for a single tree in Table 3.1.

PHENOLOGY

Phenological measurements give an idea of seasonal changes in productivity of the forest and food abundance for primates. The plant parts measured are young and mature leaves, fruits, unripe and ripe fruits, and you will need binoculars for large trees. Data on seasonal variation in the production of exudates can be useful for specialized species

---

Caption for Fig. 3.3 (cont.)
   (a) Sampling design (after Gautier *et al.*, 1994; figure drawn by
C. Chatelain; reproduction with permission from C. Chatelain)
   (b) Samples of three different forest types in eastern Madagascar. Each black square marks a point in space where the real or virtual pole touched the vegetation (after Messmer *et al.*, 2000; with permission from the author).

Table 3.1 *Variables measured to characterize trees as elements of forest structure*

Variable	Details	Source
DBH	Measuring all trees ≥ 10 cm diameter at breast height is standard; DBH is related to many important tree characteristics that are difficult to measure (tree height, biomass, age, crop size).	Whitten, 1982
Crown width	You can measure crown width (crown diameter) as the mean of the distances between the opposite extremes of the crown measured along two perpendicular axes. Similarly to DBH, crown diameter can be related to biomass and food patch size. You can use crown diameters instead of DBH when trees are highly buttressed.	Wright, 1999
Crown volume	Crowns can take many different shapes that are important to quantify biomass. However, accurate measurement of crown shapes is beyond the scope of primatological studies because the efforts required far exceed the possible benefits.	Chapman *et al.*, 1992
Tree height	Tree height has sometimes been measured but seems of limited value in primatological studies. You can measure height with a calibrated telelens, with a range finder (clinometer) or by measuring the distance at which the top of the crown is seen at an angle of 45° from the observer. Then calculate tree height as: distance between the observer and the tree + the height of the observer. The procedure looks nice in textbooks until you try it in the forest.	Whitten, 1982
Bole height	This is the height of the first branch and provides an idea of the open space below the canopy that can be used by vertical clingers and leapers.	Whitten, 1982
Solar exposure of crowns	Solar exposure influences tree productivity and food quality but there are no data yet that link primate behaviour to this variable in an evolutionarily significant sense.	Ganzhorn, 1995

Table 3.1 (*cont.*)

Variable	Details	Source
Support for locomotion	Very detailed information on the aspects and size of supports for primates might be required for analyses of locomotion.	Warren, 1997

(lemurs, galagos, tamarins; Génin *et al.*, 2010), but standardized quantitative measurement of gum production is problematic.

Ideally, phenological records should characterize (1) seasonal changes in the productivity of the forest and (2) seasonal changes in food availability for primates. You can achieve the first aspect by random or stratified sampling. The second aspect is much more difficult because it requires detailed knowledge about the diet of the study species before the phenological sampling is initiated. Also, some seasonally very important resources may be scarce and might even be located outside the normal range of the study group. If previous studies are unavailable, these so-called keystone resources are identified only during the course of the study and may not have been included in the phenological sample at the start.

Other problems are: it is sometimes difficult to distinguish between young and mature leaves; flowers and fruits can be very small, green and difficult to detect as long as they are not visited by animals; ripeness of fruits is difficult to define, and, if tree species are dioecious, half of the trees (males) will never produce fruit.

Phenological transects

Phenological transects are supposed to describe fruit, flower and leaf availability of the forest based on random samples. You can use phenological data in conjunction with the abundance and biomass of the different tree species that are present at a site. For a close match between phenological records and vegetation descriptions, take random samples from the same plots or transects that are used to describe the floristic and structural characteristics of the forest. Since leaf and fruit production vary widely within and between tree species and sites, take additional phenological samples along transects stretching over several kilometres. Phenological trails or plots therefore cover larger areas or span larger distances than the vegetation samples used to describe structural aspects of the habitat.

If you can classify and delineate vegetation formations, you can also record phenological samples within plots that might allow extrapolation to the whole area covered by the vegetation formation in question. Phenological patterns of tropical forests may not follow environmental seasonality as tightly as in temperate climates (Wright, 1999). Therefore, if possible, record phenological data over several years to account for year-to-year variation in individual trees, in individual species and on the community level. Chapman *et al.* (1994) review the application of different techniques to record phenological data, van Schaik *et al.* (1993) provide a review of phenological studies in ecological contexts, and Kay *et al.* (1997) summarize productivity for South American sites.

Phenological records and analyses

Label trees for phenology using numbered aluminium tags and nails available from forestry suppliers. Aluminium does not rust and does not cause as much damage to trees as other material. In addition, aluminium nails do not destroy saws, which may be relevant for work in timber plantations. Record phenological data every two weeks (or twice per month if data will be presented on a monthly basis). Although biased, you can rate the abundance of plant parts on relative scales from 0–4 or 0–10. Alternatively, you can estimate abundances on a logarithmic scale by actually counting fruits, for example, in a defined sector of the crown and extrapolating this figure to the total crown volume. This results in approximate absolute figures (0–1 fruits = 0; about 10 fruits = 1; about 100 fruits = 2, etc.).

You can record productivity and analyse it as simple presence/absence data. Alternatively, you can weight production by the DBH or basal area of the tree, by crown volume and by the measure of abundance. Surprisingly, the different measurements and different ways of weighting productivity yield highly correlated results (Chapman *et al.*, 1994). Therefore, simply adding the number of trees with a given plant part (fruits) is satisfactory. Using this method, you can consider each tree as an independent unit. This facilitates statistical analysis, which becomes more complicated once productivity is based on weighted data.

Fruit trails and fruit traps

Fruit trails consist of individuals of selected tree species that are of special interest (e.g. known food tree species). Tree species are sampled

in replicate (about 5 individuals per species; the more the better), allowing the study of intraspecific variation in fruit production. For the
sake of easy access, and since these trees are not supposed to represent
a random sample, you can choose these trees along existing trails.

Fruit traps consist of a plastic bag suspended from a frame, which
can be made of poles or bent wire. Measure the collecting area of the
surface (e.g. 0.08–0.25 m^2; Chapman *et al.*, 1994). Raise the whole structure off the ground so that the plastic bag does not touch the ground,
and is not perforated by rodents or eaten by ants. Make small holes in
the plastic bag to allow water to drain out, but take care that small
seeds are not flushed out.

Fruit traps have been used, for example, in Peru (Terborgh, 1983),
Uganda (Chapman *et al.*, 1994) and Madagascar (Wright, 1997). Set traps
at random or predefined intervals, e.g. 1 m off the side of existing trails.
This method is supposed to represent random sampling of fruit availability and is equivalent to phenological transects. Although data based
on phenological transects and fruit trails were significantly correlated,
fruit trap data were uncorrelated with the results of either method
(Chapman *et al.*, 1994). You can also set fruit traps under the crown
of individual trees of special interest to measure variability of fruit
production or consumption rates within and between tree species
(Böhning-Gaese *et al.*, 1999). Collect fruits, seeds and leaves as well as
caterpillar faeces at regular intervals (e.g. once per week) and identify,
count and weigh the material collected.

A method similar to fruit traps consists of sweeping a trail and
recording fruit and litter fall on the trail in regular intervals. This
method covers a larger area but exaggerates fruit and seed removal
from the trails by secondary consumers (Butynski, 1990).

Advantages: Fruit traps represent the most objective measurement
for fruit and litter fall. They can be standardized and therefore allow
comparisons between sites that are independent from observer bias.

Disadvantages: Fruit traps sample a very small surface, despite the
substantial investment required to construct and set the fruit traps.
Only fruit and litter fall is recorded. This means that fruits available but
consumed or remaining on the tree are not measured. Fallen fruits can
also start to rot or can be eaten by rodents and insects in the trap.

CONCLUSIONS

Having described all these wonderful methods, keep in mind that new
technologies are available that have the potential to catapult the

analyses of animal–habitat relationships to new standards of resolution. A plethora of GIS-based environmental variables are available free of charge. You can measure primary productivity and forest structure remotely with high resolution (Kozak *et al.*, 2008; Müller *et al.*, 2009; Chapter 4). Furthermore, near-infrared reflectance spectra in combination with other remote sensing methods permit the measurement of habitat characteristics at very fine scales over much larger areas than will ever be possible on the ground (Stolter *et al.*, 2006; Chapter 4). In a recent study, we applied all the quantitative methods described here to identify vegetation characteristics that might limit the distribution of lemurs along an environmental gradient. No correlation of these single factors was very convincing, but the distribution of species was nicely predicted by the different colours from Landsat images. Obviously, the colour pattern integrated the vegetation characteristics in a different, but more meaningful way, than our measures on the ground. Furthermore, animals do not care about Latin binomials but are more interested in the chemical composition and the availability of nutritional resources (Chapter 14). These remote sensing techniques have extremely high potential that has not yet found its way into primatology.

SELECTED REFERENCES FOR GENERAL ASPECTS (ASTERISKED) AND OTHER REFERENCES CITED

APG. (1998). An ordinal classification for the families of flowering plants. *Ann. Miss. Bot. Gard.* 85, 531–53.

Böhning-Gaese, K., Gaese, B. H. & Rabemanantsoa, S. B. (1999). Importance of primary and secondary seed dispersal in the Malagasy tree *Commiphora guillaumini. Ecology* 80, 821–32.

*Brower, J. E., Zar, J. H. & von Ende, C. N. (1990). *Field and Laboratory Methods for General Ecology.* Dubuque, IA: Wm. C. Brown Publishers.

Butynski, T. M. (1990). Comparative ecology of blue monkeys (*Cercopithecus mitis*) in high- and low-density subpopulations. *Ecol. Monogr.* 60, 1–26.

Chapman, C. A., Chapman, L. J., Wrangham, R. *et al.* (1992). Estimators of fruit abundance of tropical trees. *Biotropica* 24, 527–31.

Chapman, C. A., Wrangham, R. & Chapman, L. J. (1994). Indices of habitat-wide fruit abundance in tropical forests. *Biotropica* 26, 160–71.

Cody, M. L. (1974). Optimization in ecology. *Science* 183, 1156–64.

Ganzhorn, J. U. (1989). Niche separation of the seven lemur species in the eastern rainforest of Madagascar. *Oecologia* 79, 279–86.

Ganzhorn, J. U. (1993). Flexibility and constraints of *Lepilemur* ecology. In *Lemur Social Systems and their Ecological Basis*, ed. P. M. Kappeler & J. U. Ganzhorn, pp. 153–65. New York: Plenum Press.

Ganzhorn, J. U. (1995). Low level forest disturbance effects on primary production, leaf chemistry, and lemur populations. *Ecology* 76, 2084–96.

Gautier, L., Chatelain, C. & Spichiger, R. (1994). Presentation of a relevé method for vegetation studies based on high resolution satellite imagery. In *Proceedings of the XIIIth Plenary Meeting AETFAT, Malawi*, ed. J. H. Seyani & A. C. Chikuni, Vol. 2, pp. 1339–50. Zomba: National Herbarium and Botanic Gardens of Malawi.

Génin, F. G. S., Masters, J. C. & Ganzhorn, J. U. (2010). Gummivory in cheiroga-leids: primitive retention or adaptation to hypervariable environments? In *Evolution of Exudativory in Primates*, ed. A. Burrows & L. T. Nash. New York: Springer. (In press.)

Gentry, A. (1993). *A Field Guide to the Families and Genera of Woody Plants of Northwest South America*. Washington, D.C.: Conservation International.

Gentry, A. H. (1995). Diversity and floristic composition of neotropical dry forests. In *Seasonally Dry Tropical Forests*, ed. S. H. Bullock, H. A. Mooney & E. Medina, pp. 146–94. Cambridge: Cambridge University Press.

Kay, R. F., Madden, R. H., vanSchaik, C. & Higdon, D. (1997). Primate species richness is determined by plant productivity: implications for conservation. *Proc. Natl. Acad. Sci. USA* 94, 13023–7.

Keller, R. (1996). *Identification of Tropical Woody Plants in the Absence of Flowers and Fruits*. Basel: Birkhäuser Verlag.

*Kent, M. & Coker, P. (1992). *Vegetation Description and Analysis*. Boca Raton, FL: CRC Press.

Kozak, K. H., Graham, C. H. & Wiens, J. J. (2008). Integrating GIS-based environmental data into evolutionary biology. *Trends Ecol. Evol.* 23, 141–8.

*Krebs, C. J. (1998). *Ecological Methodology*. Menlo Park, CA: Addison Wesley Longman.

Ludwig, J. A. & Reynolds, J. F. (1988). *Statistical Ecology*. New York: John Wiley & Sons.

Messmer, N., Rakotomalaza, P. J. & Gautier, L. (2000). Structure and floristic composition of the Parc National de Marojejy, Madagascar. In *A Floral and Faunal Inventory of the Parc National de Marojejy, Madagascar: With Reference to Elevational Variation*, ed. S. M. Goodman, New Series 97, *Fieldiana Zoology*, pp. 41–104. Chicago, IL: Field Museum of Natural History.

Müller, J., Moning, C., Bassler, C., Heurich, M. & Brandl, R. (2009). Using airborne laser scanning to model potential abundance and assemblages of forest passerines. *Basic Appl. Ecol.* 10, 671–81.

Oates, J. F., Whitesides, G. H., Davies, A. G. et al. (1990). Determinants of variation in tropical forest primate biomass: new evidence from West Africa. *Ecology* 71, 328–43.

Schatz, G. E. (2001). *Generic Tree Flora of Madagascar*. Kew, St. Louis, MO: Royal Botanical Garden and Missouri Botanical Garden.

Schwarzkopf, L. & Rylands, A. B. (1989). Primate species richness in relation to habitat structure in Amazonian rainforest fragments. *Biol. Cons.* 48, 1–12.

*Southwood, T. R. E. & Henderson, P. A. (2000). *Ecological Methods*. (3rd edn.) Oxford: Blackwell Science.

Stohlgren, T. & Chong, G. (1997). Assessment of biological diversity and long-term monitoring plan for the Lower Urubamba Region. In *Biodiversity Assessment and Monitoring of the Lower Urubamba Region, Peru*, ed. F. Dallmeier & A. Alonso, Vol. 1, *SI/Monitoring and Assessment of Biodiversity Program*, pp. 41–4. Washington, D.C.: Smithsonian Institution.

*Stolter, C., Julkunen-Tiitto, R. & Ganzhorn, J. U. (2006). Application of near infrared reflectance spectroscopy (NIRS) to assess some properties of a sub-arctic ecosystem. *Basic Appl. Ecol.* 7, 167–87.

Terborgh, J. (1983). *Five New World Primates: A Study in Comparative Ecology*. Princeton, NJ: Princeton University Press.

van Schaik, C. P., Terborgh, J. W. & Wright, S. J. (1993). The phenology of tropical forests: adaptive significance and consequences for primary consumers. *A. Rev. Ecol. Syst.* 24, 353–77.

Warren, R. D. (1997). Habitat use and support preference of two free-ranging saltatory lemurs (*Lepilemur edwardsi* and *Avahi occidentalis*). *J. Zool., Lond.* 241, 325–41.

*Whitmore, T. C. (1990). *An Introduction to Tropical Rain Forests*. Oxford: Oxford University Press.

Whitten, A. J. (1982). A numerical analysis of tropical rain forest using floristic and structural data and its application to an analysis of gibbon ranging behaviour. *J. Ecol.* 70, 249–71.

Wright, P. C. (1997). Behavioral and ecological comparisons of Neotropical and Malagasy primates. In *New World Primates: Ecology, Evolution, and Behavior*, ed. W. G. Kinzey, pp. 127–41. New York: Aldine de Gruiter.

Wright, P. C. (1999). Lemur traits and Madagascar ecology: coping with an island environment. *Yb. Phys. Anthropol.* 42, 31–72.

PATRICK E. OSBORNE AND LOUISE GLEW

4

Geographical information systems and remote sensing

INTRODUCTION

A Geographic Information System (GIS) is a collection of hardware and software for the storage, retrieval, mapping and analysis of geographic data (Table 4.1). The spatial data held in a GIS refer to a particular place on earth, represented in a coordinate system such as latitude and longitude (i.e. the data are geo-referenced). GIS data are of two general types: vector, with points, lines (arcs) and polygons as fundamental components; and raster, where the data are held as a grid of pixels. One of the most common sources of raster data for a GIS is satellite remote sensing. We may define remote sensing in a variety of ways but it refers to the process of acquiring information about the Earth's surface without being in contact with it (e.g. from aircraft or satellites). Remote sensing involves recording reflected or emitted radiation and processing, analysing and applying the output to a particular problem.

In this chapter we consider how GIS and remote sensing may be applied to problems in primatology, a topic that appears to have a relatively small body of literature, although the fields of GIS and remote sensing are themselves vast. Where possible, we cite examples from the primate literature to illustrate the range of applications where GIS is invaluable but also draw on wider examples where primate work could benefit from techniques that are employed on other groups of animals.

The organization of this chapter follows a simple logic. To use GIS in primate research, you need data on the locations of primates, a set of environmental data layers in either vector or raster format, and some means to integrate the two. We therefore discuss in turn sources of primate location data, sources of environmental data layers (including remote sensing) and then examples of analyses that integrate the two.

Field and Laboratory Methods in Primatology: A Practical Guide, ed. Joanna M. Setchell and Deborah J. Curtis. Published by Cambridge University Press. © Cambridge University Press 2011.

Table 4.1 *Geographic Information Systems and digital image processing desktop applications*

Application	Source
Commercial software products	
ArcGIS	http://www.esri.com/software/arcgis
ENVI	http://www.ittvis.com/ProductServices/ ENVI
ERDAS ER mapper	http://www.erdas.com
Idrisi	http://www.clarklabs.org
Manifold System	http://www.manifold.net
MapInfo	http://www.mapinfo.co.uk
Open-source desktop applications	
DIVA-GIS	http://www.diva-gis.org
Geographic Resources Analysis Support System (GRASS)	http://grass.osgeo.org
Google Earth	http://earth.google.com
gvSIG	http://www.gvsig.gva.es
JUMP GIS	http://www.jump-project.org
MapWindow GIS	http://www.mapwindow.org
Quantum GIS	http://www.qgis.org/
SAGA GIS	http://www.saga-gis.org
Orfeo toolbox (OTB)	http://smsc.cnes.fr/PLEIADES/lien3_vm. htm
uDIG	http://udig.refractions.net/
Open-source processing/analysis	
'R'	http://www.r-project.org/
Biomapper	http://www2.unil.ch/biomapper/
Fragstats	http://www.umass.edu/landeco/research/ fragstats/fragstats.html
JMatrixNet	http://www.ecology.su.se/JMatrixNet

COLLECTING PRIMATE LOCATION DATA

Details of field techniques for working on primates are covered elsewhere in this book, as are the technical details of specialist methods such as radio-tracking (Chapter 10). Here we focus on sources of location data that form an essential input to any geographic study.

Broadly speaking, you can gather data on the locations of primates through field sightings, trapping and radio-tracking in some form (see, for example, Scholz & Kappeler 2004). In all cases, you need to capture the locations of sightings or traps, either by marking them accurately on maps from which coordinates may be read later, or by using a Global Positioning System (GPS) unit. You can often track approachable or habituated primates using hand-held GPS receivers to capture accurate location data, either as spot observations (e.g. Hope *et al.*, 2004) or by tracking groups on a continual basis (Asensio *et al.*, 2009). Where the subject is less approachable and you cannot assume that your location is the same as the animal's, you will need to correct the GPS locations using trigonometry, for example by using a laser rangefinder and electronic compass to obtain the distance and bearing between observer and subject (Grueter *et al.*, 2008). You can also use conventional VHF radio-telemetry to locate focal subjects, then record their precise locations by using a hand-held GPS (see, for example, Biebouw, 2009).

Modern GPS units are capable of spatial accuracies better than 30 m in most situations and often better than 10 m. If necessary, you can achieve greater spatial accuracy by using differential GPS (DGPS, see, for example, Pochron, 2001). DGPS uses data from a fixed base station at a known location to correct errors in your hand-held GPS, yielding locational accuracies better than 1 m. The key question is: what spatial accuracy do you need? Analyses within GIS are only as good as the data layer with the least spatial precision and it is questionable whether many environmental data sources are reliable at the sub-metre scale.

The alternative to holding the GPS yourself is to mount it on the animal as a GPS collar (Markham & Altmann, 2008). Always go for the newest GPS technology you can afford because units are continually improving in terms of signal reception and battery requirements. Remember too that a GPS mounted on a primate 'tells the animal where it is', not you! The system must therefore either send the data to you via a VHF or satellite link (e.g. via the Argos system) or you will have to retrieve the collar and download the stored data. Costs are rapidly changing but expect to pay several hundred US$ per collar, plus signal reception fees if you use the Argos system.

In general, GPS technology is advancing rapidly (i.e. monthly) so get yourself up-to-date before starting. For example, the concerns over poor GPS signal reception expressed by Phillips *et al.* (1998) probably no longer apply in many situations because they used a three-channel receiver with low sensitivity compared with modern high-sensitivity receivers using 12 channels (e.g. Garmin Oregon 400t), although dense canopies may still present problems.

In some cases it may not be necessary to go to the field to collect new data. You can perform worthwhile analyses on assembled location records from mixed sources including museum specimen localities and other historical records (see, for example, Baumgarten & Williamson, 2007).

COLLECTING ENVIRONMENTAL DATA

Three main sources of environmental data can be used in a GIS: digital map data; remotely sensed data; and field data gathered with a GPS. Comprehensive GIS databases often integrate data from all these sources (e.g. Smith *et al.*, 1997).

Digital map data

Digital map data include all forms of maps (e.g. showing vegetation or roads) that are geo-referenced, i.e. that show changing coordinates as a computer mouse is moved across the screen (for examples, see Table 4.2). Beware of map images that are not geo-referenced (e.g. scanned JPEG or TIFF files of aerial photographs or maps): while useful for project planning, such images cannot be used in a GIS unless they are first geo-referenced (see, for example, Anderson *et al.*, 2007). You can only do this if it is possible to identify features on the image of a fine enough resolution such that you can locate them in the field and capture their coordinates by using a GPS. You can use these coordinates within a GIS as control points to geo-rectify the image. Digital map data are typically in vector format (consisting of lines and polygons) rather than raster format (in pixels) but topographic layers may have been converted from contours to pixels already.

Since the aim of most studies involving GIS is to relate the presence of a primate to the environmental layers, it is crucial to consider the extent to which this is possible from the beginning of the project. When using vector data, there are two main issues to consider: is the mapped information accurate; and at what resolution were the data captured? For example, when using a vegetation data layer, ask yourself whether the methods used to assess the vegetation are likely to have revealed the true vegetation types at the resolution you want (Chapter 3). The vegetation class 'savanna' may be unhelpful because the primate species of interest may only use the tree component of the habitat rather than the grassland. Secondly, if the original field mapping was done at coarse resolution, the polygons will have

Table 4.2 *Examples of GIS and remotely sensed data products*

Note that the list is by no means exhaustive and is only intended to be used as a guide to the kinds of data available to researchers. Pedro J. Leitão kindly provided the list of satellite imagery.

Data	Summary	Website
1. Global data sources		
(a) Base layers		
Digital chart of the world	Ageing but comprehensive data layers for political boundaries, main topographic features together with human settlement and infrastructure	www.maproom.psu.edu/dcw
GeoCommunity	Collection of national-level geospatial data from a variety of sources, some free to download	http://data.geocomm.com
(b) Climate		
Climatic Research Unit datasets	A series of high resolution gridded datasets for multiple climatic variables covering various spatial extents	www.cru.uea.ac.uk/cru/data
WORLDCLIM	Grids representing different climatic variables at various resolutions	www.worldclim.org
(c) Soils, geology and topography		
Harmonized world soil database	Soils database with global coverage. Data and a simple mapping application are free to download	www.iiasa.ac.at
Global 30 Arc-Second Elevation (GTOPO30) dataset	Global elevation dataset at 1 km resolution	http://eros.usgs.gov

Table 4.2 (*cont.*)

Data	Summary	Website
(d) Habitat, land cover and land use		
Global land cover characterization (GLCC) dataset	Land cover dataset derived from 1 km AVHRR (Advanced Very High Resolution Radiometer) imagery taken between 1992 and 1993	http://eros.usgs.gov
(e) Species data		
Ape Populations, Environments, and Surveys (APES) database	Geospatial database of ape distribution, occurrence and threats. Data free to download on application	http://apes.eva.mpg.de/
(f) Protected areas and priority areas for conservation		
World database of protected areas	Free to download global database containing vector data layers of protected areas, with periodic updates available.	www.wdpa.org
(g) Socioeconomic data		
United Nations Environment Programme/Global Resource Information Database (UNEP/GRID)	Spatially referenced population databases by continent together with various base map layers	www.na.unep.net
Measure DHS	Collection of datasets, some geo-referenced on human health and disease incidence	www.measuredhs.com/
(h) Satellite imagery		
GOES	Optical and thermal meteorological sensor with imagery available at multiple resolutions	www.goes.noaa.gov
Meteosat (MVIRI/SEVIRI)	Optical and thermal meteorological sensor with imagery available at multiple resolutions	www.eumetsat.int

Table 4.2 (*cont.*)

Data	Summary	Website
SPOT VGT	Optical sensor producing imagery for use in vegetation and natural resources monitoring with a resolution of 1150 m	www.spotimage.com
NOAA AVHRR	Optically sensed images with a resolution of 1090 m with applications in climatology and vegetation monitoring	http://earth.esa.int/ NOAA-AVHRR
TerraMODIS	Remotely-sensed images available at 250 m, 500 m, and 1000 m for use in land cover and vegetation monitoring	http://modis.gsfc.nasa. gov
ENIVSAT MERIS	Optical sensor producing images at a resolution of 3000 m, with a variety of applications	http://earth.esa.int/ dataproducts
ENVISAT ASAR	Radar images with a resolution of 30 m, 150 m or 300 m for use in the analysis of topography and vegetation	http://earth.esa.int/ dataproducts
Landsat TM/ETM+	Optical and thermal imagery available at spatial resolutions ranging between 15 and 120 m. The entire archive is now available for download free of charge	http://landsat.gsfc.nasa. gov
RADARSAT-1 & -2	High resolution (3 m to 100 m) radar imagery with applications in vegetation monitoring	http://gs. mdacorporation.com

Table 4.2 (*cont.*)

Data	Summary	Website
NASA SRTM	A radar-based acquisition satellite, the images of which form the basis of many digital elevation models (DEM)	http://www2.jpl.nasa.gov/srtm/dataprod.htm
Terra ASTER	Optical and thermal sensors acquire images with resolutions <90 m for use in land cover monitoring	http://asterweb.jpl.nasa.gov
IRS LISS-III & -IV	High-resolution optical sensor, the output of which may be used to analyse land cover and vegetation dynamics	http://earth.esa.int
ERS SAR	Radar-based sensor providing imagery at 30 m resolution, commonly used in vegetation monitoring	http://earth.esa.int/ers/eeo4.128
EO-1 Hyperion	Optical sensor acquiring imagery used in land cover and natural resource monitoring at a spatial resolution of 30 m	http://eo1.usgs.gov/
TerraSAR-X	High resolution (<16 m) radar sensor used to monitor vegetation and natural resources	www.infoterra.de
Ikonos	Very high resolution (<3.5 m) optical sensor use to monitor natural resources	www.geoeye.com
Quickbird	Very high resolution (<3 m) imagery with applications in land cover, vegetation and natural resources monitoring	www.digitalglobe.com

Table 4.2 (*cont.*)

Data	Summary	Website
GeoEye-1	Similar to Quickbird, this very high resolution optical sensor provides imagery at resolutions of <2 m. Commonly used in fine-scale vegetation monitoring	www.geoeye.com

2. Regional data sources
(a) Base layers

Data	Summary	Website
International Livestock Research Institute	Open access data layers for eastern and southern Africa, with particular resources for Kenya and Tanzania. Data include national base layers together with economic datasets	www.ilri.org
SAFARI2000	Repository of base and environmental data layers for southern Africa, including a selection of remotely sensed imagery.	http://daac.ornl.gov/ S2K/safari.html
Environment Information Centre	A geo-referenced database covering India. Data available upon request	www.eicinformation. org
AfricaMap	A database of free to download data layers covering Africa and Madagascar	http://africamap. harvard.edu/

(b) Climate

Data	Summary	Website
Africa rainfall estimates	High resolution precipitation grids for continental Africa	www.cpc.ncep.noaa. gov/products/fews/rfe. shtml
South America daily gridded precipitation analysis	Daily downloadable precipitation grids for South America	www.cpc.noaa.gov/ products/GIS/ GIS_DATA/sa_precip/ index.php

Table 4.2 (cont.)

Data	Summary	Website
(c) Soils, geology and topography		
Soil and terrain database for Latin America and the Caribbean	Georeferenced soils database at a scale of 1:5 million for South America	www.isric.org/UK/About +ISRIC/Projects/Track +Record/SOTERLAC. htm
(d) Habitat, land cover and land use		
Africover	Detailed land cover datasets derived from LandSat satellite imagery available for ten African nations together with a selection of base layers (e.g. political boundaries, roads and rivers)	www.africover.org
(e) Species data		
African mammal databank	Modelled species extent of occurrence and distribution datasets for all medium- and large-bodied African mammals	www.gisbau.uniroma1. it/amd
Digital distribution maps of the mammals of the western hemisphere	Distribution maps and data products for mammals in Latin and South America	www.natureserve.org/ getData/ mammalMaps.jsp
(f) Protected areas and priority areas for conservation		
Central American vegetation/land cover classification and conservation Status	Conservation gap analysis data based on AVHRR imagery covering Central America	http://sedac.ciesin. columbia.edu/ conservation/proarca. html
Central African Regional Programme for the Environment	Data mapper with limited analysis functions and links to downloadable data layers relating to forest resources in Central Africa	http://carpe.umd.edu

Table 4.2 (*cont.*)

Data	Summary	Website
(g) Socioeconomic data		
China country-level data on population and agriculture	Dataset containing demographic and agricultural data for China	http://sedac.ciesin. columbia.edu/china/ popuhealth/popagri/ census90.html

uncertain boundaries when examined at a fine scale. There is always a scale limit to sensible use of vector data.

Seemingly perfect topographic and climate data represented in pixel form may also disguise problems. For example, temperature surfaces are usually derived by interpolating data from ground stations (Chapter 5), which may only be sparsely distributed in more remote regions and may suffer from gaps in data records, particularly in developing countries. The reliability of such surfaces may depend on the distance of a pixel to the nearest ground station, the algorithms used in the interpolation and whether altitude has been taken into account (see, for example, Hijmans *et al.*, 2005). Topographic data interpolated from contour maps may also be problematic and it may be preferable to use the global product available from the Shuttle Radar Topography Mission (SRTM: http://www2.jpl.nasa.gov/srtm/) which has uniform coverage of most of the Earth (see, for example, Di Fiore & Suarez, 2007).

Remotely sensed data

There is an increasing range of remotely sensed data available that may be of interest to primatologists (Table 4.2). All remotely sensed data comes in raster format consisting of pixels. It is important to choose a product of an appropriate spatial resolution to match the primate data (Technical Note 4.1). For example, if the primate data consist of radio-tracking locations accurate to 50 m, it would be wise to choose a product with pixels of 100 m × 100 m or larger to minimize errors due to mis-location of data (Osborne & Leitao, 2009). You often need to compromise between spatial resolution, the number of pixels that make up the study area, and the number of spectral bands in the imagery, as it is easy to over-burden a GIS analysis with data. Both

---

Technical Note 4.1: Spatial resolution

Spatial resolution refers to the size of the smallest unit (i.e. the pixel) that is recorded in a raster image. It is therefore also a measure of the ability of an imaging system to separate closely adjacent features on the ground. Pixels are typically square areas ranging in size from 410 m × 410 m (for GeoEye-1 data) to 8 km × 8 km (for GOES imagery). Not all raster data comes in true squares, however. For example, the WorldClim dataset is available as a 30 arc second product (nominally referred to a 1 km × 1 km). At the equator, an arc second of longitude is approximately equal to an arc second of latitude (30.87 m). Whereas arc seconds of latitude remain nearly constant, arc seconds of longitude decrease towards the earth's poles. For example, at a latitude of 49° north, one arc second of longitude equals only 20.25 m.

---

hyper-spectral and multi-spectral imagery can be very useful for distinguishing vegetation types because they have several bands around the 'red edge' (boundary of visible red and infra-red). However, hyper-spectral data sets are generally composed of about 100–200 spectral bands of relatively narrow bandwidths (5–10 nm), whereas multi-spectral data typically involve 5–10 bands of relatively large bandwidths (70–400 nm). Such considerations are becoming less important as computer power increases, but could be important if you are thinking of running analyses in the field on a laptop.

Satellite imagery typically comprises a reflectance value per band per pixel and you need to decide how to process the data before analysis. All imagery requires atmospheric, radiometric and geometric correction before use (covered in any remote sensing textbook; see, for example, Lilliesand *et al.*, 2007), but assuming that this has been done, there are usually three choices: use the raw image values; calculate spectral indices from the image values; or use a classification algorithm to derive land cover classes. Our search of the literature revealed no primate examples using raw image data in analysis, but see Hepinstall and Sader (1997) for a study involving birds. Use of spectral indices and band combinations is also rare, but an excellent example is provided by Stickler and Southworth (2008). They used a suite of indices and band combinations to detect differences in green biomass and soil signatures in Uganda. Specifically, they used the Normalized Difference Vegetation Index (NDVI), Tasseled Cap Analysis (TCA),

---

Technical Note 4.2: Spectral indices and texture measures

Spectral indices are mathematical combinations of the reflectance values of selected spectral bands on an image. They are often used in vegetation studies to detect green standing biomass (productivity). The most common index is the Normalized Difference Vegetation Index (NDVI), which is defined as (Near IR band – Red band) divided by (Near IR band + Red band). For a Landsat TM image, this would be (Band 4 – Band 3)/(Band 4 + Band 3). NDVI and variants such as the Soil Adjusted Vegetation Index (SAVI) make use of information on only few bands. The tasselled cap transformation, on the other hand, uses all six Landsat TM bands to derive a moisture index, a soil brightness index and a green vegetation index. Importantly, these indices are orthogonal (i.e. zero correlated) as the tasselled cap procedure is similar to a principal components analysis.

Variability in reflectance values among neighbouring pixels can be caused by horizontal variability in plant growth forms (St-Louis *et al.*, 2009) and this variability may be quantified through so-called texture measures. First-order texture measures include the coefficient of variation and range of reflectance values among a group of pixels. Second-order texture measures are calculated from the grey-level co-occurrence matrix and account for the spatial arrangement of pixel values (Haralick *et al.*, 1973). There is a wide variety of measures with technical-sounding names including the angular second moment, difference entropy, inverse difference moment and sum of squares variance. See Haralick *et al.* (1973) for technical details and St-Louis *et al.* (2009) for an application to avian biodiversity.

Principal Components Analysis (PCA) and textural analysis (Technical Note 4.2) to derive predictor variables for the occurrence of redtail monkeys, *Cercopithecus ascanius*. There are many examples of the use of satellite imagery to derive land cover maps (e.g. Wilkie *et al.*, 1998 – Landsat Multi-Spectral Scanner) and to detect vegetation change (Srivastava *et al.*, 2001 – Indian Remote Sensing Satellite, IRS-1D; Miller *et al.*, 2004 – Landsat Thematic Mapper). These studies employed standard classification procedures (either supervised or unsupervised) which are described in remote sensing textbooks (e.g. Lilliesand *et al.*, 2007). Adventurous researchers should note, however, that traditional classification algorithms such as Maximum Likelihood are falling out of favour because of the unrealistic assumption of normally distributed data. Primatologists should make far greater use of machine-learning techniques such as Support Vector Machines (SVM: Huang *et al.*, 2002) in the future. Although studies employing classifiers have been successful, it is important to realize that land cover mapping has limited use in more homogeneous or continuous landscapes where within-class variability is of greater interest than the variability associated with clearly discrete landscape units (Stickler & Southworth, 2008). In such cases, vegetation and texture indices (Technical Note 4.2) are likely to provide better discriminators of primate habitats.

Although most remotely sensed data are satellite-derived, it may be possible to acquire airborne imagery (which includes aerial photographs) in some areas. Airborne imagery is captured by specially equipped aircraft, often owned by national environmental agencies or consulting firms. Costs of commissioning flights are high but it might be worth asking what data are already available for your study area. Be aware though, that airborne imagery is often challenging to process and you will almost certainly need the assistance of professional remote sensing scientists to use it. Aerial photos are a further source of remotely sensed data, although they are far less straightforward to use in analyses than satellite imagery since they require geo-referencing and may be limited to three RGB (red, green, blue) layers. Three layers are rarely sufficient to provide enough spectral information for automated classifications of habitat or land use.

Field data gathered with a GPS

A final source of environmental data is field data gathered with a GPS. Although it is highly labour-intensive, this may be the only way to capture fine-scale features that do not exist on maps and cannot be seen on satellite imagery. Good examples might be rough tracks through the bush, or point features such as water holes. Modern GPS devices permit you to record 'tracks' (i.e. continuous point records) that allow you to drive or walk routes while logging your position. You can download the tracks to computer via a USB cable and overlay them onto a base map with the correct coordinate system for the study area. This approach is very useful for updating or supplementing the features on an existing map (see, for example, Scholz & Kappeler, 2004; Anderson *et al.*, 2007).

INTEGRATING PRIMATE AND
ENVIRONMENTAL DATA

Although the use of satellite imagery and GIS data has increased in studies on primates, full integration of species location data with environmental data is rare. The examples that follow are broadly ordered from more straightforward approaches to complex techniques involving advanced statistics and satellite image analysis, including methods rarely used on primates. We would strongly advise anyone considering the complex approaches to consult specialists before embarking on data collection.

Simple GIS overlays

You can use remotely sensed data to build habitat maps that then direct field studies but serve little part in subsequent analysis. For example, you can use Landsat Enhanced Thematic Mapper (ETM) images to identify field sites for habitat and population studies or to divide habitat into areas depending on factors known to influence primate density (see, for example, Lahm *et al.*, 1998). You can use GIS to build distribution maps for primates without complex statistical techniques. For example, you can divide primate habitat into classes (based on forest cover maps, topography, conservation areas and other data sources) and create distribution maps by selecting the species-specific vegetation-class polygons that contain presence records for each species. You can then edit these maps using personal knowledge of features that are avoided, such as mountain ranges or rivers (see, for example, Meijaard & Nijman, 2003 for Borneo). It is far better, however, to produce maps more objectively using species distribution modelling tools that have rarely been applied in primate research (see below). Once you have distribution maps you can overlay them to find hotspots of primate diversity or endemism (e.g. Meijaard & Nijman, 2003) or across multiple taxonomic groups. For example, Hopkins and Nunn (2007) overlaid pre-existing primate range maps with data on the distribution of parasites from the Global Mammal Parasite Database (Nunn & Altizer, 2005) to perform a 'gap analysis', i.e. identify geographic areas that have been under-sampled for parasites in relation to primate ranges.

There are several other examples of use of simple GIS overlays in primate research. Banks *et al.* (2007) used a time-series of Landsat 7 images to assess how habitat changes might have affected a critically endangered lemur, Perrier's sifaka (*Propithecus perrieri*), in Madagascar, but did not model habitat use through locations (see below). Baumgarten and Williamson's (2007) study of howler monkeys (*Alouatta pigra* and *A. palliata*) overlaid species coordinates on GIS data layers but did not make use of the armoury of statistical tools available to integrate the two. Similarly, although Di Fiore and Suarez (2007) assessed the concordance between travel paths of sympatric spider monkeys (*Ateles belzebuth*) and woolly monkeys (*Lagothrix poeppigii*) in Amazonian Ecuador and topography visually by superimposing routes on a digital elevation model, they did not undertake any formal analysis. These are all examples of studies that could be improved through a better appreciation of GIS tools and modern statistics using the approaches below.

Analysing habitat use with basic statistics

Zinner *et al.* (2001) assessed the habitat characteristics of hypothetical home ranges (3 km radius) around baboon (*Papio hamadryas*) sleeping sites. Rather than doing this laboriously in the field, they used a GIS to analyse the values of NDVI (derived from AVHRR – Advanced Very High Resolution Radiometer – data; see Technical Note 4.2) and a vegetation classification (based on Landsat Multispectral Scanner imagery) to compare the characteristics of sleeping sites with background values from the entire study site, so inferring selection. In a further study, Zinner *et al.* (2002) reported that the home ranges of vervet monkeys (*Cercopithecus aethiops*) in central and eastern Eritrea had significantly higher average NDVI values than the entire area surveyed. Willems and Hill (2009) used the same index at a finer scale to study within-range habitat use. They found that within the annual home range area, monthly NDVI values were related to field measurements of leaf cover and vervet monkey food availability. The amount of time the monkeys spent on the ground decreased with monthly NDVI, interpreted as an anti-predatory response to changes in habitat visibility, associated with leaf cover. This study illustrates that simple spectral indices such as NDVI may be useful for understanding fine-scale habitat use within ranges in contrast to the more usual use in remote sensing at coarser spatial scales.

Analysing habitat use with complex statistics

Rather than distinguishing sites with and without the target species, Smith *et al.* (1997) related the abundance of lemur species to a suite of environmental variables at 64 sites in western Madagascar using regression and decision tree analysis. This paper is an early example of species distribution modelling, a subject which has exploded since the year 2000 with seminal reviews by Guisan and Zimmermann (2000), Scott *et al.* (2002) and the papers in collections by Guisan *et al.* (2002), Lehmann *et al.* (2002), Guisan *et al.* (2006), Moisen *et al.* (2006) and Franklin (2010). Species distribution modelling encompasses a range of statistical techniques for predicting the range of a species over a given area from incomplete location data. The statistics employed may be quite simple (e.g. linear regression analysis) but are usually complex modern approaches such as Generalized Linear or Additive Models, Boosted Regression Trees, genetic algorithms, neural networks and Maximum Entropy modelling (see Hastie *et al.*, 2001 and Phillips *et al.*, 2006 for an explanation of these methods).

Very few primatologists have applied species distribution models to their data, although Stickler and Southworth (2008) present a habitat suitability model for the redtail monkey (*Cercopithecus ascanius*), built using logistic regression analysis. Importantly, their study compared the relative performance of two high-resolution satellite sensors (2.5 m resolution Quickbird and 30 m resolution Landsat ETM+) in capturing patterns of habitat selection. Their work suggests that a model combining data from both the Quickbird and ETM+ sensors predicts monkey presence–absence best, and that individually, ETM+ data provide better predictors than Quickbird data. Moreover, model fit was best at larger spatial scales (ETM+ 120 m, 150 m and 240 m pixels). One practical implication of this study might be that it is not worth the additional cost of analysing habitat selection at the fine scale offered by the Quickbird sensor, an important consideration for any researcher.

Although species distribution modelling studies often employ advanced statistics, few consider the biasing effects of spatial autocorrelation and non-stationarity using methods specifically designed to handle these problems (Osborne *et al.*, 2007). A unique exception in the primate world is Willem and Hill's (2009) work on spatial range use by vervet monkeys. They used a mixed regressive–spatial regressive (or lagged predictor) model (Florax & Folmer, 1992) to assess the relative influence of resource availability and fear of predators in determining spatial range use. They found that the study animals spent most time in the two habitat types in which food was most abundant and also stayed close to sleeping trees and surface water. However, the models suggested that the effects of fear caused by predators may exceed the attraction of food resources, emphasizing how important it is to conduct integrative studies to achieve a good understanding of animal behaviour.

Landscape analysis

A useful product of any GIS and remote sensing approach that defines suitable habitat patches is the ability to derive landscape metrics such as patch sizes and distribution. Simple landscape metrics may routinely be calculated using Fragstats (Table 4.1) but more is possible. Bodin and Norberg (2007) provide a more sophisticated example using an agricultural landscape with scattered dry-forest patches in southern Madagascar, inhabited by ring-tailed lemurs (*Lemur catta*). They apply network-centric methods from the social sciences to identify (among other things) individual habitat patches that are disproportionately high in importance in preserving the ability of organisms to move across fragmented landscapes. The authors provide a free Java-based computer program named JMatrixNet (Table 4.1) to compute networks of habitat patches in landscape map images.

CONCLUDING REMARKS

It is at large spatial extents that GIS technology really comes into its own. Thus Devos *et al.* (2008) note in their study of ape densities and habitats in northern Congo that a combination of satellite imagery and ground surveys can effectively be used to assess, quantify and interpret large and remote tracts of forest, something that would be impossible using ground surveys alone. Other examples that would benefit from use of GIS include studies of human impacts (usually carried out through field surveys over relatively small areas; see, for example, Pinto *et al.*, 2003) and habitat assessments (for example, Anderson *et al.*'s 2007 field study of 46 forest patches in Kenya would have been so much easier if remotely sensed data had been used). If you are alert to these opportunities and aware of the latest technological developments, GIS and remote sensing promise a bright future for primatological research.

REFERENCES

Anderson, J., Cowlishaw, G. & Rowcliffe, J. M. (2007). Effects of forest fragmentation on the abundance of *Colobus angolensis palliatus* in Kenya's coastal forests. *Int. J. Primatol.* 28, 637–55.

Asensio, N., Korstjens, A. H. & Aureli, F. (2009). Fissioning minimizes ranging costs in spider monkeys: a multiple-level approach. *Behav. Ecol. Sociobiol.* 63, 649–59.

Banks, M. A., Ellis, E. R., Wright, A. & Wright, P. C. (2007). Global population size of a critically endangered lemur, Perrier's sifaka. *Anim. Conserv.* 10, 254–62.

Baumgarten, A. & Williamson, G. B. (2007). The distributions of howling monkeys (*Alouatta pigra* and *A. palliata*) in southeastern Mexico and Central America. *Primates* 48, 310–15.

Biebouw, K. (2009). Home range size and use in *Allocebus trichotis* in Analamazaotra Special Reserve, Central Eastern Madagascar. *Int. J. Primatol.* 30, 367–86.

Bodin, O. & Norberg, J. (2007). A network approach for analyzing spatially structured populations in fragmented landscape. *Landscape Ecol.* 22, 31–44.

Devos, C., Sanz, C., Morgan, D. *et al.* (2008). Comparing ape densities and habitats in northern Congo: Surveys of sympatric gorillas and chimpanzees in the Odzala and Ndoki regions. *Am. J. Primatol.* 70, 439–51.

Di Fiore, A. & Suarez S. A. (2007). Route-based travel and shared routes in sympatric spider and woolly monkeys: cognitive and evolutionary implications. *Anim. Cogn.* 10, 317–29.

Florax, R. & Folmer, H. (1992). Specification and estimation of spatial linear regression models: Monte-Carlo evaluation of pretest estimators. *Reg. Sci. Urban. Econ.* 22, 405–32.

Franklin, J. (2010). *Mapping Species Distributions*. Cambridge: Cambridge University Press.

Grueter, C. C., Li, D. Y., van Schaik, C. P. *et al.* (2008). Ranging of *Rhinopithecus bieti* in the Samage Forest, China. I. Characteristics of range use. *Int. J. Primatol.* 29, 1121–45.

Guisan, A. & Zimmermann, N. E. (2000). Predictive habitat distribution models in ecology. *Ecol. Model.* 135, 147–86.

Guisan, A., Edwards, J., Thomas, C. & Hastie, T. (2002). Generalised linear and generalised additive models in studies of species distributions: setting the scene. *Ecol. Model.* 157, 89–100.

Guisan, A., Lehmann, A., Ferrier, S. *et al.* (2006). Guest editorial: making better biogeographical predictions of species' distributions. *J. App. Ecol.* 43, 386–92.

Haralick, R. M., Shanmugam, K. & Dinstein, I. (1973). Textural features for image classification. *IEEE Trans. Syst. Man Cybernetics SMC* 3, 610–21.

Hastie, T., Tibshirani, R. & Friedman, J. (2001). *The Elements of Statistical Learning*. New York: Springer-Verlag.

Hepinstall, J. A. & Sader, S. A. (1997). Using Bayesian statistics, Thematic Mapper satellite imagery, and breeding bird survey data to model bird species probability of occurrence in Maine. *Photogramm. Eng. Rem. S.* 63, 1231–7.

Hijmans, R. J., Cameron, S. E., Parra, J. L., Jones, P. G. & Jarvis, A. (2005). Very high resolution interpolated global terrestrial climate surfaces. *Int. J. Climatol.* 25, 1965–78.

Hope, K., Goldsmith, M. L. & Graczyk, T. (2004). Parasitic health of olive baboons in Bwindi Impenetrable National Park, Uganda. *Vet. Parasitol.* 122, 165–70.

Hopkins, M. E. & Nunn, C. L. (2007). A global gap analysis of infectious agents in wild primates. *Diversity Distrib.* 13, 561–72.

Huang, C., Davis, L. S., & Townshend, J. R. G. (2002). An assessment of support vector machines for land cover classification. *Int. J. Rem. Sens.* 23, 725–49.

Lahm, S. A., Barnes, R. F. W., Beardsley, K. & Cervinka, P. (1998). A method for censusing the greater white-nosed monkey in northeastern Gabon using the population density gradient in relation to roads. *J. Trop. Ecol.* 14, 629–43.

Lehmann, A., Overton, J. M. & Austin, M. P. (2002). Regression models for spatial prediction: their role for biodiversity and conservation. *Biodivers. Conserv.* 11, 2085–92.

Lilliesand, T. M., Kiefer, R. W. & Chipman, J. W. (2007). *Remote Sensing and Image Interpretation.* (6th edn.) New York: John Wiley & Sons.

Markham, A. C. & Altmann, J. (2008). Remote monitoring of primates using automated GPS technology in open habitats. *Am. J. Primatol.* 70, 495–9.

Meijaard, E. & Nijman V. (2003). Primate hotspots on Borneo: predictive value for general biodiversity and the effects of taxonomy. *Conserv. Biol.* 17, 725–32.

Miller, L., Savage, A. & Giraldo, H. (2004). Quantifying remaining forested habitat within the historic distribution of the cotton-top tamarin (*Saguinus oedipus*) in Colombia: implications for long-term conservation. *Am. J. Primatol.* 64, 451–7.

Moisen, G. G., Edwards Jr, T. C. & Osborne, P. E. (2006). Further advances in predicting species distributions. *Ecol. Model.* 199, 129–31.

Nunn, C. L. & Altizer S. (2005) The Global Mammal Parasite Database: an online resource for infectious disease records in wild primates. *Evol. Anthropol.* 14, 1–2.

Osborne, P. E. & Leitao P. J. (2009). Effects of species and habitat positional errors on the performance and interpretation of species distribution models. *Diversity Distrib.* 15, 671–81.

Osborne, P. E., Foody, G. M. & Suárez-Seoane, S. (2007). Non-stationarity and local approaches to modelling the distributions of wildlife. *Diversity Distrib.* 13, 313–23.

Phillips, K. A., Elvey, C. R. & Abercrombie, C. L. (1998). Applying GPS to the study of primate ecology: A useful tool? *Am. J. Primatol.* 46, 167–72.

Phillips, S. J., Anderson, R. P. & Schapire, R. E. (2006). Maximum entropy modelling of species geographic distributions. *Ecol. Model.* 190, 231–59.

Pinto, A. C. B., Azevedo-Ramos, C. & de Carvalho, O. Jr., (2003). Activity patterns and diet of the howler monkey *Alouatta belzebuth* in areas of logged and unlogged forest in eastern Amazonia. *Anim. Biodiv. Conserv.* 26, 39–49.

Pochron, S. T. (2001). Can concurrent speed and directness of travel indicate purposeful encounter in the yellow baboons (*Papio hamadryas cynocephalus*) of Ruaha National Park, Tanzania? *Int. J. Primatol.* 22, 773–85.

Scholz, F. & Kappeler P. M. (2004). Effects of seasonal water scarcity on the ranging behavior of Eulemur fulvus rufus. *Int. J. Primatol.* 25, 599–613.

Scott, J. M., Heglund, P. J., Samson, F. *et al.* (2002). *Predicting Species Occurrences: Issues of Accuracy and Scale.* Covelo, Washington, D.C.: Island Press.

Smith, A. P., Horning, N. & Moore, D. (1997). Regional biodiversity planning and lemur conservation with GIS in western Madagascar. *Conserv. Biol.* 11, 498–512.

Srivastava, A., Biswas, J., Das, J. & Bujarbarua, P. (2001). Status and distribution of golden langurs (*Trachypithecus geei*) in Assam, India. *Am. J. Primatol.* 55, 15–23.

St-Louis, V., Pidgeon, A. M., Clayton, M. K. *et al.* (2009). Satellite image texture and a vegetation index predict avian biodiversity in the Chihuahuan Desert of New Mexico. *Ecography* 32, 468–80.

Stickler, C. M. & Southworth, J. (2008). Application of multi-scale spatial and spectral analysis for predicting primate occurrence and habitat associations in Kibale National Park, Uganda. *Remote Sens. Environ.* 112, 2170–86.

Wilkie, D. S., Curran, B., Tshombe, R. & Morelli, G. A. (1998). Modeling the sustainability of subsistence farming and hunting in the Ituri forest of Zaire. *Conserv. Biol.* 12, 137–47.

Willems, E.P. & Hill, R.A. (2009). Predator-specific landscapes of fear and resource distribution: effects on spatial range use. *Ecology* 90, 546–55.

Zinner, D., Pelaez, F. & Torkler, F. (2001). Distribution and habitat associations of baboons (*Papio hamadryas*) in Central Eritrea. *Int. J. Primatol.* 22, 397–413.

Zinner, D., Pelaez, F. & Torkler, F. (2002). Distribution and habitat of grivet monkeys (*Cercopithecus aethiops aethiops*) in eastern and central Eritrea. *Afr. J. Ecol.* 40, 151–8.

JULIAN C. MAYES AND NICHOLAS PEPIN

5

Monitoring local weather and climate

INTRODUCTION

Weather and climate have a profound influence on ecosystems. The climate of a region can act as a resource for an ecosystem as a whole, influencing the phenology of an area (Chapter 3), and as a determinant of food supply for primates, which can in turn affect reproduction, ranging and social interaction. Whereas climate provides the background conditions, the real-time monitoring of local weather can provide a much more direct link to primate field studies at any given time. For example, daily temperature cycles can influence activity patterns (Chapter 18).

Local weather information can be straightforward to collect and the value of local observations at field study sites has been enhanced in recent years by the increasing use of automated observations. Weather conditions also vary within a field study area as a result of the microclimate (see, for example, Geiger *et al.*, 1995; Rosenberg *et al.*, 1983); as a result, local conditions can differ significantly from larger-scale climatological surveys such as those of Buckle (1996) and McGregor and Nieuwolt (1998).

This chapter describes the types of local weather information that you can collect; shows how weather can vary over both space and time, enabling the interpretation of point observations in the context of wider conditions; and shows how you can measure local atmospheric conditions easily during fieldwork using different types of portable instrumentation.

Key terms for the main subjects covered are:

Weather: the state of the atmosphere at a given place and time.

Climate: the aggregation of weather conditions over several decades (typically at least 30 years) comprising both averages and extremes of particular climatic parameters (temperature, rainfall, etc.).

Field and Laboratory Methods in Primatology: A Practical Guide, ed. Joanna M. Setchell and Deborah J. Curtis. Published by Cambridge University Press. © Cambridge University Press 2011.

Meteorology: the science of the atmosphere; concerned with explaining variations in weather in both horizontal and vertical planes (the latter extending from the soil to the upper atmosphere).

Microclimate: the variation of weather and climate over small areas – from tens of metres up to a few hundred (Oke 1987).

Meteorological variation can be investigated at two distinct scales: local and regional. The local scale corresponds with variations in microclimate within a small study area; key influences are the type of land cover and the nature of the soil and ground surface. Most meteorological elements (such as temperature and wind speed) change abruptly between the surface and the first few metres of the atmosphere, within a zone easily accessed by human observers. Although primates may often live above this level, the near-surface conditions may influence food supply potential, especially in forested environments in which microclimate is heavily modified by such factors as light attenuation. At the regional scale, variations in altitude and topography (land shape) exert a greater influence on local weather conditions. Measurement of weather variables therefore allows you to place the findings from a specific study area within the context of a larger region or to compare different sites. The climatology of a site is expressed in terms of climatic averages, usually derived from daily measurements. It is important that these data are published in a consistent form throughout the world; standard terminology for the main climatic variables is defined as follows.

Mean daily maximum (minimum) temperature: the average of the highest (lowest) temperatures recorded over each 24 h period within a specified period (e.g. one month).

Mean temperature: usually the average of the mean daily maximum and minimum temperature; with automatic weather stations it is more likely to be the mean of the hourly temperatures. It is often misinterpreted as referring to day temperature. Though frequently quoted in published data, the concept of mean temperature may have less climatic relevance than mean daily maximum and minimum. For example, the minimum temperature stimulates flowering of some species the fruits of which are eaten by primates (Edwards & White, 2000).

Mean monthly precipitation: the average sum of the daily precipitation totals over a month.

Sunshine: usually measured in terms of duration of bright sunshine in hours.

Solar radiation: typically expressed as a daily mean in watts per square metre (W/m^2). Because the attenuation of light and solar radiation is such an important facet of habitat modification in forest microclimates, measuring solar radiation can provide highly relevant data.

Relative humidity: the actual estimated water vapour content of the air, expressed as a proportion of the maximum amount that air can hold at that temperature. The latter is temperature-dependent: as air cools, its capacity to hold water vapour diminishes (owing to changes in molecular bonding). Condensation of water droplets can occur when air approaches saturation (this can start at slightly below 100% relative humidity because of condensation nuclei, which attract water). If you know the relative humidity, you can calculate the temperature at which saturation will take place – the dewpoint temperature. This is the temperature below which dew, fog or low cloud may start to form. The high humidity of tropical rainforests means that dewpoint temperatures often reach 20 °C. You can estimate dewpoint temperature using a three-stage process (Kuemmel, 1997):

1. Calculate the saturation vapour pressure (the maximum amount of water vapour that air can hold at the recorded air temperature T_{db})

$$es_{db} = \exp\left(\frac{16.78T_{db} - 116.9}{T_{db} + 237.3}\right) \qquad (5.1)$$

2. Calculate the amount of moisture actually held in the atmosphere

$$e_d = \frac{RH \times es_{db}}{100} \qquad (5.2)$$

3. Calculate the dewpoint temperature

$$T_{dew} = \frac{116.9 + 237.3\ln(e_d)}{16.78 - \ln(e_d)} \qquad (5.3)$$

where T_{dew} = dewpoint temperature, T_{db} = air temperature, RH = % relative humidity, es_{db} = saturation vapour pressure, e_d = vapour pressure.

Airflow: Wind direction is expressed as the direction from which the wind is blowing to the nearest 10 degrees from North. North is recorded as 360 degrees and calm conditions as 0. Table 5.1 shows the relationships among the most common units of wind speed. Scientific surveys frequently use metres per second but, for example, in the British Isles wind speed is officially recorded in knots. Take care

Table 5.1 *Conversion factors between units of wind speed*

Unit	Conversion
1 m/s	= 2.237 miles per hour
	= 3.6 km/h
	= 1.94 knots (nautical miles per hour)
1 knot	= 1.15 miles per hour

when averaging successive wind direction readings. N should be logged as 36, NNE as 02, etc.; it would clearly be erroneous to average 36 and 02 over successive sampling intervals and assume the mean direction was 190°.

Wind speed is usually measured over at least 1 minute, to record a range of gusts and lulls. It is usual to note both the mean over this period and the maximum gust. You can also quote the gust: lull ratio, especially in the turbulent wake downwind of obstacles to the airflow (such as a stand of trees or buildings) where the contrast between gusts and lulls typically increases. Similarly, a 'gust factor' expresses the 3 second maximum gust as a proportion of the mean speed (Linacre, 1992).

Cloud cover is estimated by eye as the number of eighths (oktas) of sky that are covered with cloud.

Evaporation from a wet vegetated surface can be estimated by using the following formula (Linacre, 1992):

$$Et = [0.015 + 4 \times 10^{-4}T + 10^{-6}z]$$
$$\times [380(T + 0.006z)/(84 - A) - 40 + 4u(T - T_d)]\text{mm/d} \quad (5.4)$$

where T = dry bulb air temperature; T_d = dewpoint temperature; A = latitude; z = altitude in m; u = wind speed in m/s.

METEOROLOGICAL VARIATIONS AT THE LOCAL SCALE

Microclimate variation occurs according to (among others) time of day, the effects of vegetation, altitude and topography. The fundamental element is temperature, which has an intricate relationship with the surroundings, depending on the nature of the local terrain and the role of land cover, radiation, airflow and humidity.

Some definitions

Active surface

The active surface is the principal plane of meteorological activity in a system, where the majority of the solar energy is absorbed, reflected and emitted. It is where radiant energy is transformed into thermal energy (through absorption), where rainfall is intercepted and where drag on airflow is maximized. In the absence of vegetation, the active surface will be at the ground. Figure 5.1 shows the typical temperature profiles above and below the active surface of bare ground. During the day the ground is the focus of surface heating, but it is also the focal point of heat loss at night since it is where loss of long-wave radiation takes place. The concept of the 'active surface' identifies the energy transformations that create an area's microclimate. Most field surveys of temperature are carried out near the 'active surface', an area of sharp variations in temperature and other meteorological elements.

Albedo

The active surface warms as it absorbs a proportion of the solar radiation. Although cloud cover determines the amount of direct solar

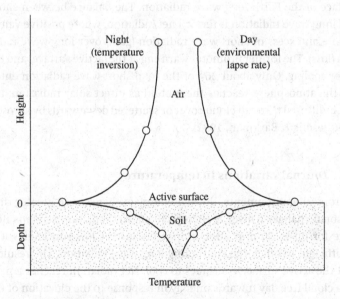

Fig. 5.1 Diurnal temperature profiles adjacent to an active surface (assuming light winds and clear skies).

radiation, the proportion of incoming radiation reflected by the surface also influences the degree of warming; this reflectance is expressed as the albedo of the surface. This can vary from less than 0.1 for dark tarmac or black sand to over 0.9 for fresh snow.

Environmental lapse rate (ELR)

This is the drop in temperature that occurs as height increases in the free atmosphere away from the immediate influence of the active surface. The average lapse rate in the global atmosphere is around −0.65 °C / 100 m, with some variation with altitude and latitude (Linacre, 1992). It is steeper when the surface below is being heated by sunshine. At night, unless winds and cloud cover are high, temperature often rises with increasing height (a temperature inversion, or positive lapse rate) (Fig. 5.1).

Net radiation

Heat energy is transferred through the atmosphere by radiation and is gained when a surface absorbs solar radiation. However, all surfaces also emit radiation according to the surface temperature; the higher the temperature, the shorter the wavelength of radiation. Solar radiation is therefore termed short-wave radiation and that emitted by the surface of the Earth long-wave radiation. The balance between short- and long-wave radiation is termed net radiation, where positive values denote an excess of short-wave radiation (gain) over long-wave radiation (loss). The former promotes warming at the 'active surface' and the latter cooling. Only about 26% of the total short-wave radiation entering the atmosphere reaches the ground as direct solar radiation; the rest is diffused through cloud cover or scattered downwards by aerosols (Stoutjesdijk & Barkman, 1992).

Diurnal variations in temperature

In tropical environments, diurnal energy exchanges are greater than seasonal changes (Oliver & Hidore, 2002). Net radiation remains negative for up to 30 min after sunrise because at low sun angles it is insufficient to offset the continuing long-wave radiation. The result is that the active surface continues to cool. Net radiation reaches a peak on a cloud-free day towards midday in response to the elevation of the sun. Lapse rates in sunny conditions may rise well above average levels at this time; −10 °C/km is not unusual and rates can be as much as

− 100 °C/km or steeper in the zone immediately above the active surface. Surface temperature will peak after a short delay; there is a further delay (perhaps an hour or two) before the temperature peak is experienced further away from the active surface, such as at the height at which air temperature is officially measured, 1.25 m (Fig. 5.2).

Vegetation effects

Vegetation raises the height of the active surface, typically to around two-thirds of the height of the vegetation or towards the height of maximum foliage density (tree canopy in the case of forest). This has a major effect on the vertical profile of temperature, wind speed and moisture above (and below) the ground. The principles of this modification apply to most vegetation types and environments.

Influence on temperature and radiation balance

The obvious shading effect of vegetation arising from the vertical attenuation of solar radiation often results in a temperature inversion below the height of the active surface. As the active surface is approached, highest day temperatures tend to be recorded at around the height of the highest leaf density. The proportion of short-wave radiation reaching the ground through woodland is often less than 20% of that above the canopy. At the ground, the diurnal range in temperature is reduced and this dampening of the temperature extremes is carried down into the soil.

Night-time temperatures are elevated under a vegetation canopy, owing to reduced emission of long-wave radiation. You can quantify this effect by means of the Sky View Factor (SVF): a completely unobstructed sky hemisphere is said to have a SVF of 1.0 and a site having a completely obscured sky has a SVF of 0.0.

Albedo varies with both vegetation density and time of day. The albedo of an individual leaf is typically around 0.30, but a whole stand of crops or trees will generally have a value closer to 0.18–0.25 owing to trapping of solar radiation between leaves (Oke, 1987). This effect is greater at midday when the sun angle is higher.

The thermal microclimate of individual leaves shows a sharp contrast to ambient air temperatures since sunlit leaves take on the function of active surfaces. The upper surface of a sunlit leaf can be 5–10 °C warmer than the adjacent air (with undersides being a few degrees cooler). At night, leaves are an effective radiator of energy and

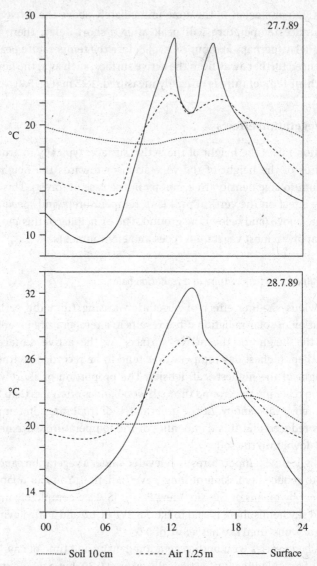

Fig. 5.2 Typical diurnal variations in air, ground and soil temperature measured by simple sensors on fine days during the UK summer (27 and 28 July 1989). Note the larger temperature range at the surface compared with 1.25 m and the sharp reduction of surface temperature during a cloudy period just after midday on 27 July.

can be up to 10 °C cooler than their surroundings, with possible implications for arboreal primates. The same applies to primates themselves, that will experience ambient temperatures much more extreme than those in the ambient environment as they absorb and exchange radiation with their surroundings, and act as an active surface.

Influences on airflow

Vegetation has two effects on airflow. First, the vertical profile of wind speed is modified in relation to vegetation density, owing to frictional drag. The vertical profile of average wind speed shows a peak above canopy level, a sharp decrease below this – in the zone of maximum leaf density – and an increase closer to the ground if there is a substantial zone of trunks and stems clear of leaves. Values decrease again at the ground owing to frictional effects. Both wind speed and direction in the stem zone can be independent of ambient (free-air) conditions above the canopy.

Second, the regularity of wind speed and direction is disturbed, with eddies forming, especially in clearings. Frictional effects of whole stands of trees disturb the wider pattern of airflow, creating swirling eddies downwind of woodland. This can lead to a variety of wind directions in forest clearings and an increased gust:lull ratio. Both of these changes have important implications for the accurate measurement of wind speed and direction around forest edges and in clearings.

Influence on moisture and relative humidity

Large forests are significant moisture stores; the leaf canopy intercepts incoming rainfall but a proportion reaches the ground either as stem-flow (water flow down vegetation stems), throughfall (flow directly to ground falling between leaves and stems) or leaf drip (flow bouncing and dripping off vegetation). Interception rates are highest in short-lived showers, when forest can be a temporary refuge from heavy rainfall. As rainfall duration increases, the infiltration ratio decreases and drip can continue well after rainfall has ceased. Forests also act as a source of evapotranspiration and humidify the air. Relative humidities are thus consistently higher in forested areas in comparison with clearings and open landscapes.

Regional variations

Meteorological variations over tens of metres upwards are influenced by additional factors, most fundamentally altitude and topography. Temperature variations at the surface are accompanied by density changes of the air close to the surface. In the absence of significant general (gradient) winds caused by pressure systems (a common occurrence in the tropics), daytime heating results in upslope (anabatic) winds and nocturnal cooling produces a downslope counterpart (the katabatic wind). Height variations of five metres or more within a study area may be sufficient to initiate such motion. This will be accompanied by ponding of cold air in hollows or valley bottoms at night, contributing to the nocturnal temperature inversion shown in Fig. 5.1. This is explored comprehensively in Geiger *et al.* (1995).

FIELD TECHNIQUES FOR MEASURING METEOROLOGICAL VARIABLES

This section introduces the key principles of meteorological measurement; you can achieve these either by traditional instruments that are read manually or by automated sensors (sometimes assembled as an automatic weather station). Electronic sensors work either in conjunction with a data-logger to record observations at a pre-determined frequency (Strangeways, 1997, 2003, 2004), or independently with an internal logging mechanism. Portable manual instruments can serve as a backup if automated systems fail (Crump, 1996).

Key principles

Whatever type of instrumentation you choose, it is important to understand the principles on which official observations are made. This allows comparison with official observing sites, and also informs the exact locations chosen for exposing the instruments (in the context of microclimatic variations considered above) and the timing of observations.

The basic assumptions of meteorological observation include:

- Air temperature is measured at a height of 1.25–1.5 m in shade (the temperature indicated by any thermometer exposed to direct sunlight is a function of the colour and construction of the thermometer itself as well as the amount of sunlight).
- Precipitation (the generic term for rain, hail, sleet, snow and dew) is measured as a depth that accumulates in a straight-sided

container, or as a volume (mm^3 or ml) divided by the area over which it fell (mm^2).

- Wind speed is usually measured officially at a height of 10 m above the local ground surface. Anemometers measuring wind speed and direction erected at lower heights can provide readings having relative if not absolute validity.
- In the absence of data-logging, observations should be made at the same time each day.

Measuring temperature

If time and finance permit, expose thermometers at a semi-permanent measurement site within a Stevenson screen: a ventilated wooden box, painted white to maximize reflection of direct radiation. Include dry- and wet-bulb thermometers (to measure air temperature and humidity) together with those for recording maximum and minimum temperature (Fig. 5.3). This standard for exposing thermometers is adapted in automatic weather stations with the use of the smaller and more portable Gill screen. It is important that thermometers exposed in screens are well ventilated in order to be representative of

Fig. 5.3 The interior of a Stevenson screen, showing dry-bulb and wet-bulb thermometers (vertical, in the centre), maximum and minimum thermometers (horizontal, in the centre), hygrograph (left) and thermograph (right).

Fig. 5.4 Cheap radiation shield constructed to provide protection for temperature/humidity data-loggers against radiative fluxes. The tube is open to provide ventilation and painted white to reflect radiation load. The orientation of the shield is explained in the text.

their surroundings. An additional refinement in poorly ventilated environments, such as forests, is an aspirated screen (fitted with a fan) if power is available.

Temperature measurements are often required from locations other than a base station. You can also construct highly portable home-made radiation shields for minimal cost. Figure 5.4 shows an example. A tube of white PVC (open at both ends) is orientated with its longest axis perpendicular to the mean solar path (i.e. the sun does not shine directly into the tube). Calculate the optimal angle of installation (measured as displacement from the vertical) by subtracting the latitude from 90°. At high latitudes the angle of the tube should be more vertical (the upper end facing the pole); at the equator a horizontal instalment is required (open ends facing north and south).

In unshaded terrain (where the active surface is at ground level) the ground surface is a zone of temperature extremes. To measure ground temperature, electrical sensors have the advantage of compactness and can be shaded more easily for daytime use than thermometers. Soil temperatures are most easily obtained by burying the small sensors of electrical thermometers to the required depth.

To measure the microclimate of a forest, you will gain more information by traversing in and out of the woodland, recording temperatures in contrasting environments over a short period of time than by *in situ* measurements (an alternative is to establish separate observing sites). There are two main methods of obtaining representative measurements at many locations when you only have one or two observers. The first is the transect method (Chandler, 1965) whereby you traverse the same route twice and space out the timing of the measurements evenly so that when you take a mean at each location it is representative of the 'middle time'. This makes observations from different sites comparable. The second, or control, method is to have one observer (or an automated system) taking consistent readings at a fixed location (the control) while the other observer moves about to the required locations (the rover) and to map the differences between the two simultaneous observations.

There are two simple, low-cost solutions to temperature measurement:

1. A whirling psychrometer (whirling hygrometer) consists of a thermometer that records the air temperature (the dry-bulb thermometer) and another that records the temperature that results from evaporation at the thermometer bulb (the wet-bulb thermometer). The whirling action shades and ventilates the thermometer bulbs (ventilation increases the rate at which thermometers adjust to the ambient temperature after storage). When whirling stops, the temperatures recorded may increase, so you need to read the thermometers quickly.

2. Simple battery-powered electronic sensors based on measurement of electrical resistance. These are often used if a spatially extensive survey is required. Most sensors are now automated and have a typical memory of up to 16 000 readings, with their own power supply (batteries) and data storage. You define the sampling interval (and hence the length of time for which the sensor can be deployed without becoming full). This can be years. In addition, most sensors allow wrap-around (the

sensor overwrites the oldest data when it is full and continues logging) or stop when full. Data can be downloaded in the field by using small portable hand-held devices.

Thermometers: accuracy, resolution and response time

Officially approved thermometers have an **accuracy** judged against reference instruments to within roughly ± 0.3 °C across a wide range of temperatures. The **resolution** of observations refers to the smallest change in a variable that can be measured or read, i.e. the precision rather than the accuracy of the reading. It is customary to read to one decimal place on the Celsius scale. High resolution combined with low accuracy can result in spurious accuracy. However, you should also consider whether it is essential for readings to be accurate in an absolute sense or merely to be internally consistent (relative calibration between instruments). The latter is more important in most extensive field surveys, and can be achieved by calibrating all sensors before and after field use, in the lab.

Response time is a critical factor in most field surveys. This is the time taken for the displayed temperature to come into equilibrium with the ambient temperature. Most electrical resistance thermometers used with sensors have a slightly faster response time than mercury thermometers. During a mobile survey, when carrying thermometers between observing sites, it is important to know the response time and allow for it before making readings.

Measuring humidity

Humidity is measured by a set of wet-bulb and dry-bulb thermometers (a psychrometer), visible as the vertical thermometers in Fig. 5.3. The difference in temperature recorded by the two thermometers is related to the humidity of the air because if a thermometer bulb is wet, evaporation will increase as the surrounding air becomes drier. This has a cooling effect as latent heat is absorbed from the surroundings with the change of state from liquid to vapour. If air is saturated, no evaporation will take place and both thermometers will give the same reading. The wet-bulb reading is not the same as the dewpoint temperature (which is lower) but you can calculate the dewpoint from hygrometric tables or slide rules supplied with a whirling psychrometer (or estimate it by using formula 5.3).

A surrogate for direct measurement of humidity involves the use of a hair hygrograph. Human hair expands in length as humidity increases. This movement can be shown by a moving arrow on a simple instrument (a hair hygrometer); a humidity recorder is created by connecting this to a rotating drum and a pen arm (Fig. 5.3). Finally, many electronic temperature sensors, such as those described in the temperature section, also record humidity (since air temperature is one of the required measurements for relative humidity).

Measuring precipitation

Precipitation is measured in units of length (conventionally in millimetres) since it is expressed as the depth that would be measured in any straight-sided cylinder. It is conventional to measure precipitation every 24 h. In general, the water collected (or melted) for a given diameter of funnel is measured in an appropriately calibrated measuring cylinder, but you can achieve a home-made calibration by dividing the amount of water collected (mm^3 or ml) by the area of the funnel (mm^2).

The funnel has a narrow tube to minimize evaporation from the inner collecting vessel. Standard rain gauges in the British Isles have a 5" (12.7 cm) diameter funnel; those in the United States have an 8" (20 cm) diameter funnel. A standard 5" gauge can hold at least 75 mm of precipitation, depending on the size of collecting vessel. The top of the funnel should be 30 cm above ground level. Lower gauges may record more due to in-splash and for this reason gauges should ideally be situated in short grass. Although it may seem convenient in field surveys to mount a gauge on a mast, any gauge higher than 30 cm will tend to record lower totals due to over-exposure in strong winds. This should be avoided, even if readings are not intended for comparison with nearby official gauges, because of a lack of consistency between days of strong and light winds. The resolution of all measurements should be to 0.1 mm (the measuring cylinder is narrower than the funnel to lengthen the vertical scale by a factor of more than 10). The top of the water column will form a meniscus around the edge of the cylinder and you should judge the level from the bottom of the meniscus.

Several types of recording rain gauge have been developed, of which the tipping bucket system is ideally suited to integration into an automatic weather station. A pulse is transmitted for each filling of a small cup corresponding to a known amount of precipitation (typically

0.2 mm). By recording the time of each tip, the intensity of rainfall can be deduced, although the precise duration of rainfall is easier to ascertain from traditional clock-driven tilting-siphon or natural-siphon gauges. In intense rainfall or hailstorms, erroneous results can be obtained from tipping bucket gauges owing to over-tipping; conversely, drizzle often fails to fill the bucket, and may evaporate before it is recorded.

Measuring wind speed and direction

Wind speed is officially measured by an instrument called an anemometer on a mast at an effective height of 10 m. The effective height is the height above nearby obstructions such as trees or buildings. This is often difficult in the field, and automatic weather stations usually have much shorter masts. It is important to site the anemometer as far as possible from local features that might disturb wind speed and direction. Adequate estimates of wind speed can be obtained from handheld anemometers held at arm's length at right angles to the wind, although a 10% correction is sometimes applied to give an estimate of the 10 m wind speed. However, this method makes estimation of wind speed in near calm conditions more difficult.

The maximum gust will often be 50%, or more, higher than the mean and the resulting gust:lull ratio is an important facet of the wind climatology of any region; it will tend to be higher in the vicinity of obstructions. Anemometer cups accelerate faster in gusts than they decelerate in lulls (Strangeways, 2003), leading to a risk of small overestimation of mean speed.

Automated measurement

A major advantage of automated wind speed and direction instruments is the facility for automatic calculation and display of scalar and vector means, avoiding the need for observer estimation. This is a useful function of even inexpensive hand-held digital anemometers. The number of revolutions is measured by the closure of a switch contact generated by the revolution of the cups. The total number of contacts divided by time can be used to give a display of the mean wind speed and the shortest time between pulses can indicate the maximum gust. Logging anemometers usually have a sampling interval (the period between instantaneous readings) and a recording interval (the period over which the individual samples are averaged). This allows you to calculate mean vector speeds and directions (for the case of wind direction a scalar mean can be very misleading).

Measuring radiation

The development of automated weather observation has made it possible to measure the intensity of solar and net radiation in the field. These variables are more relevant for many purposes than sunshine duration since they influence light levels and energy transfer. You can measure solar and net radiation by using pyranometers and net radiometers, respectively. Both work on the principle that receipt of radiation will result in a temperature difference between adjacent black and white panels due to contrasting surface albedo. Net radiation can be measured by exposing discs to both solar and terrestrial radiation (i.e. facing up and down). The spectral response of a net radiometer is wider than a solarimeter since it responds to both short-wave and long-wave radiation. Two solarimeters mounted up and down make an albedometer, which measures the reflected solar radiation as a proportion of the total incident input. Light-sensitive diodes can provide a cheaper alternative source of solar radiation or light data (Strangeways, 2003). If you take observations manually at a series of different sites and/or times, note the following: presence of sunlight on the observed area, state of ground (dry, wet, nature of surface) and sky view factor. The exact orientation of the instruments is crucial, as radiation receipt is sensitive to the slope of the surface. Most long-term climate studies are interested in a horizontal surface, but studies of microenvironments may involve surfaces and measurements at differential orientations (slope gradient and azimuth).

DATA-LOGGING AND AUTOMATED DATA ACQUISITION

Data-loggers can provide flexible round-the-clock monitoring of atmospheric conditions, and are invaluable at remote field sites or where regular fixed-point observations do not provide sufficient monitoring of meteorological changes. The principles of data-logging and details of the technical characteristics of data-loggers are outlined in Strangeways (2003). There are two main types of sensor: those that plug in to a central data-logger (as in an automatic weather station array) and stand-alone miniature sensors with in-built logging capacity. The choice between the two depends on the application. The major limitation of the former is that all sensors have to be in a small area, so this is the obvious choice for a climate base station at a fixed location. If the focus of study is spatial variability, then unless the spatial scale is small (<50 m) multiple stand-alone sensors are usually preferred. Stand-alone sensors also limit the need for long wires, which can be an additional advantage.

Modern data-loggers

Data-loggers are housed in an insulated sealed box, up to the size of a brick, connected by cable to the appropriate sensors, and with an internal or external power supply. The latter may be replaceable alkaline cells, lithium batteries or solar panels (charging lead-acid batteries). It is important that the data-logger is not subjected to extremes of temperature. Operating ranges typically span –30 °C to +60 °C but they have been used successfully in extreme environments such as Antarctica. Battery lifetime is maximized when a data-logger is sited in shelter (ideally indoors). Shade data-loggers from solar radiation and excess moisture in hot climates. Modern data-loggers are highly programmable. Two key variables are the sampling and recording intervals (see 'Airflow', above). Battery lifetime and period of operation depend on the sampling interval and recording interval, respectively. Data are downloaded to a laptop PC, or to a 'shuttle', removing the need for a laptop in the field. Real time downloading via the Internet is possible in advanced applications, even in remote locations, allowing you to interrogate data-loggers from afar.

CONCLUSIONS

Weather observation provides a rich source of local environmental data relevant to primate activity. Although methods of instrumentation may vary, the principles of meaningful climate observation remain the same. Individual field observations may be highly specific to particular times and locations but, collectively, observations made today also provide a potential source of data for future generations of scientists who may be interested in investigating the degree to which global climatic change has altered local environments.

REFERENCES

Buckle, C. (1996). *Weather and Climate in Africa*. Harlow: Addison Wesley Longman Limited.
Chandler, T. J. (1965). *The Climate of London*. London: Hutchinson.

Crump, M. L. (1996). Climate and environment (in Chapter 4: Keys to a successful project: associated data and planning). In *Measuring and Monitoring Biological Diversity - Standard Methods for Mammals*, ed. D. E. Wilson, F. R. Cole, J. D. Nichols, R. Rudran & M. S. Foster, pp. 52–56. Washington, London: Smithsonian Institution Press.

Edwards, A. & White, L. (2000). Methods for recording the weather. In *Conservation Research in the African Rain Forests: A Technical Handbook*, ed. L. White & A. Edwards, pp. 85–92. New York: Wildlife Conservation Society.

Geiger, R., Aron, R. H. & Todhunter, P. (1995). *The Climate Near the Ground*. (2nd edn.) Braunschweig: Vieweg Press.

Kuemmel, B. (1997). *Temp, Humidity and Dew Point ONA*. http://www.faqs.org/faqs/meteorology/temp-dewpoint/ accessed 7 Jan 2010.

Linacre, E. (1992). *Climate Data and Resources: A Reference and Guide*. London: Routledge.

McGregor, G. R & Nieuwolt, S. (1998). *Tropical Climatology*. (2nd edn.) Chichester: Wiley.

Oke, T. R. (1987). *Boundary Layer Climates*. London: Routledge.

Oliver, J. E. & Hidore, J. J. (2002). *Climatology: An Atmospheric Science*. (2nd edn.) New Jersey: Prentice Hall.

Rosenberg, N. J., Blad, B. L. & Verma, S. B. (1983). *Microclimate: The Biological Environment*. Chichester: John Wiley & Sons.

Stoutjesdijk, P. & Barkman, J. J. (1992). *Microclimate, Vegetation and Fauna*. Knivsta, Sweden: Opulus Press.

Strangeways, I. (1997). Ground and remotely sensed measurements. In *Applied Climatology: Principles and Practice*, ed. R. D. Thompson & A. H. Perry, pp. 13–21. London: Routledge.

Strangeways, I. (2003). *Measuring the Natural Environment*. (2nd edn.) Cambridge: Cambridge University Press.

Strangeways, I. (2004). Back to basics: The 'met. enclosure': Part 10 - Data loggers. *Weather* 59, 185–9.

LIST OF SUPPLIERS

http://www.campbellsci.com - Campbell Scientific Ltd. (automated logging equipment).

http://www.casellameasurement.com - Casella Measurement (traditional and automated meteorological and environmental monitoring equipment).

http://www.geminidataloggers.com - Gemini Data Loggers (suppliers of the *Tiny Tag* series of sensors).

www.munro-group.co.uk - RW Munro (traditional and automated meteorological monitoring equipment).

http://www.onsetcomp.com - Onset Computer Corporation (micro-datalogger company)

www.benmeadows.com/weather/default.htm - The Weather Place at Ben Meadows Co. (automated meteorological observing sensors and equipment)

6

Survey and census methods: population distribution and density

INTRODUCTION

A population study of a wild primate typically involves a considerable investment of time and resources (i.e. money, equipment, labour) and it is vital to ensure that such effort is well targeted. When designing your study, a key issue is whether your study objectives genuinely demand an **absolute estimate** of the population density from either a **census** (a total count) or a **survey** (in which density is estimated from statistically valid samples), or whether less information will suffice. **Relative estimates** of density using data from methods such as 'catch per unit effort' from trapping or systematic searching do not provide absolute densities but, as long as the sampling methods and other conditions are standardized, can allow reliable comparisons between locations and monitoring of population change over time. **Population indices** are based on indirect indicators that can be correlated with population density, such as the density of faeces or other characteristic signs. Such methods may be a more practical alternative to searching for secretive, hard-to-find animals.

In practice, no population survey or census is completely bias-free and many studies may find that a reliable relative population estimate or index is more achievable than a reliable absolute estimate of the population size (Bibby *et al.*, 1992; Greenwood, 1996; Krebs, 1999). There is a trade-off between the depth of the data gathered and the number of replicate samples that can be obtained. The choice of method in the field will also be affected by the degree to which data from individuals are required (e.g. sex, age, reproductive status, individual recognition) and whether capture and identification tagging is feasible (Chapters 7 and 10). Figure 6.1 should help you to choose the appropriate method for your own study.

Field and Laboratory Methods in Primatology: A Practical Guide, ed. Joanna M. Setchell and Deborah J. Curtis. Published by Cambridge University Press. © Cambridge University Press 2011.

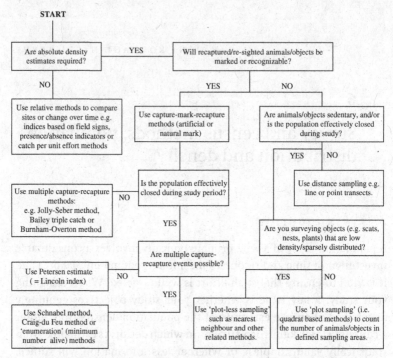

Fig. 6.1 Choice-chart giving a general guide to measures of abundance most likely to be useful in primate-related studies of abundance. Note that the precise application of each method, and variants of those methods, will be governed by specific assumptions too detailed to be included in this chart.

To review this entire topic here is impossible. We aim to provide a straightforward, non–mathematical guide to some key issues to consider in primate population studies and to direct the reader to detailed descriptions of the analysis methods and useful further reading.

BEFORE YOU START

- Decide what resources you can commit to the population study.
- Learn about the study site and target species (Chapter 2) and prepare for handling and trapping animals if needed (Chapters 7 and 8).
- Think about the pros and cons of different census methods for your particular study. It is vital that you start thinking about statistical analysis at this stage! Your time, energy and money will be wasted if you cannot interpret your data.

- Practise using different techniques, perhaps in a local park where you could carry out a plot sample of birds' nests or spiders' webs or a line transect counting squirrels (or people). Play around with your data and do a variety of analyses.
- Use a pilot study to try out different methods of data collection, for example indirect versus direct observation methods, line transects in different places, walking transects at different speeds, working alone (if it is safe) or with others, familiarizing yourself with traps or other equipment. If possible, collect enough data for preliminary analysis – it may have a major effect on the type of study you eventually decide to carry out.

WHAT TO COUNT AND RECORD

Always record the following: date and time, observer(s), weather conditions and other special circumstances (e.g. recent hunting). When you detect animals, you may also record activity, height above ground, age/sex class etc. Assessing age requires study of the age classes used by previous researchers and some practice (Chapter 9). You can count group size and composition. If groups are stable, repeated estimations should yield increasingly accurate counts (Fig. 6.2). However, these records may be inaccurate if some classes behave more conspicuously, avoid humans (e.g. mothers with infants) or the group is widely dispersed and not all animals are located. Estimate biomass if you can make accurate estimates of group numbers and compositions – this will be more accurate than estimates based on absolute numbers only, if good body mass data are available for different age/sex classes.

SPECIES PRESENCE/ABSENCE RECORDS

Recording presence or absence can help to establish the geographical range or habitat requirements of primate species (see, for example, Singh et al., 1999). Such data can also contribute to species lists, biodiversity evaluations, and the identification of suitable sites for further ecological or behavioural studies (see, for example, van Krunkelsven et al., 2000; McGraw, 1998). These surveys are comparatively simple to carry out, but the data yield is relatively low. Of course, proving absence conclusively may require a prolonged study or repeated investigations.

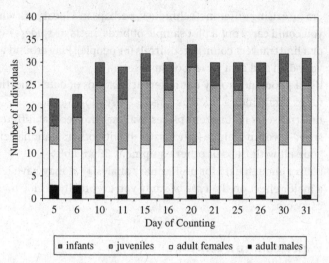

Fig. 6.2 Repeated counts of a single group of Hanuman langurs (*Presbytis entellus*) taken over a month in Southern India. Early counts have clearly missed some animals whereas later counts show between 25 and 30 animals. On most days, one adult male is present. The 'extra' males present on days 5 and 6 were peripheral, chased repeatedly by the third male and once seen in association with another group. Juveniles were difficult to count as they were frequently playing and chasing on the edge of the group (for example, on day 26 when juveniles were playing with those from another group the number recorded was probably too high).

The presence of primates in an area may be clear from both direct and indirect evidence, such as:

- Direct observation of live animals.
- Indirect observation of characteristic signs (tracks, faeces, nests, carcasses, feeding signs, calls, etc.). See Wemmer *et al.* (1996) and Parnell (2000) for summaries of field signs used in mammal surveys.
- Observation of animals captured or killed (e.g. in markets or local villages). In this case, take care to ascertain where the animals were obtained.
- Reports from local people. The value of such reports will depend on the accuracy of identification of both the species of interest and the exact location of the sighting.

For the sake of both internal consistency and comparability with other studies, standardize survey techniques and sample all areas in the locality equally.

ABSOLUTE ESTIMATES OF POPULATION DENSITY

The choice of method for absolute density estimates depends on several factors, but two fundamental questions are:

- Are natality, mortality, emigration and immigration *during the study period* likely to be significant (open population) or can they be discounted as negligible (closed population)?
- Can you identify the animals by using natural or artificial markers (Chapter 10)?

Total counts

A total count of all animals in the study area (a census) is rarely feasible but a virtually complete count of all animals may be possible if the population is effectively closed, and in the following circumstances:

- Few animals will be missed, for example because of good visibility or a very high chance of trapping all animals. Detection of animals may be enhanced by well-developed field skills and knowledge of predictable daily or seasonal activity patterns – animals may be more visible at certain times of the day or year. You could target counts on aggregations at known key locations such as at sleeping sites or water holes.
- Individuals can be identified to prevent double-counting (Chapter 10). If you can identify stable social groups (by location or presence of key animals) and count them accurately, then individual identification may not be needed to ensure a good estimate.

In long-term studies, you can combine census work with other research to give an overall picture of population demography and ecology. For example, research on mountain gorillas (*Gorilla beringei*) in the Virunga volcanoes has included both repeated censusing (of animals, nests, other signs) and the study of a number of focal groups. This has allowed researchers to count almost all individuals, map their distribution into social groups and estimate home range sizes (Fossey & Harcourt, 1977; Harcourt *et al.*, 1981). However, even short studies can yield almost complete counts, if an area can be covered rapidly, for example by using teams of workers (see, for example, McNeilage *et al.*, 2001 for mountain gorillas in Uganda).

POPULATION ESTIMATES FROM SAMPLES

When you cannot make a total count for the whole study area, estimate population density from a survey of a statistically robust, representative sample of smaller areas within the main study area. Ideally, these areas should be randomly located but stratified random sampling is often advisable. Stratification in this case involves identification of the different habitat types and *separate* random sampling of these to ensure each is sampled sufficiently well (Krebs, 1999). Three common sampling methods used in primate studies are: sampling of habituated groups, plot sampling and distance sampling. Less common methods include plot-less sampling, capture–mark–recapture methods and estimates from indirect observation.

Sampling habituated groups

You can obtain accurate density estimates if you know the group size, home range and group overlap for one or more groups in an area. Here, the area being censused is the group home range area of one or more well-studied groups. Such density estimates are likely to be very accurate if only needed for this local area, and many studies assume this to be the most accurate measure that can be obtained (Brugiere & Fleury, 2000). However, accurate estimates of group overlap may be impossible if non-habituated neighbouring groups avoid the area when observers are present, resulting in an under-estimate of density. This scenario is more likely if the study relies on a relatively small number of habituated groups that overlap with several non-habituated groups.

Another problem arises if you extrapolate from data from well-studied groups to estimate population density over a wider area, a form of plot sampling (see below), where the plot is the home range of these groups. This assumes that estimates of home range, group size and group overlap from studied groups are representative of the population as a whole. Obtaining good estimates of these parameters almost always relies on long-term data from habituated animals, but these groups may not be representative of the population as a whole and may have been chosen because other factors make them amenable for study, for example proximity to roads, ease of habituation, visibility, large group size. In such situations, you will need to sample larger areas and estimate population size from representative samples.

Plot sampling

Plot sampling is effectively the census (total count) of a representative set of replicate quadrats (standard subsamples) of the study area. The population density estimate for the area is the mean of those obtained for the plots. Plots are typically either square (e.g. trapping grids) or rectangular (e.g. strip transects), but can be circular (e.g. point transects). Plot sampling is a good method for censusing inanimate objects (e.g. nests, field signs or plants) that are relatively easy to find and to mark, thus preventing double-counting (Chapter 3).

The major problems with plot sampling are related to difficulties in finding all the animals (or objects) within the area. If animals are hard to find, then you may need extensive replication. In areas of good visibility, you can use strip transects and count animals within a defined distance from a central line – each transect can be a separate plot or you can include several replicate transects within a large plot. If you are working with several people then you can traverse a series of parallel transects to carry out a sweep of the general area, although you need to take care that you don't double-count.

Strip transects require knowledge of the area that can be accurately sampled, i.e. you need to determine the distance over which all animals can be seen from the line. This is derived from the distribution of the distances of the animals from the transect line. In Fig 6.3a, the distribution suggests that all animals are visible up to a certain cut-off point, in this case 40 m on either side of the transect line. The population density in this example is simply derived from the number of animals within a strip 80 m × 1000 m (8 ha). Because the strip width is determined by factors affecting visibility, there may be differences between transects in different habitats, times of day, seasons and weather conditions. If results of transects in different locations or seasons are significantly different, you may need to standardize strip width to the width of the transect with lowest visibility.

Using a line transect and counting all animals may yield results as in Fig. 6.3b, in which animals are increasingly difficult to see the further they are from the observer. If plot sampling were used to analyse these data, the area sampled would be an extremely thin strip transect (18 m wide in Fig. 6.3b) and very few of the observed animals would be included in the data. This is problematic because density estimates based on such a small, unrepresentative sample would be inaccurate. However, distance sampling seeks to overcome

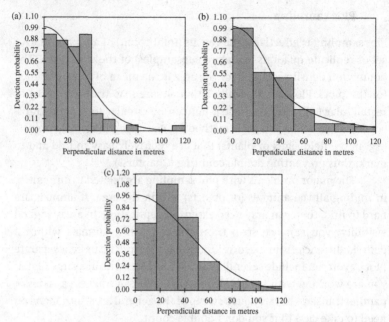

Fig. 6.3 Examples of data recorded from line transect sampling. In all cases we used a total of 62 observations to generate the graphs shown. We used the program Distance to generate the detection functions (g), which is shown as a curve on each graph (when using the program you can fit a number of models to the data and choose the best fit).

(a) Here a high proportion of animals are seen up to 40 m away. If the data used are truncated at 40 m it could be used to give a total count of all animals within the area 40 m either side of the transect line. Alternatively, a higher proportion of the data could be used if a model were used to estimate the proportion of animals not seen at greater distances, using g.

(b) The detection function falls off more rapidly with distance than in (a).

(c) The data suggest that relatively few animals have been detected on the line (see text).

this problem by allowing you to include a greater sample of animals in the analyses.

Distance sampling

Distance sampling counts as many animals (or signs of an animal's presence) as possible from a point or line within an area. Whereas plot sampling assumes that all individuals or signs within the area are detected, distance sampling assumes that only a proportion will be seen. This proportion (which declines with distance from the observer) is then estimated and used to calculate an estimate of the total number of animals present.

There are two forms of distance sampling: **line transect sampling** in which the observer walks (or drives) along a line and records animals seen on either side of the line and **point sampling** in which the observer stands at one or more selected viewpoints and records animals seen from that location (effectively a transect of zero length). Point samples, often used in bird surveys, have certain design advantages. They are easier to locate randomly or systematically (especially in woodland or other dense habitats where the need for access may bias the location of transects) and detection is not compromised by the noise of walking, obstacles or uneven ground (Bibby *et al.*, 1992). However, primates are typically wide-ranging and occur at relatively low densities, requiring a large number of point samples over a wide area and travel between dispersed points, whereas line transects allow you to cover a wider area while recording data simultaneously (Buckland *et al.*, 2001). Point sampling may be preferred, for example, because of difficult terrain or when movement is restricted for political reasons. Bibby *et al.* (1992) and Buckland *et al.* (2001) provide further information on point sampling; see also Bollinger *et al.* (1988) for a comparison with line sampling.

If you plan to rely heavily on distance sampling, it is important to look at this subject in detail, including the mathematical basis of population density estimations, which we do not discuss in depth here (Anderson *et al.*, 1985; Buckland *et al.*, 2001). The comprehensive guide by Buckland *et al.* (2001) is supported by the PC software 'Distance' (Thomas *et al.*, 1998), which allows relatively easy analyses of data obtained from distance sampling, can be downloaded free (see Useful Internet sites) and is used widely (see, for example, Blom *et al.*, 2001). Notation and abbreviations used here follow those used in Buckland *et al.* (2001) and in the software.

The basic procedure for line transect sampling is as follows:

- Locate the transect line (L) randomly in the area.
- Traverse the line, recording the observed number (n) of objects on either side of the line.
- Each time an object is seen record *either* (i) its perpendicular distance from the line (x) *or* (ii) the sighting distance from the observer (r) *and* the sighting angle (θ) (Fig. 6.4). If you choose (ii), both r and θ are then used to estimate the detection function.

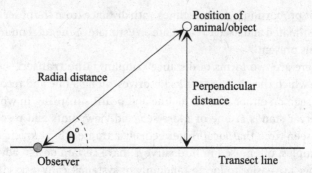

Fig. 6.4 Measuring the distance of sighted animals/objects from the transect line. Ideally you should measure the perpendicular distance directly, but you can also calculate it from the radial distance and the bearing (θ = angle from the transect line).

- Calculate the 'detection function' (g) (Fig. 6.3), using data for a number of objects detected at different distances, and use this to estimate the total number of animals in the area. Estimation of g is best done using 'Distance' (Buckland *et al.*, 2001). White and Edwards (2000) discuss alternative methods of estimating the detection function and hence population density if you do not have access to a computer.

For line distance sampling to be an effective means of estimating population density, a number of assumptions are made:

- *Objects directly on the transect line have never been missed.* For arboreal species this assumption may be violated, as animals above the transect line might be missed. Buckland *et al.* (2001) emphasize the importance of ensuring that this condition is met by paying particular attention to the area on and directly above the line. If objects on the transect line are missed then the distribution of animals against distance may reflect this, as shown in Fig. 6.3c.
- *Objects do not move before being detected.* If animals move before detection, the results will be biased towards far or near distance sightings.
- *Objects are not counted twice in a single transect walk.* Double counting is most likely to occur if animals flee by moving ahead of the observer or the transect is not straight and the same area is seen twice from different parts of the line. The second problem can be avoided by careful placing of transects and mapping of areas visible from two parts of the same line. The first problem is less

easily overcome, but avoidable if you can identify individuals. It is less important if objects are detected twice in two different walks of the same transect or in two walks of different transects, providing detection in the first walk is independent of detection in the second. In such cases, record each sighting and treat it as a separate independent sighting for analysis.

- *Distances and angles are measured accurately.* Ideally, use a tape measure or laser rangefinder and carry an alternative means of measurement. You can measure the angle of sighting with an angleboard or compass. Visual estimation of distances or pacing are less accurate. Avoid putting distances into bands (e.g. 0–5 m, 6–10 m, etc.) as this decreases the accuracy of the analyses. Check the accuracy of estimated distances frequently, to quantify measurement error, to prevent 'drift' of estimations over time and to ensure intra-observer reliability. Normally, the simplest measure to take is the perpendicular distance of the object from the transect line, but if this is difficult to do take the radial distance and the angle (θ) (Fig. 6.4).

- *Sightings are independent events.* Analysis assumes that sightings are both randomly distributed and independent. For social species, such as most primates, individuals are likely to be clustered and the group is treated as the single object for analysis rather than each individual animal (see, for example, Fashing & Cords, 2000), with distances measured either to the group centre or to the nearest individual.

- *Sufficient sightings are made for an accurate estimate of the detection function.* To calculate an accurate estimate usually requires at least $n = 60$–80 individual sightings, although $n = 40$ may sometimes be sufficient (Buckland *et al.*, 2001). For group-living animals n = the number of groups and the data are used to calculate mean group density. You can then estimate population density using the mean number of animals per group – data that can be obtained either during line transect sampling or from reliable group counts achieved in other situations.

There has been some debate about the best way to carry out line transects and the methods for analysis. This focuses on three questions (Marshall *et al.*, 2008): (i) whether to use radial (animal–observer) distance or perpendicular distance from track to the transect, (ii) whether to estimate individual densities by only using individuals seen on transect walks or by assuming a mean group size estimated during census walks or from an independent census, and (iii) whether or not to use information on group spread to estimate the group centre. Before you embark on any such work, refer to Marshall *et al.* (2008) and comparisons of different methods (Brockelman & Ali, 1987; Brugiere & Fleury, 2000; Peres, 1999; Plumptre, 2000). These papers emphasize that the accuracy of different methods will depend on how well the data fit with the assumptions of the model being used and this, in turn, will vary with species, habitat, and the time and money available. Marshall *et al.* (2008) provide a very useful flow chart that helps with choice of method.

Traversing the area and detecting animals

Most primate population surveys are carried out by walking through a study site, looking and listening for animals. Vehicles permit you to travel longer distances but usually limit the survey to trails, roads or waterways. Aerial surveys are rarely appropriate (but see Ancrenaz *et al.*, 2004). Whatever your mode of transport, stop, look and listen at regular intervals. You may need to leave the transect line to count and detect animals or objects, although search effort should concentrate on areas near to the line (Buckland *et al.*, 2001). Take care to return to the line and not to miss any part of the transect.

All nocturnal primates except for owl monkeys (*Aotus* spp.) and tarsiers (*Tarsius* spp.) have a reflective layer (the tapetum) at the back of the eye – creating 'eyeshine' in torchlight. Eyeshine has been used successfully to detect nocturnal primates in several studies, and it is particularly useful in species that freeze when threatened (Charles-Dominique, 1977; Singh *et al.*, 1999). Thermal imaging, especially of nocturnal species, is increasingly being used to aid detection; hand-held devices are now available. Thermal imaging has its limitations, especially if the thermal contrast between the animal and its environment is poor or forest cover is dense. Nevertheless, it has been used successfully in both aerial and ground surveys of a number of bird and mammal population studies (e.g. Kinzel *et al.*, 2006; Garner *et al.*, 1995; Havens & Sharp, 1998), and in difficult situations, e.g. semi-fossorial ground squirrels (*Spermophilus parryii*, Hubbs *et al.*, 2000) and forest deer (Cervidae, Gill *et al.*, 1997), as well as in comparison to spotlighting for a range of mammals (Focardi *et al.*, 2001).

Minimizing observer bias and maximizing the reliability of detection is particularly important. Key factors to standardize in the survey protocol include:

- Observer technique, vigilance, skill, motivation and other inter-individual differences.
- Sampling effort (e.g. length of transects and time spent travelling them). Walking speed needs to be slow enough to detect enough animals while allowing a sufficiently large area to be surveyed. Most forest surveys of primates have found that a speed of 1–2 km/h is appropriate.
- Habitat conditions (e.g. seasonal variation) can drastically alter the detection function.
- Weather conditions – differences in visibility and environmental noise will alter the detection function.

Plot-less sampling methods

Several methods can be used to estimate the density of immobile objects (e.g. field signs), based on measuring the distance between randomly chosen objects and their nearest neighbour or between randomly chosen locations and the nearest object. These can allow you to establish not only the density of objects, but also the nature of their spatial distribution (e.g. random, clumped or regular). These techniques are also often referred to as 'distance' methods (see, for example, Krebs, 1999; Southwood & Henderson, 2000) creating the potential for confusion with the distance sampling described above. A range of plot-less sampling methods, such as the T–square method, are explained well in Krebs (1999).

Capture–mark–recapture methods

These methods are among the most common ways of providing absolute mammalian population density estimates, but normally require that animals are caught (usually by systematic trapping), distinctively marked (Chapter 10), released and recaptured later on one or more regularly timed occasions. Trapping is labour-intensive, but creates opportunities for additional data collection (physical, physiological, genetic, parasitological, etc.), radio-tagging and identification marking (Chapters 7, 8, 9 and 10). You can use capture–mark–recapture methods without trapping (e.g. using plot or distance sampling) if animals can be systematically detected and identified at a distance. Artificial marks may not be required if individuals can be distinguished by their natural features.

A real strength of these methods is the detailed information that can be obtained from longitudinal studies, allowing not only reliable absolute population estimates to be made but also determination of demographic variables, such as birth and mortality rates and population age structure. However, you must be alert to the detailed assumptions of the technique you use.

Choice of capture–mark–recapture analysis method

There are many methods, and variants of methods, to choose from and a vast primary literature covering their application to a variety of taxa. Begon (1979), Greenwood (1996), Krebs (1999) and Southwood and Henderson (2000) are very useful overview texts that detail the assumptions and calculations for a wide range of methods. For a good review of techniques for mammals in general, see Nichols and Conroy (1996). A number of free software packages are also available (see Southwood & Henderson, 2000).

Table 6.1 summarizes the basic characteristics of the principal, most commonly used methods (see also Fig. 6.1). Some of the key issues to bear in mind when planning your capture–mark–recapture study are:

- Is your study population open or closed (see above)?
- Can you assume that all animals in the population (irrespective of sex, age, social status, etc.) are equally trappable? This assumption is fundamental to most methods except the Burnham–Overton method (Greenwood, 1996).
- Is the probability of recapture affected by trapping and marking? Animals may learn to avoid capture or to exploit the traps for food and/or shelter (Chapter 7).
- Marks must be permanent (at least during the study) and date-specific. Marks can be used either just to distinguish all animals caught at a particular time, or to identify certain subgroups or individuals.
- Will multiple recapture events be possible? How many are appropriate and can be resourced? The simple Petersen estimate requires only one capture and subsequent recapture event, but more sophisticated methods will require several regularly timed recapture events.
- Trapping/catching effort must be enough to initially catch and recapture animals in sufficient numbers for good statistical power; you may need to mark as much as 50% of the population to obtain accurate estimates (for details see Krebs, 1999).

Table 6.1 *Summary of some of the principal methods judged most likely to be useful in the estimation of primate population abundance from capture–mark–recapture data*

All methods, unless otherwise stated, require that: (i) marks are not lost during the study, (ii) marks are always correctly recognized, (iii) all animals (whatever age, sex or status, marked or not) have an equal probability of capture, (iv) the duration of sampling itself is brief in relation to the total study time. Multiple capture methods usually assume strictly regular spacing of sampling periods.

Method	Application, assumptions, key points
Closed populations	
Petersen estimate (Lincoln index)	Single capture/marking and single recapture events. Good for short studies, but restrictive assumptions and requires high levels of catchability.
Schnabel method	Multiple capture/marking and recapture events. Successive Petersen estimates used to calculate a weighted mean population estimate. Graph plot of the proportion of marked animals per catch against total previously marked (each estimate is a point on the graph) is linear if assumptions are met.
Craig du Feu method	Multiple capture/marking and recapture events. Successive, cumulative population estimates are made until the sufficient samples have been taken to stabilize the value with an acceptable standard error (Greenwood, 1996).
Enumeration (minimum no. alive methods)	Multiple capture/marking and recapture events. Based on a log of successive captures of individually marked animals. Those known to have been alive throughout any specified period are counted. Always an underestimate of the true population – a bias that worsens with low overall catchability (Krebs, 1999).
Open populations	
Jolly–Seber method	Multiple capture/marking and recapture events. Normally requires individually unique marks to determine when individuals were last captured. Calculates losses and recruitment to population. Equal catchability is a key assumption. Method

Table 6.1 (*cont.*)

Method	Application, assumptions, key points
	of choice when assumptions are met and numbers captured are not too low.
Bailey triple-catch	Three capture/marking and recapture events. Effectively a special case version of the Jolly–Seber method, allowing population losses and recruitment to be estimated quickly, but with less power (Begon, 1979).
Burnham & Overton method	At least four capture/marking and recapture events. Uses counts of number of animals caught exactly 1, 2, 3 and 4 times. Sensitive to change in capture probabilities over time and trapping effort must be constant. Allows individuals to differ in probability of capture (Greenwood, 1996).

Indirect observation

If you cannot observe animals directly, you may be able to use calls to identify species and to estimate population density (Brockelman & Ali, 1987). If species identification is in doubt, then always make recordings of calls to allow later verification (Chapter 16).

Observation of signs such as faeces, nests and tracks is useful for indicating presence, but you can also use the density of the objects (assuming a known probability of detection) to calculate population density if you can determine the following factors:

- Creation rate of the object per animal or group (e.g. number of nests built per week or number of defecations per day).
- Duration, i.e. how long the object persists in the environment after its creation.

Thus: population density = object density / (duration × creation rate).

You can use nest counts to estimate great ape population numbers (Tutin *et al.*, 1995), and the size of associated faeces to estimate the age class of the animals in each nest (McNeilage, 2001). Great apes usually build a new nest each night and only unweaned animals share a nest, hence the creation rate per weaned individual is relatively easy to calculate. However, animals may build more than one nest per night, resulting in over-estimates, as suggested for estimates of gorilla numbers in Bwindi Impenetrable National Park, Uganda (Guschanski et al., 2009). One way of ameliorating this is to use faecal DNA analyses to determine the number of individuals that have used nest groups (Guschanski et al., 2009). Nest decay rates can be estimated by using marked nests and vary with nest material and environment both between and within study sites (Tutin et al., 1995; Mathewson et al., 2008). Decay rates are highly variable and you must ensure you use good estimates (Walsh & White, 2005; Marshall & Meijaard, 2009). Detection rates are calculated as for detection rates of animals (Fig. 6.3). As with animal counts, count groups of nests (found in gorillas and chimpanzees) as a single object and measure to the centre of the group. Nest-counting methods are discussed in Blom et al. (2001), Furuichi et al. (2001) and Hashimoto (1995); aerial counting of orang-utan nests is discussed by Payne (1987); and problems with these nest counting methods are discussed by Walsh and White (2005) and Marshall and Meijaard (2009).

OBTAINING A REPRESENTATIVE SAMPLE

Whichever sampling method you select, ensure that the locations of your sample areas represent the habitat of the study area in an unbiased way, and that sampling effort is sufficient for a robust population estimate. Pragmatic decisions on the size of the study area, sampling methods and level of replication must be balanced against, but not override, the need to avoid bias and ensure adequate statistical power. A pilot study is vital if the area and/or species have not been studied previously. Chapter 2 discusses familiarization with your study species.

You can achieve the random location of a plot or transect simply by using a map and random numbers (from tables or computer-generated) to give grid references. Transects can be sited within the study area (or within an area of specific habitat if stratification is required) along a line joining two randomly generated grid references. However, plot or transect location may not be entirely random if your study aims dictate that it should cover a particular range of habitats (Chapter 3). For example, if you were interested in how population density of one or more species varied in relation to the proximity of human disturbance, you might place transects perpendicular to roads or fields (Fig. 6.5). Transects are often placed in regularly spaced parallel lines. This does not seriously violate the assumption of randomness,

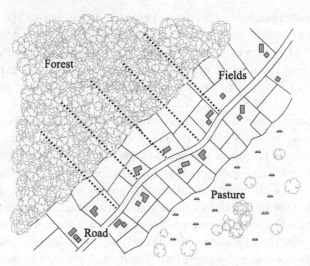

Fig. 6.5 A series of parallel transects (dashed lines) designed to investigate how species composition changes with variation in human disturbance.

provided the first transect was chosen randomly, the distance between lines does not replicate a feature of the landscape, and lines are sufficiently separated to ensure independence.

Transects pose particular problems when attempting to avoid bias and represent the range of different vegetation types, altitudes or other features in the study area properly. Make sure transects do not follow contour lines or water courses or run along established paths or tracks, as these are unlikely to be representative. In thick vegetation, you may need to cut trails, but minimize disturbance. If natural features prevent cutting of trails or movement along the planned line, decide whether to truncate the transect, allow a detour, or abandon the line and start another. Whatever you decide, consider all possible biases you may be introducing to your survey; for example, you might lose data on an important habitat by avoiding marshy ground. If no practical solution is found, note that your data exclude marshy areas.

Transect length will depend on population density and factors affecting detection rates – where detection rates are low the minimum n of 60 will take longer to accumulate. In multi-species surveys, the species with the lowest detection rate sets the standard. A single transect through one particular habitat might be appropriate in some cases, but, if you are investigating differences between several habitat types, either a long transect through different habitats or a number of shorter transects may be needed (N.B. $n = 60$ is required for each habitat type).

It is important to map plots, transect lines and trap positions carefully to allow later replication. This can be done using traditional means, artificial markers and/or photographs of key features or GPS (Edwards & White, 2000; Chapter 4). Note the locations of observed animals on maps as this may allow analysis of habitat use or ranging patterns.

REFERENCES

Ancrenaz, M., Goossens, B., Gimenez, O., Azri Sawang, A. & Lackman-Ancrenaz, I. (2004). Determination of ape distribution and population size using ground and aerial surveys: a case study with orang-utans in lower Kinabatangan, Sabah, Malaysia. *Anim. Conserv.* 7, 375–85.

Anderson, D. R., Burnham, K. P. & Crain, B. R. (1985). Estimating population size and density using line transect sampling. *Biom. J.* 7, 723–31.

Begon, M. (1979). *Investigating Animal Abundance, Capture–Recapture for Biologists.* London: Edward Arnold.

Bibby, C. J., Burgess, N. D. & Hill, D. A. (1992). *Bird Census Techniques.* London: Academic Press.

Blom, A., Almasi, A., Heitkonig, I. M. A., Kpanou, J. B. & Prins, H. H. T. (2001). A survey of the apes in the Dzanga-Ndoki National Park, Central African Republic: a comparison between the census and survey methods of estimating the gorilla (*Gorilla gorilla gorilla*) and chimpanzee (*Pan troglodytes*) nest group density. *Afr. J. Ecol.* 39, 98–105.

Bollinger, E. K., Gavin, T. A. & McIntyre, D. C. (1988). Comparison of transects and circular-plots for estimating bobolink densities. *J. Wildlife Manag.* 52, 777–86.

Brockelman, W. Y. & Ali, R. (1987). Methods of surveying and sampling forest primate populations. In *Primate Conservation in the Tropical Rain Forest*, ed. C. W. Marsh & R. A. Mittermeier, pp. 23–62. New York: Alan R. Liss.

Brugiere, D. & Fleury, M. C. (2000). Estimating primate densities using home range and line-transect methods: a comparative test with the black colobus monkey *Colobus satanas*. *Primates* 41, 373–82.

Buckland, S. T., Anderson, D. R., Burnham, K. P. et al. (2001). *Introduction to Distance Sampling: Estimating Abundance of Biological Populations.* Oxford: Oxford University Press.

Charles-Dominique, P. (1977). *Ecology and Behaviour of Nocturnal Primates.* London: Duckworth.

Edwards, A. & White, L. (2000). Maps, compasses, GPS units and principles of navigation. In *Conservation Research in African Rain Forests: A Technical Handbook*, ed. L. White & A. Edwards, pp. 31–51. New York: Wildlife Conservation Society.

Fashing, P. J. & Cords, M. (2000). Diurnal primate densities and biomass in the Kakamega Forest: an evaluation of census methods and a comparison with other forests. *Am. J. Primatol.* 50, 139–52.

Focardi, S., De Marinis, A. M., Rizzotto M. & Pucci, A. (2001). Comparative evaluation of thermal infrared imaging and spotlighting to survey wildlife. *Wildl. Soc. Bull.* 29, 133–9.

Fossey, D. & Harcourt, A. H. (1977). Feeding ecology of free-ranging mountain gorillas (*Gorilla gorilla beringei*). In *Feeding Ecology*, ed. T. H. Clutton-Brock, pp. 415–47. London: Academic Press.

Furuichi, T., Hashimoto, C. & Tashiro, Y. (2001). Extended application of a marked-nest census method to examine seasonal changes in habitat use by chimpanzees. *Int. J. Primatol.* 22, 913–28.

Garner, D. L., Underwood, H. B. & Porter, W. F. (1995). Use of modern infrared thermography for wildlife population surveys. *Environ. Manag.* 19, 233–8.

Gill, R. M. A., Thomas, M. L. & Stocker, D. (1997). The use of portable thermal imaging for estimating deer population density in forest habitats. *J. Appl. Ecol.* 34, 1273–86.

Greenwood, J. J. D (1996). Basic techniques. In *Ecological Census Techniques: A Handbook*, ed. W. J. Sutherland, pp. 11–110. Cambridge: Cambridge University Press.

Guschanski, K., Vigilant, L., McNeilage, A. *et al.* (2009). Counting elusive animals: comparing field and genetic census of the entire mountain gorilla population of Bwindi Impenetrable National Park, Uganda. *Biol. Conserv.* 142, 290–300.

Hashimoto, C. (1995). Population census of the chimpanzees in the Kalinzu Forest, Uganda: comparison between methods with nest counts. *Primates* 36, 477–88.

Harcourt, A. H., Fossey, D. & Sabater Pi, J. (1981). Demography of *Gorilla gorilla. J. Zool. Lond.* 195, 215–33.

Havens, K. J. & Sharp, E. J.. (1998). Using thermal imagery in the aerial survey of animals. *Wildl. Soc. Bull.* 26, 17–23.

Hubbs, A. H., Karels, T. & Boonstra, R. (2000). Indices of population size for burrowing mammals. *J. Wildl. Manag.* 64, 296–301.

Kinzel, P. J., Nelson, J. M., Parker, R. S. & Davis, L. R. (2006). Spring census of mid-continent sandhill cranes using aerial infrared videography. *J. Wildl. Manag.* 70, 70–7.

Krebs, C. J. (1999). *Ecological Methodology.* (2nd edn.) Menlo Park, CA: Addison Wesley Longman.

Marshall, A. J. & Meijaard, E. (2009). Orang-utan nest surveys: the devil is in the details. *Oryx* 46, 413–8.

Marshall, A. R., Lovett, J. C. & White, P. C. L. (2008). Selection of line-transect methods for estimating the density of group-living animals: lessons from the primates. *Am J. Primatol.* 70, 452–62.

Mathewson, P. D., Spehar, S. N., Meijaard, E. *et al.* (2008). Evaluating orang-utan census techniques using nest decay rates: implications for population estimates. *Ecol. Appl.* 18, 208–21.

McGraw, W. S. (1998). Three monkeys nearing extinction in the forest reserves of eastern Cote d'Ivoire. *Oryx* 32, 233–6.

McNeilage, A., Plumtre, A. J., Brock-Doyle, A. & Vedder, A. (2001). Bwindi Impenetrable National Park, Uganda: gorilla census, 1977. *Oryx* 35, 39–47.

Nichols, J. D. & Conroy, M. J. (1996). Techniques for estimating abundance and species richness. In *Measuring and Monitoring Biological Diversity: Standard Methods for Mammals*, ed. D. E. Wilson, F. R. Cole, J. D. Nichols, R. Rudran &. M. S. Foster, pp. 177–234. Washington, D.C.: Smithsonian Institute Press.

Payne, J. (1987). Surveying orang-utan populations by counting nests from a helicopter: a pilot survey in Sabah. *Prim. Conserv.* 8, 92–103.

Parnell, R. J. (2000). Information from animal tracks and trails. In *Conservation Research in African Rain Forests: A Technical Handbook*, ed. L. White & A. Edwards, pp. 157–89. New York: Wildlife Conservation Society.

Peres, C. A. (1999). General guidelines for standardizing line transect surveys of tropical forest primates. *Neotrop. Primates* 7, 11–6.

Plumptre, A. J. (2000). Monitoring mammal populations with line-transect techniques in African forests. *J. Appl. Ecol.* 37, 356–68.

Singh, M., Lindburg, D. G., Udhayan, A., Kumar, M. A. & Kumara, H. N. (1999). Status survey of slender loris *Loris tardigradus lydekkerianus* in Dindigul, Tamil Nadu, India. *Oryx* 33, 31–7.

Southwood, T. R. E. & Henderson, P. A. (2000). *Ecological Methods*. (3rd edn.) Oxford: Blackwell Science.

Thomas, L., Laake, J. L., Derry, J. F. *et al.* (1998). *Distance 3.5. Research Unit for Wildlife Population Assessment.* St. Andrews, UK: University of St. Andrews.

Tutin, C. E. G., Parnell, R. J., White, L. J. T. & Fernandez, M. (1995). Nest building by lowland gorillas in Lope Reserve, Gabon: environmental influences and implications for censusing. *Int. J. Primatol.* 16, 53–76.

van Krunkelsven, E., Inogwabini, B.-I. & Draulans, D. (2000). A survey of bonobos and other large mammals in Salonga National Park, Democratic Republic of Congo. *Oryx* 34, 180–8.

Walsh, P. D & White, L. J. T. (2005). Evaluating the steady state assumption: simulations of gorilla nest decay. *Ecol. Appl.* 15, 1342–50.

Wemmer, C., Kunz, T. H., Lundie-Jenkins, G. & McShea, W. J. (1996). Mammalian signs. In *Measuring and Monitoring Biological Diversity: Standard Methods for Mammals*, ed. D. E. Wilson, F. R. Cole, J. D. Nichols, R. Rudran &. M. S. Foster, pp. 157–76. Washington, D.C.: Smithsonian Institute Press.

White, L. & Edwards, A. (2000). Methods for assessing the status of animal populations. In *Conservation Research in African Rain Forests: A Technical Handbook*, ed. L. White & A. Edwards, pp. 225–73. New York: Wildlife Conservation Society.

USEFUL INTERNET SITES

www.ruwpa.st-and.ac.uk/distance/ – for Distance sampling.

CLIFFORD J. JOLLY, JANE E. PHILLIPS-CONROY
AND ALEXANDRA E. MÜLLER

7
Trapping primates

INTRODUCTION

There are many reasons to capture study animals. They include marking or radio-collaring (Chapter 10), taking morphological measurements (Chapter 9), or biological samples (Chapters 1 and 8), and
estimating age and condition. For small nocturnal primates, capture
is essential to radio-tag animals for direct observation, the most effective method of determining the spatial distribution and social interactions of individuals and estimating population densities (Chapter 6;
Sterling et al., 2000). Historically, studies in which wild, larger-bodied,
primates are habituated for long-term observation have rarely included
capture, perhaps because researchers have been understandably wary
of its effects on subsequent behaviour and habituation (Chapters 2
and 11). However, a survey of more than 120 studies that combined
observation with capture, and which involved about 65 primate
species, showed that a careful capture–release programme using trapping will not cause a previously habituated population to change its
behaviour towards human observers, and will not be associated with
excess mortality or serious injury (Jolly & Phillips-Conroy, 1993 and
unpublished data). Changes in ranging habits will be temporary at
worst, and basic social organization and structure will not be affected.
The survey also provided a comparison between capture methods.
Trapping has been used most often to catch diurnal-terrestrial and
nocturnal-arboreal species. Diurnal-arboreal primates (apart from callitrichines) have generally been captured by darting (Chapter 8), a bias
that seems unjustified. Although in skilled hands either trapping
or darting can produce good results, where trapping is possible it is
much more productive, far less hazardous to the subjects, and less
likely to cause them to become frightened of an observer. It is true

Field and Laboratory Methods in Primatology: A Practical Guide, ed. Joanna M. Setchell and Deborah
J. Curtis. Published by Cambridge University Press. © Cambridge University Press 2011.

that some species are notoriously difficult or even impossible to trap – either because they will not approach a strange object such as a trap (e.g. owl monkeys, *Aotus*, and titis, *Callicebus*: Silveira *et al.*, 1998), or because they are not attracted to foods offered as bait (most folivores and gummivores) – but trapping should certainly be attempted for arboreal and semi-arboreal, frugivorous-omnivorous primates such as lemurids, platyrrhines such as capuchins (*Cebus*), and catarrhines such as forest cercopithecines and some colobines.

Here we outline trapping methods, drawing both on the results of our survey and on our experience catching cercopithecines (CJJ & JP-C) and nocturnal lemurs (AEM). Although details vary greatly with the biology of the species and the study's objectives, all live-trapping projects involve similar considerations: when to capture, which animals to capture, trap design, number of traps, trap placement, baiting, trapping and processing the animals, and a protocol of holding, recovery and release.

WHEN TO CAPTURE

Since trapping is unlikely to prejudice subsequent observation, it is best to trap early in the study programme, immediately after the animals have been habituated to human presence (Chapter 2). This gives the study the immediate advantage of marked and radio-collared individuals, and the support of a comprehensive dataset that includes individuals that emigrate or die during the study. Data for animals that immigrate or are born later on must of course be added by individual trapping or by non-invasive sampling of 'bio-detritus' (Chapter 21). Although trapping can be carried out at any time of year, it is generally most productive when natural food is less abundant (Charles-Dominique, 1977).

WHICH ANIMALS TO CAPTURE

There are obvious economies of scale to be derived by sampling the population comprehensively, and no sex or age category needs to be excluded. Some capture programmes have avoided pregnant females and young infants, because these are felt to be particularly at risk from trapping and handling. This restriction is unnecessary, however, as long as you are careful to avoid separating dependent infants from their caregivers – usually their mothers, but (especially among some platyrrhines) other group members too. We (CJJ & JP-C) have successfully captured, processed, and observed after recovery pregnant female baboons that gave birth the next day, and at least one infant with its umbilical cord still attached. Contrary to expectation, the most vulnerable animals among cercopithecines seem to be the adult males, which are most likely to sustain minor trauma, and are also most susceptible to overheating under ketamine sedation (Chapter 8).

Even if your observation is to focus on a single social group, it is advantageous to trap neighbouring groups as well. Not only will this document the wider demographic and genetic context, it will also allow animals that are likely to immigrate into the primary study group to become accustomed to traps and bait. Unless previously familiarized in this way, immigrants tend to be excluded from bait and traps by more confident, acclimatized, residents.

All studies report that the vast majority of individuals simply resume their previous social roles after trapping and release. There is, however, a possibility that, unless the group is captured in its entirety at one time, the temporary absence of key individuals may cause changes in the status of other group members. In particular, consort-ships may be disrupted long enough for mating to be affected, and in species such as gelada (*Theropithecus gelada*) and hamadryas (*Papio hamadryas, sensu stricto*) baboons, 'harems' may disperse or be taken over by rival males. Such events can be considered analogous to natural injuries or predation events that similarly affect individual life histories, but do not alter the overall structure of society. By timing the trapping season as early as possible, you will ensure that any capture-induced changes in status occur before individual identities have become a crucial component of your observational data.

TRAP DESIGN

Trapping primates depends upon conditioning them to recognize traps as desirable and 'friendly' feeding sites, which means that traps can be entirely visible and do not have to be disguised. Two major kinds are used. Gang traps are designed to catch several animals at a time; individual traps are meant to catch one animal per closure, although in practice multiple captures often occur. In theory, gang traps have the advantage of a high capture rate and of catching whole social units together, thus minimizing the chance of social disruption. Although several investigators (T. Shotake, personal communication, 1999; J. Bert, personal communication, 1970) have reported using gang traps successfully to catch macaques (*Macaca* spp.) and baboons

(*Papio* and *Theropithecus*), our experience of them has not been positive. Large traps are difficult to transport and set up in the field; removing animals from the trap is much more difficult, and correspondingly more stressful to both the operator and the animals. There is also a greater risk that smaller individuals will be bitten when trapped *en masse*. An alternative is to use a multi-chambered trap to capture entire social groups (see, for example, Garber *et al.*, 1993; Savage *et al.*, 1993).

Most traps used to catch monkeys consist of a simple, custom-built, individual cage with a hinged or (more usually) a vertically sliding door (Brett *et al.*, 1982; de Ruiter, 1992). An alternative design catches the animal under a shallow tray-like box that is propped on edge when set (Brett *et al.*, 1982). Commercially available Tomahawk traps, ideally wrapped with 5 mm^2 wire mesh to prevent bait stealing, have been used for callitrichines (Albernaz & Magnusson, 1999; Dawson, 1977; J. Dietz, unpublished data).

Custom-built cage traps, probably the design of choice for most species, will obviously vary greatly in size. For convenience and for the animal's safety, they should be as small and light as possible. For example, a cage that is a cube with sides of about 90 cm and a door measuring 50×50 cm^2 can catch and hold a 30 kg baboon. The height should be sufficient to allow the animal to sit upright, but need not be much greater than that. Traps are best constructed with a rigid frame but flexible mesh walls, using materials as light in weight as is compatible with necessary strength. If securely tied down to prevent tipping, a trap of chain link, or even good quality chicken wire, on a light metal frame will hold any primate up to baboon size. Screen the bait with fine mesh netting to prevent animals reaching it from outside.

Most traps are designed so that they can be closed either automatically (tripped by the animal itself) or manually (released from a distance by an operator). An automatic trigger can be a treadle, operated by the animal's weight, or a simple catch that the animal releases by pulling on a bait. Manual closure usually involves the operator pulling a string attached to the catch, from a distance (commonly *c.* 25 – 100 m in open habitats, less in forest) that must be determined empirically in the field. The operator can be concealed in a hide (blind) (Garber *et al.*, 1993), but this restricts vision and movement, and we have not found it necessary for catching baboons or vervet monkeys (*Chlorocebus aethiops*).

For nocturnal species, the classic ('Chardonneret') trap is a $30 \times 35 \times 50$ cm^3 box made of wire mesh stapled to a wood frame, with a large, automatically triggered entrance on top (Charles-Dominique &

Bearder, 1979; Müller, 1999). It has proved effective for most galago species, pottos (*Perodicticus potto*) and small cheirogaleids, and has also caught pygmy marmosets (*Cebuella pygmaeus*; Zingg, 2001) and brown lemurs (*Eulemur fulvus*; A. Müller, unpublished data). Appropriately sized commercial traps such as Sherman and Tomahawk Live Traps are also commonly used to catch cheirogaleids (see, for example, Atsalis 1999; Ehresmann, 2000; Fietz, 1999; Fietz & Ganzhorn, 1999; Kappeler, 1997; Lahann, 2007; Wright & Martin, 1995). They have the advantage of being portable, convenient and comparatively light, but importation can cause bureaucratic problems and expense.

Whatever its design, it is important that a trap has no sharp edges inside; the use of plastic-covered wire can prevent some injuries. If the animal's tail is likely to lie across the threshold, incorporate a device to prevent guillotining when the door shuts. Trapped animals often try to enlarge any small break in the netting, so it is very important to fix the netting securely to the frame (binding it with wire is usually more reliable than welding) and to mend breaks promptly. Escape efforts are usually concentrated on the door, which must be rigid enough not to spring out of its runners when impacted vigorously from inside. Many primates become adept at raising the door manually. You can prevent this with a simple catch that locks when the door drops.

NUMBER OF TRAPS

The number of traps employed will obviously vary with the demography of the target species, and anticipated capture rates. For group-living species, more traps will produce a larger per-day yield, and may make it easier to catch the shyest individuals. Processing the captives then becomes the limiting step, especially if they are to be released immediately. About 20 traps per group will probably be a practical maximum. Some studies of nocturnal strepsirrhines have used up to 200 portable, commercial traps, processed animals in the camp and released them the following night (see, for example, Fietz & Ganzhorn, 1999; Schwab, 2000). Other studies in which up to 80% of the traps were successful, and animals were processed at the capture site, have used 20–40 traps (Müller, 1999; A. Müller, unpublished data).

SELECTION OF THE TRAP SITE

The prime consideration here is that the animals should visit the trap site frequently, so that they can be acclimatized to taking bait – a site near a

favoured sleeping-place is ideal. It should be close enough that it is usually traversed as the animals enter and leave their sleeping-place, but not
so close that a visit at dusk or dawn by the trappers is likely to scare the
animals. If a cluster of manually closed traps are to be used, choose the
site so that they can all be scanned and operated from a single position.

Arboreal species must be trapped in the trees, at a height where
the animals are accustomed to feed. Traps should be positioned so as to
afford easy access, for example fixed between two or more small trees,
in a fork of a branch, or wired securely to a horizontal branch or vine.
For group-living species, it is often advantageous to construct an elevated baiting platform ($1 \times 2\,\mathrm{m}^2$) and place traps on it and in the surrounding trees (J. Dietz, unpublished data; Dietz *et al.*, 1994; Savage
et al., 1993). When trapping nocturnal strepsirrhines that forage alone,
you can disperse a large number of traps in a grid (see, for example,
Fietz, 1999; Fietz & Ganzhorn, 1999), or group fewer traps at a feeding
site, so as to minimize the risk of trapping animals other than the
target species (Müller, 1999).

HABITUATION AND PREPARATORY BAITING

This is probably the most crucial phase in determining the success of a
trapping campaign; obviously, if no animals take the bait none will
be caught. With a new species or a naïve population, plan a month
or more of patient and careful preliminary observation, non-interactive
habituation (Chapter 2), and evaluation of potential baits. It may
take some time for naïve animals to try an unfamiliar food, and it is
a common mistake to abandon a potential bait after insufficient trials.

Sweet fruits such as bananas are a highly attractive bait for many
species, including most nocturnal strepsirrhines and callitrichines
(J. Dietz, unpublished data; Dietz *et al.*, 1994; Müller, 1999; Savage
et al., 1993; Wright & Martin, 1995) but they are also perishable and
attract insects such as ants. For species (especially macaques, baboons
and some guenons) that find dried grains attractive, loose maize (corn),
rice or wheat kernels are very convenient to handle and store, and also
hold the animals' attention to the trap site, since they cannot easily be
carried off for consumption elsewhere. Some studies have used a
'caller' – a captive of the target species – to attract wild groups to the
baiting area, but this raises issues of logistics and animal welfare
(see, for example, Savage *et al.*, 1993), and it has been shown repeatedly
to be unnecessary with careful pre-baiting (J. Dietz, unpublished data;
Garber *et al.*, 1993; Savage *et al.*, 1993).

Once you have identified a viable bait, start baiting at least one to two weeks before attempting trapping. If you concentrate traps at a feeding site or platform, offer bait at this locality without traps present at first. When animals visit the site regularly, introduce unarmed (fixed open) traps and place bait around, on and in them. Lock multiple traps on a grid open and bait them individually. If possible – especially for group-living, diurnal species – the later phases of preparatory baiting should be monitored by an observer, to ensure that all target animals are entering the traps to feed. Individuals that do not enter traps during preparatory baiting are unlikely to be caught when traps are set.

CAPTURE, TRANQUILLIZATION AND RELEASE

In most reported studies of nocturnal strepsirrhines, portable, automatic traps have been set at dusk and checked at dawn. Captives are carried to the field lab in the traps, processed, and released at dusk at the site of their capture. Alternatively, and preferably, you can monitor traps regularly during the night and process captives immediately, at the capture site (Müller, 1999). As most animals are trapped soon after nightfall, this minimizes holding time, does not prevent lactating females from attending to their 'parked' infants, and allows the animals the rest of the night for foraging. Small nocturnal strepsirrhines can be taken out of the trap easily and tranquillization is necessary only when radio-collaring (if at all). When removing larger nocturnal strepsirrhines from a trap, one person can push the animal to the side of the trap with a cloth or glove while a second person gives a subcutaneous injection (O. Schülke, personal communication, 2001). After processing, nocturnal strepsirrhines are usually held in their individual traps (covered with a cloth) or in dark cotton bags until release.

When trapping callitrichids, set traps before dawn. Traps can be closed manually, by pulling a string (see, for example, Garber et al., 1993; Savage et al., 1993), or automatically (Albernaz & Magnusson, 1999; J. Dietz, unpublished data). In the latter case, check traps in the late morning. At dusk, close the traps and bring trapped animals to the field lab for processing. Put them in a cloth-covered trap, cloth bag or cage for the night, and release them at their capture site the next morning (see, for example, J. Dietz, unpublished data; Savage et al., 1993).

You can use much the same protocol for cercopithecine monkeys. Capture animals in the morning as they leave their sleeping place. Close traps manually or automatically, and continue catching until all traps are filled and closed, the group has left, or (if closure is manually controlled) when no 'new' animals are entering the traps. Do not catch animals in excess of the processing team's daily capacity, because if released without being tranquillized they will be very difficult to re-trap. It is unclear why tranquillization should make monkeys easier to re-trap, but ketamine and phencyclidine hydrochloride are known to curtail short-term memory (Jansen, 2000) and so may reduce or eliminate the adverse memories of capture. Tranquillize captives with a pole- or hand-syringe (Jolly, 1998) at intervals determined by the pace of processing, examine them at a field lab close to the capture site, then place them in a recovery cage and release them individually as soon as the effects of sedation wear off. Do not tranquillize small, nursing infants, which usually cling to their mother while she is sedated and processed. As a precaution against an infant jumping off and running away, secure it to its mother's wrist by a cord tied around its waist or ankle. Remove the cord when you place the mother and infant in the recovery cage.

Immediate release has some drawbacks if complete ascertainment is important, and if your target species (like many terrestrial and semi-terrestrial cercopithecines) lives in groups consisting of more animals than can be trapped and examined in a single day. Daily capture rates rapidly dwindle as untrapped animals are outnumbered by those that have already been released (and who then tend to 'hog' the traps). A residue of wary or subdominant individuals will probably never be caught.

For monkey species that live in large groups, and are known to tolerate captivity, a hold–release strategy may be preferable (Brett *et al.*, 1982; Melnick *et al.*, 1984). In this case, set traps to close automatically. After the group has moved away from the trapping area, tranquillize all captives, transport them to a processing site, examine them, and put them into holding cages where they will live until trapping of their group is completed. A major advantage is that no operator need be present during capture, allowing you to trap minimally habituated groups. On the other hand, extended holding involves expenditures on cages and food, and possibly increased stress and health risks to the captives. Another potential problem is that as a group is progressively deprived of its most powerful members – inevitably trapped first – it may be displaced from the trap site by neighbours, resulting in 'mixed bags' drawn from more than one social group.

If you adopt a hold–release strategy, you will need to build dedicated holding cages. These should be as small as is consistent with comfort, and should be placed so that group members are able to see each other and communicate. To avoid injury, co-housing arrangements should be compatible with the species' social behaviour. Often, adult males are best housed singly, while juveniles and compatible females should be co-housed. The holding area should, of course, be sanitary and provide protection against weather, snakes and predators, and should be off-limits to all non-essential human visitors.

Whether released singly or *en masse*, animals must be released in their home range, usually at their capture site. In a mass release, liberate the youngest animals first to reduce the chance of adults leaving them behind (J. Dietz, unpublished data). With wide-ranging species – such as baboons – the group may move a considerable distance before captives are released. We have not, however, found it necessary – or practicable – to keep animals under sedation and carry them back to their group before release. Once recovered from sedation, they are much more efficient at finding their troop-mates than we could be.

DATA GATHERING

Because trapping is costly, it makes sense not only to mark or radio-collar animals, but also to collect as comprehensive a dataset as is compatible with the animal's well-being (Chapters 8, 9, 10 and 19). Even materials of no immediate relevance to your study objectives often prove to be useful to later investigations, some using techniques undreamed of at the time of collection.

The complexity of the data to be collected, the number of animals to be caught each day, and the size of individual animals will determine the size and composition of your field team. In the simplest cases, trapping and data collection can be carried out by one researcher and an assistant. When sampling larger and more numerous animals, it is preferable to organize separate trapping/sedating and data-gathering teams. At the examination table, assign tasks such as applying an ear-tag or other marker, making dental moulds and casts, blood sampling, weighing and measuring individually. As far as is practical in the field, observe the usual precautions against zoonotic infection (gloves, masks, goggles). Assign one team member to monitor the sedated animal's condition (Chapter 8), and another to remain 'clean' to record data and to check off procedures. An experienced team can gather an animal's full dataset in about 30 min, which imposes a practical limit of about 15–20 animals per day. A hold–release protocol affords a second opportunity to collect data when animals are tranquillized again for transportation to the release site.

Re-trapping animals in subsequent surveys is rarely problematic in the case of cercopithecines, callitrichines (J. Dietz, unpublished data; Savage *et al.*, 1993) or nocturnal strepsirrhines. Indeed, conditioning may even be reinforced by the experience of repeated trapping, tranquillization, and handling, as animals tend to become attached to traps and trap-sites as favoured feeding places. A more likely problem than trap aversion is animals that become 'trap-happy', or learn to take bait without entering or closing the trap. There is, however, much interindividual variation in reaction to traps, and some individuals are much harder to trap/re-trap and do not enter the same trap twice.

CATCHING SMALL, UN-TRAPPABLE PRIMATES

Some species (typically those with a natural diet composed largely of small animals, exudates, or leaves) apparently cannot be lured into baited traps, and many are also too small to be darted easily, but can be caught in other ways. Slender lorises (*Loris* spp.), for instance, seem to be relatively easy to catch by hand if the animal is low in the tree (Petter & Hladik, 1970). You can catch hairy-eared dwarf lemurs (*Allocebus trichotis*) using hand-held noose poles or by fitting a net in front of their tree hole, if their sleeping location is known (Biebouw, 2009). Mist-netting captures tarsiers (*Tarsius* spp.) safely, but recapture is difficult (Gursky, 2000). Slow lorises (*Nycticebus* spp.) can be caught in commercial traps but success is low, and you need many traps (*c.* 200) and creative tactics (Wiens, 2002). Aye ayes (*Daubentonia madagascariensis*) can be captured from their nests during the day (Ancrenaz *et al.*, 1994; Petter & Peyrieras, 1970). O. Schülke (personal communication, 2001; Schülke & Kappeler, 2003) succeeded in capturing fork-marked and sportive lemurs (*Phaner* and *Lepilemur*, respectively) by leading them into a Tomahawk trap via a tunnel of netting stapled around the entrance to their nest hole in a hollow tree. Inventiveness is the name of the game, and it would also be worth while experimenting with lures other than conventional baits – natural or artificial scents, for example, live insect bait, recorded calls or images of conspecifics – to attract animals into traps.

CLOSING THOUGHTS

A trapping programme in a populated area always attracts the attention of local people, and it is important to convey the study objectives clearly to them – in particular, to stress that tranquillized animals are sleeping, rather than dead or injured, and that animals will be released unharmed, will not be eaten, and will not be killed (even at the request of local farmers). Often, people who have never seen a non-human primate close up are amazed by its human-like anatomy and physiognomy, and this can be turned to advantage in stimulating empathy. Such demonstrations should, of course, be confined to a very select audience; it is disastrous if a trapping operation becomes a public spectacle. Between trapping seasons, store traps securely to prevent illicit uses.

Although trapping primates is generally safe, it can be stressful for both the captive and the captor. Primatologists accustomed to observing their subjects with minimum contact and disturbance are seldom psychologically prepared for the kind of interactions required by trapping. A trapped animal (and sometimes its group mates, if present) will scream blood-curdlingly when you approach the trap, and will certainly bite in self-defence if given the chance. Since it is the fear of accidentally injuring or even killing a study animal that makes capturing stressful for the novice trapper, it is sensible wherever possible to call on experienced trappers for help when beginning a project. Trapping primates can be very easy; avoiding associated problems and knowing when not to trap is the real challenge!

ACKNOWLEDGEMENTS

CJJ and JP-C thank the many colleagues who contributed unpublished data to the survey of trapping, and the numerous colleagues, students and volunteers who have assisted in trapping seasons. AEM thanks Simon Bearder, Jim Dietz, Joanna Fietz, Leanne Nash, Oliver Schülke and Frank Wiens for their information on trapping techniques, Urs Thalmann for his assistance on trapping dwarf lemurs during her own field study and Beno Schoch for his help with preparing the traps. Their fieldwork has been supported by the NSF, the NIH, the Harry Frank Guggenheim Foundation, Washington University, New York University, and Earthwatch (CJJ and JP-C) and the A. H. Schultz Foundation, Family Vontobel Foundation, Goethe Foundation, G. & A. Claraz Donation and the Swiss Academy of Natural Sciences (AEM).

REFERENCES

Ancrenaz, M., Lackman-Ancrenaz, I. & Mundy, N. (1994). Field observations of aye-aye (*Daubentonia madagascariensis*) in Madagascar. *Folia Primatol.* 62, 22–36.

Albernaz, A. L. & Magnusson, W. E. (1999). Home-range size of the bare-ear marmoset (*Callithrix argentata*) at Alter do Chão, Central Amazonia, Brazil. *Int. J. Primatol.* 20, 665–77.

Atsalis, S. (1999). Seasonal fluctuations in body fat and activity levels in a rain-forest species of mouse lemur, *Microcebus rufus. Int. J. Primatol.* 20, 883–910.

Biebouw, K. (2009). Home range size and use in *Allocebus trichotis* in Analamazaotra Special Reserve, Central Eastern Madagascar. *Int. J. Primatol.* 30, 367–86.

Brett, F. L., Turner, T. R., Jolly, C. J. & Cauble, R. G. (1982), Trapping baboons and vervet monkeys from wild, free-ranging populations. *J. Wildlife Manag.* 46, 164–74.

Charles-Dominique, P. (1977). *Ecology and Behaviour of Nocturnal Primates*. London: Duckworth.

Charles-Dominique, P. & Bearder, S. K. (1979). Field studies of lorisid behavior: Methodological aspects. In *The Study of Prosimian Behavior*, ed. G. A. Doyle & R. D. Martin, pp. 567–629. New York: Academic Press.

Dawson, G. A. (1977). Composition and stability of social groups of the tamarins (*Saguinus oedipus geoffroyi*) in Panama: ecological and behavioral implications. In *The Biology and Conservation of the Callitrichidae*, ed. D. G. Kleiman, pp. 23–8. Washington, D.C.: Smithsonian Press.

de Ruiter, J. (1992). Capturing wild long-tailed macaques (*Macaca fascicularis*). *Folia Primatol.* 59, 89–104.

Dietz, J. M., Baker, A. J. & Miglioretti, D. (1994). Seasonal variation in reproduction, juvenile growth, and adult body mass in golden lion tamarins (*Leontopithecus rosalia*). *Am. J. Primatol.* 34, 115–32.

Ehresmann, P. (2000). Ökologische Differenzierung von zwei sympatrischen Mausmaki-Arten (*Microcebus murinus* und *M. ravelobensis*) im Trockenwald Nordwest-Madagaskars. PhD thesis, University of Hanover.

Fietz, J. (1999). Mating system of *Microcebus murinus. Am. J. Primatol.* 48, 127–33.

Fietz, J. & Ganzhorn, J. U. (1999). Feeding ecology of the hibernating primate *Cheirogaleus medius*: how does it get so fat? *Oecologica* 121, 157–64.

Garber, P. A., Encarnación, F., Moya, L. & Pruetz, J. D. (1993). Demographic and reproductive patterns in moustached tamarin monkeys (*Saguinus mystax*): implications for reconstructing platyrrhine mating systems. *Am. J. Primatol.* 29, 235–54.

Gursky, S. (2000). Sociality in the spectral tarsier, *Tarsius spectrum. Am. J. Primatol.* 51, 89–101.

Jansen, K. L. (2000). A review of the nonmedical use of ketamine: use, users and consequences. *J. Psychoactive Drugs*. 32, 419–33.

Jolly, C. J. (1998). A simple and inexpensive pole syringe for tranquilizing primates. *Lab. Primate Newsl.* 37, 1–2.

Jolly, C. J. & Phillips-Conroy, J. E. (1993). The use of capture in field primatology. *Am. J. Phys. Anthropol.* Suppl.16, 158.

Kappeler, P. M. (1997). Intrasexual selection in *Mirza coquereli*: evidence for scramble competition in a solitary primate. *Behav. Ecol. Sociobiol.* 45, 115–27.

Lahann, P. (2007). Biology of *Cheirogaleus major* in a littoral rainforest in southeast Madagascar. *Int. J. Primatol.* 28, 895–905.

Melnick, D. J., Pearl, M. C. & Richard, A. F. (1984). Male migration and inbreeding avoidance in wild rhesus monkeys. *Am. J. Primatol.* 7, 229–43.

Müller, A. E. (1999). Aspects of social life in the fat-tailed dwarf lemur (*Cheirogaleus medius*): inferences from body weights and trapping data. *Am. J. Primatol.* 49, 265–80.

Petter, J. J. & Hladik, C. M. (1970). Observations sur le domaine vital et la densité de population de *Loris tardigradus* dans les forêts de Ceylan. *Mammalia* 34, 394–409.

Petter, J. J. & Peyrieras, A. (1970). Nouvelle contribution à l'étude d'un lémurien malgache, le aye-aye (*Daubentonia madagascariensis* E. Geoffroy). *Mammalia* 34, 167–93.

Savage, A., Giraldo, H., Blumer, E. S. *et al.* (1993). Field techniques for monitoring cotton-top tamarins (*Saguinus oedipus oedipus*) in Columbia. *Am. J. Primatol.* 31, 189–96.

Schülke, O. & Kappeler, P. M. (2003). So near and yet so far: territorial pairs but low cohesion between pair partners in a nocturnal lemur, *Phaner furcifer*. *Anim. Behav.* 65, 331–43.

Schwab, D. (2000). A preliminary study of spatial distribution and mating system of pygmy mouse lemurs (*Microcebus* cf. *myoxinus*). *Am. J. Primatol.* 51, 41–60.

Silveira, G., Bicca-Marques, J. C. & Nunes, C. A. (1998). On the capture of titi monkeys (*Callicebus cupreus*) using the Peruvian method. *Neotropical Primates* 6, 114–15.

Sterling, E. J., Nguyen, N. & Fashing, P. J. (2000). Spatial patterning in nocturnal prosimians: a review of methods and relevance to studies of sociality. *Am. J. Primatol.* 51, 3–19.

Wiens, F. (2002). Behavior and ecology of wild slow lorises (*Nycticebus coucang*): social organization, infant care system, and diet. PhD thesis, University of Bayreuth.

Wright, P. C. & Martin, L. B. (1995). Predation, pollination and torpor in two nocturnal prosimians: *Cheirogaleus major* and *Microcebus rufus* in the rain forest of Madagascar. In *Creatures of the Dark: The Nocturnal Prosimians*, ed. L. Alterman, G. A. Doyle & M. K. Izard, pp. 45–60. New York: Plenum Press.

Zingg, J. (2001). Nahrungsökologie und Sozialverhalten von Zwergseidenäffchen (*Cebuella pygmaea*) in Ecuador. Diploma thesis, University of Zurich.

USEFUL INTERNET SITES

www.shermantraps.com – for Sherman traps.
www.tomahawklivetraps.com – for Tomahawk Live Traps.

STEVE UNWIN, MARC ANCRENAZ AND WENDI BAILEY

8

Handling, anaesthesia, health evaluation and biological sampling

INTRODUCTION

Primatologists capture wild primates for many reasons, including medical screening (Chapter 1), sampling for endocrinology, genetics (Chapter 21) and physiology (Chapter 19), or for marking and radio-telemetry (Chapter 10). Since capture and handling is always accompanied by the risk of injury or mortality, it is ethically important to maximize the information gathered during these procedures (Karesh *et al.*, 1998). You should therefore collect biological samples every time wild primates are handled. New techniques are also emerging where biological and disease data can be obtained remotely (Jensen *et al.*, 2009; Leendertz *et al.*, 2004) and we encourage researchers to obtain data by using the least invasive methods available.

Infectious agents affect population dynamics, ecology, behaviour and reproductive success, and disease and health issues are thus widely recognized as important factors in wildlife conservation (Daszak *et al.*, 2000; Deem *et al.*, 2001). Primates deserve special attention for general health issues because of the potential for disease exchange between non-human primates and humans (zoonoses) (Chapter 1). The health status and diseases of free-living primates, the inter-relationships between diseases of wild primates and other biological parameters, and more generally the baseline biological parameters of primates under natural conditions, are rapidly growing areas of research. Moreover, emerging infectious diseases are a cause for concern in primates (Chapter 1; Leendertz *et al.*, 2006) and primate researchers themselves have also been the origin of disease in wild primates (Kondgen *et al.*, 2008). Therefore, we recommend a disease risk analysis approach to biological investigations, to minimize the risks of such disease transmission from researchers to wild animals (Beck *et al.*, 2007).

Field and Laboratory Methods in Primatology: A Practical Guide, ed. Joanna M. Setchell and Deborah J. Curtis. Published by Cambridge University Press. © Cambridge University Press 2011.

Here, we give basic methods for handling primates, clinical examination, and collecting biological samples. Many of the topics covered, especially in relation to anaesthesia, are medically specialized, and will require wildlife veterinarians or medical personnel. However, you can gain useful basic veterinary knowledge by attending training courses at zoos or at other captive facilities prior to departure for the field. You will also need to obtain permission for the capture, anaesthesia and sampling of free-ranging animals and for the export/ import of biological samples (Chapter 1).

PHYSICAL HANDLING

Handling is a compromise between human and animal safety in a situation where stress levels must be kept as low as possible. Primates weighing less than 3 kg can be handled by one person, but most of these species are agile and will bite if threatened. Always handle primates with protective leather gloves. Rapid capture is essential. Being chased is stressful for an animal, and can lead to hyperthermia, for example, which can rapidly result in death. Methods for trapping free-ranging primates are described in Chapter 7.

Grip the smallest animals around the neck with the thumb and forefinger, or by a fold of skin on the flank (especially strepsirrhines). Prolonged handling of small primates is extremely stressful, and we recommend anaesthesia if the animal will be handled for longer than a couple of minutes. Hold larger strepsirrhines by the scruff of the neck with one hand, and by the feet or base of the tail with the other (Figure 8.1 left). Never hold an animal by only one arm, as this could result in fracture or dislocation of the humerus. Grasp primates weighing less than 10 kg just under the arms, with your fingers encircling the upper chest. This position is convenient for intramuscular or intraperitoneal injection. However, for more effective immobilization, hold the arms together behind the animal's back with one hand, and the two feet together with the other (Figure 8.1 right). You will always need a second person to carry out procedures, and you should keep restraint as brief as possible to minimize stress. For obvious safety reasons, you will need to chemically restrain species weighing more than 10 kg.

CHEMICAL HANDLING

Chemical immobilization of wild primates is difficult and hazardous. Besides the risks related directly to capture, an anaesthetic event can

Fig. 8.1 Left: Manual restraint of a ring-tailed lemur (source: Chester Zoo). Right: Correct monkey hold, with a guenon (source Sofie Meilvang, taken at Limbe Wildlife Centre, Cameroon, handler Jonathan Kang).

disrupt the behaviour of a group, resulting, for example, in a change in the social status of a darted individual, and alter its response to the human observer (Karesh *et al.*, 1998). In addition to guidance provided below, procedures for darting small arboreal primates are described by Glander *et al.* (1991), Jones and Bush (1988) and Karesh *et al.* (1998); guidelines for chemical capture of large terrestrial primates can be found in Sapolsky and Share (1998) and Sleeman *et al.* (2000); practical tips for chemical restraint of mammals and a list of the equipment necessary for field anaesthesia are given in Osofsky and Hirsch (2000). The anaesthetic chapter in West *et al.* (2007) provides a thorough overview of primate anaesthesia. Finally, practical information on chemical immobilization, emergency medicine and biological sampling of wild primates is available from various websites, in particular the Pan African Sanctuary Alliance website (see 'Useful Internet sites').

Although modern drugs have a wide safety margin in primates, none is absolutely safe and the reaction of any individual primate to anaesthesia can be unpredictable. Anaesthesia should therefore never be undertaken by untrained personnel. We recommend that a veterinarian be involved with the field anaesthesia of primates, at least to provide advice, but preferably to conduct the anaesthesia themselves, particularly if it will be prolonged. You will need a plan of action in case an anaesthetic does not go as planned. Use a standard **A**irway **B**reathing **C**irculation **D**rug therapy protocol (Unwin *et al.*, 2009), or a variant of it adapted to your own situation.

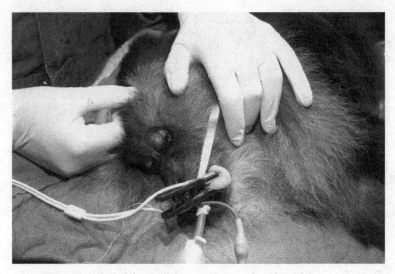

Fig. 8.2 Intubated lion-tail macaque (source Steve Unwin).

Is anaesthesia necessary?

Anaesthesia can be physiologically stressful to any animal, and the decision to sedate or anaesthetize an animal must not be taken lightly. However, with the animal immobile, risk of disease transmission (from bites and scratches) is also reduced, and the quality of samples taken will be enhanced. Ethical and welfare considerations must also be taken into account.

Airway patency during anaesthesia

This is an especially critical issue in field situations where animals may have fed before anaesthesia, increasing the possibility of vomiting. Although you can use a well-fitting mask in small primates less than 1 kg, larger monkey species, and the apes, should be intubated (Fig. 8.2). If the procedure is brief (e.g. fitting a radio-collar or blood collection), you should at least have a correctly sized endotracheal tube on hand in case of an anaesthetic emergency. You can provide oxygen easily in field situations, with the development of new ultra-light field kits. The trachea is relatively short in all primates, so only insert endotracheal tubes to just past the larynx, to prevent bronchiole intubation and consequent over-inflation of one or other lung. Spray the larynx with local anaesthetic such as lignocaine before intubation, particularly following induction with ketamine alone, as the laryngeal reflexes will still be functioning.

Choice of drugs

The choice of substance and dosage is always a compromise between safety and efficiency, and depends on the delivery system, skill and preparation of the human team, ecological conditions (temperature) and the biological features of the primate itself. An equivalent dosage of anaesthetic will have a variable effect, depending on the metabolic size of the individual: all aspects of anaesthesia are accelerated in smaller individuals and slowed in larger individuals. Also consider other biological features such as age, body condition, pregnancy and lactation. The first priority when choosing an anaesthetic is to maximize the safety margin of the drug (the difference between an efficacious and a lethal dose), since you rarely know the exact weight of the animal to be caught when anaesthetizing animals in the wild.

Injectable anaesthetics

Ketamine used alone as an injectable anaesthesia agent has long been the mainstay of primate anaesthetics. It is the induction agent of choice when pre-anaesthetic fasting is not possible, as gagging reflexes are maintained. It is easily available, relatively cheap, and has a wide safety margin. In contrast to classical anaesthetics, thermoregulation, blood pressure, respiratory and spinal reflexes are well maintained with ketamine (Sapolsky & Share, 1998) and the persistence of these reflexes can sometimes be confused with insufficient anaesthesia (Schobert, 1987). The drug can also be given on repeated occasions, although some tolerance may develop. High doses decrease leukocyte and erythrocyte values. Animals given ketamine are often adequately anaesthetized for placement of an endotracheal tube and you can maintain prolonged anaesthesia by inhalation anaesthesia for invasive or major surgery. Ketamine can also be mixed with other anaesthetic drugs to improve the anaesthetic rating.

Ketamine : Medetomidine (or medetomidine equivalent) combination is a common induction regime for primates. However, anaesthetic maintenance becomes unpredictable after 30–40 minutes without the addition of, for example, an inhalation agent. Medetomidine has a wide safety margin and induces good myorelaxation. However, it causes cardiorespiratory depression, which could be life threatening if overdosing occurs and no reversal is administered (Muir *et al.*, 1995; West *et al.*, 2007). A major advantage of medetomidine is that its effects are reversible with atipamezole. When using ketamine and medetomidine, wait 30 minutes before administering the reversal. If you reverse earlier than this, the primate may wake up still under residual ketamine effect and have a turbulent and disoriented recovery. Takako *et al.* (2001) provide details on effective medetomidine combination anaesthetics in macaques, and a transmucosal sedation regime has been developed for chimpanzees (Vidal *et al.*, 1998).

Tiletamine–zolazapam (Telazol®/Zoletil®) has been widely used in primate medicine, either alone or combined with an α-2 agonist such as medetomidine. It is a good combination for anaesthetic novices, as side effects are minimal, and the therapeutic index is wide, so potential overdosing is unlikely to cause long-term problems. Tiletamine is more potent than ketamine and only small volumes are required. In our experience, Zoletil provides excellent immobilization in chimpanzees for 60 minutes, but full recovery can take a couple of hours. Zoletil-immobilized primates tend to maintain stable cardiopulmonary parameters (Horne, 2001).

Inhalation anaesthetics

All of the common inhalation agents have been used in primates, with isoflurane being the most popular currently (Horne, 2001). All these agents cause cardiovascular depression, although minimally so with sevoflurane and isoflurane, and all depress the respiratory system. Inhalation anaesthetic field kits (see, for example, Lewis, 2004) provide an excellent option for anaesthetic maintenance.

Route and method of anaesthetic induction and maintenance

This decision will be based on the animal's species, size, age, and state of health. Try to isolate the animal from the group. Work in a clean area (e.g. on a ground sheet), and wear clothes suitable to minimize the risk of disease transmission. In primates that can be caught safely and hand-held, you can deliver the anaesthetic by direct intramuscular injection into the quadriceps, hamstrings or shoulder muscles. Crush cages also allow relatively safe intramuscular injection and transmucosal use of medetomidine might allow hand injection.

The most common approach to capturing wild primates is remote injection by using blowguns or capture rifles. These systems and their advantages and disadvantages are described by Sapolsky and Share (1998) and Karesh et al. (1998). Check whether you require a firearms licence to use darting equipment. The two most common systems used in primates are Daninject™ and Telinject™. Remote delivery systems should be used only by experienced people owing to the high risks of injury to the animal from a badly located dart, and hazards related to the pre-anaesthetic stage (e.g. falling from a tree or into water, aggression from a conspecific or a predator). Learn how to use darting equipment before attempting to use it, or recruit a skilled darter. Primates often remove the dart following impact, before the full dosage has been injected, and only a mild tranquillization stage is achieved. If this occurs you must monitor the animal until it has completely recovered. As primates are often high up, you will need suitable netting or other material to attempt to catch the anaesthetized animal as it drops.

When using a blowpipe, ensure that the anaesthetic agent is delivered intramuscularly and not subcutaneously, as the latter may prolong the induction time or fail to produce complete immobilization. It is essential that primates are not excited prior to or during anaesthetic induction or they will require higher doses and take longer to become recumbent. Both these factors, and higher adrenaline levels, increase the potential for complications during anaesthesia. It is also important to have a quiet environment to avoid disturbance during induction, as hearing is usually the last sense to be abolished by an anaesthetic. After delivering an anaesthetic agent, allow a minimum of 10–15 minutes to elapse before disturbing the primate, unless an emergency situation such as respiratory arrest develops during this period. This allows the drugs to take their full effect. Failure to allow such a period may result in partial arousal during the induction period, leading to a lighter plane of anaesthesia than would normally be expected with a given dose rate.

Anaesthetic maintenance

When you need to prolong anaesthesia beyond the time allowed by a single dose of the injectable agent, the best option is maintenance with gaseous anaesthetic agents such as isoflurane administered with oxygen via an endotracheal tube. Some injectable drugs (e.g. ketamine) can be given in incremental intravenous (IV) doses, which may significantly prolong recovery time; we recommend that you don't extend an anaesthetic by injectables alone for longer than 60 minutes.

Assessment and monitoring of primates during anaesthesia

The primary purpose of monitoring is to maximize the safety of an anaesthetic procedure, particularly by avoiding excessive anaesthetic depth or adverse reactions. Problems are easier to correct if detected early. You can find anaesthetic monitoring techniques in Horne (2001), West *et al.* (2007) and Unwin *et al.* (2009).

Assess the effectiveness of anaesthesia before carrying out any procedures (including moving the primate). This is especially important with large individuals, which can be dangerous if incompletely immobilized. When the jaw is sufficiently relaxed, check the mouth thoroughly and remove any food, excess saliva or even pieces of dart or dart needles. A member of the team should hold the primate's hand during the procedure. The grip reflex is one of the first to return, and can be used as another indicator of anaesthesia becoming too light. Cover the eyes to protect them from sunlight. Keep detailed notes of all measured physiological parameters, including respiratory and heart rates, body temperature, pulse quality, mucous membrane colour and muscle tone. Pink mucous membrane is normal, red may indicate hyperthermia, white or pale might be a sign of low blood pressure or vascular problems, and blue may indicate respiratory problems. Capillary refill time should be less than 2 seconds; more than this indicates low blood pressure. Use a pulse oximeter to measure blood oxygenation levels. Record body mass during anaesthesia to calculate accurate dose rates of agents retrospectively.

Anaesthetic recovery

Conduct recovery in as calm and quiet an environment as possible. Keep the animal warm and away from draughts in cold areas. In hot weather, choose a cool recovery area. Before leaving a primate to recover, repeat the oral cavity check to remove foreign material. If possible, leave the primate in lateral recumbency with head and neck extended and the tongue protruding from the mouth to allow saliva to drain. Although you should not extubate an endotracheal tube until the animal can swallow, this is often not possible in the field, so the timing of extubation is reliant on the anaesthetist's experience.

Primates try to climb upwards as soon as they recover consciousness and rapidly regain the ability to cling with their hands and feet, and support their own weight. They usually then remain static until fully recovered, but be aware of potential hazards in the immediate environment. Recovery takes longer in fat animals.

CLINICAL EXAMINATION

Record all physical examination data and all biological samples collected on a datasheet. Take close-up photographs or video (Chapter 17) of the primate's body, face, hands, feet, etc., and of any special findings during the clinical examination.

EXTERNAL EXAMINATION AND GENERAL BODY CONDITION

Take notes on general appearance, sex, fur colour, and presence of scars, injuries and disabilities. Record body mass (Chapter 9), abundance of subcutaneous fat deposits, and hydration level. To determine the hydration level, pinch the skin with two fingers and release. If the fold disappears instantaneously, hydration is normal; if the fold holds a few seconds after release, the animal is dehydrated and needs parenteral infusion or oral hydration.

Morphometrics, oral cavity and dentition

See Chapter 9.

Body temperature

You can use a digital thermometer to record rectal temperature in anaesthetized primates (Chapter 19). Mean body temperature of primates ranges from 37 °C to 40 °C, with wide interspecific fluctuations. For example, body temperature of mouse lemurs (*Microcebus* spp.) and dwarf lemurs (*Cheirogaleus* spp.) can fall as low as 12 °C in winter periods of inactivity (aestivation) (Charles-Dominique, 1977). Body temperature also fluctuates across the day: it is minimal around 3 a.m. and maximal around 3 p.m. in diurnal species and the opposite in nocturnal species (Chapter 18). Compared with adults, temperature is higher in neonates and lower in young and juvenile individuals. Stress and muscular activity before capture induce transient hyperthermia.

Examination of the different systems

Cardiorespiratory system: Listen to the heart and lungs with a stethoscope. Record heart and respiratory rates (heartbeats and breaths per minute, respectively). As a rule, these decrease with increased body size and they will obviously be affected by the capture and anaesthesia.

Locomotor system: Investigate muscular reflexes and the integrity of bones and articulations by using manipulation and palpation.

Digestive system: Use transabdominal palpation to examine the size and shape of internal organs (liver, spleen, kidneys, presence of abnormal mass). Use rectal palpation to obtain a faecal sample.

Reproductive system: Use palpation (animal lying in a lateral position) to explore the genital tract and obtain swabs for cytology. Abdominal palpation can be used to determine late stages of pregnancy in females.

Other: Examine the ears and eyes and apply a sterile eye protector such as Lacrilube™ or hydrocellulose to protect the cornea from desiccation in anaesthetized animals.

BIOLOGICAL SAMPLING

Table 8.1 summarizes the biological samples that you can obtain from an anaesthetized primate, the information they can yield, and storage methods. Cheeseborough (2005) is an excellent resource on field laboratory techniques. Label samples carefully (with a solvent-proof pen) with the date, location, species, individual identification, the type and number of the sample. Many biological samples require specific storage conditions (refrigeration or freezing) and rapid processing after collection. Different labs will recommend different storage conditions, so check first. You can store samples in relatively inexpensive field-compatible portable fridges, freezers and liquid-nitrogen containers for subsequent analysis, and screen several biological parameters *in situ* with diagnostic field kits.

Blood sampling

Whole blood

Collect whole blood after venipuncture from the femoral vein for the smallest species, or when large quantities are required for other species. With the animal in a supine position, feel for the femoral artery pulse in the femoral triangle (Fig 8.3). The femoral vein is parallel

Fig. 8.3 Femoral blood collection from a juvenile chimpanzee under field conditions (source Steve Unwin).

to and immediately in front of the artery. Disinfect the site with iodine solution, then insert the needle at a 45–90° angle to the body. You will see a flash of blood in the hub of the needle when you are in the vein. After sample extraction, apply strong digital pressure to the site as you withdraw the needle, and for a minute or so afterwards, to prevent haematoma (bruise) formation. Sample the smallest species (small strepsirrhines, marmosets, and tamarins) with a 0.4 mm gauge needle and 1 ml syringe, squirrel monkeys (*Saimiri* spp.) with a 0.9 mm needle and 1 or 2.5 ml syringe or small vacuum tube, and larger species with a 1.2–2 mm needle and 2.5–10 ml vacuum tube. Other sites are the jugular vein (especially for the smallest species) or tibial vein (strepsirrhines, many monkey species). The tibial vein lies just under the skin on the caudal surface of the gastrocnemius muscle, and is easy to locate after compression of the upper thigh. However, it is small and collapses easily if large quantities of blood are taken. This is the site of choice for intravenous injections for many species of primates. The cephalic vein, or its branches, can be attempted in apes. As a rule, blood sampling should never exceed 1 ml per 100 g body mass per month to prevent risk of hypotension and associated heart failures. In recently dead primates, you can obtain blood from the heart.

Haematological parameters

To prevent clotting, place whole blood in a collection tube containing sodium heparin or EDTA (ethylenediamine tetra-acetic acid) disodium salt immediately following venipuncture. Determine red and white blood cell counts (RBCC and WBCC, respectively) with a Coulter Counter or a Malassez' Cell slide after staining. Make differential WBCC by examining 100 leukocytes in smears stained by using the Wright–Giemsa method. These results provide valuable information on the individual's medical status, including anaemia, infection, parasitism, neoplasm and other haematological abnormalities. You can also determine other parameters (e.g. mean corpuscle volume, mean corpuscle haemoglobin) that explore haematological function more precisely. Haematocrit or packed cell volume (PCV) gives information on the general hydration level, and is determined by centrifuging blood-filled capillary tubes in a portable microcentrifuge. These analyses are conducted from a few hours to a few days after venipuncture (if samples are stored at 4–8 °C). Do not freeze or shake tubes, to prevent haemolysis of blood cells. You can find the physiological range for biological parameters in primates in the literature, or ISIS/MedArks system (available from all zoos and most captive facilities holding primates).

Bacteria and virus isolation

Refer to Table 8.1.

Blood smears

Blood smears are easy to make, but hard to make well, and yield information concerning haematology, blood parasites and bacterial infection. Collect peripheral or capillary blood from the tip of a finger or ear, or use part of a larger sample. Disinfect the site, slightly puncture the skin with a small needle, and collect a drop of blood. To make a thin blood smear, place a single drop of whole or peripheral blood near one end of a horizontal microscope slide. You can use a capillary tube to provide a precise, small volume of blood. Bring the end of a second slide up to the drop at 45° until the drop disperses along its edge. Then push this slide quickly and evenly towards the opposite end of the first slide. The smear will dry immediately (if not it is too thick). To make a thick smear, place a drop of blood on a slide, and spread in a small circle with

Table 8.1 *Biological samples that you can obtain from an anaesthetized primate*

Sample	Area of investigation	Examples	Equipment and methods	Storage[1]	Storage time	In situ analysis	Costs[2] Sampling	Storage	Analysis
Whole blood	Haematology	Cell counts (thin blood smear)	See text	Fridge	Max. 4 days	Possible	$	$$	$$
		Haematocrit	See text	Fridge	Max. 4 days	Easy	$	—	$
	Pathogens	Bacteria (thin blood smear)	Slide, coloration kit[3], microscope (see text)	Room temp.	Months	Easy	$	$	$
		Haemoparasites (thick blood smear)	Slide, coloration kit, microscope (see text)	Room temp.	Years	Easy	$	$	$
		Virus or bacteria isolation/PCR	Use sterile equipment and transport media (e.g. buffered glycerine) to prevent contamination	Wet ice	Few days	Possible	$$	$$	$$
			Freeze in a mixture of CO_2 and alcohol, or in liquid nitrogen, store at $-70\,°C$ or below (but tissue samples are better)	Frozen	Months	Possible	$$	$$$	$$

Table 8.1 (cont.)

Sample	Area of investigation	Examples	Equipment and methods	Storage[1]	Storage time	In situ analysis	Costs[2] Sampling	Storage	Analysis
	Toxicology	Pesticides	Dry tube[4]	Frozen	Months	Unlikely	$	$$$	$$$
	Immunology	Antibodies	Dry tube[4]	Frozen	Months	Possible	$	$$	$$
			2-3 drops on Whatman-type filter paper in a 50 ml tube over silica gel	Cool place	Months	Possible	$	$	$$
	DNA/RNA extraction	General PCR screening for genetics or investigations of various pathogens, or specific PCR for candidate pathogens	2-3 drops of blood or buffy coat[5] on Whatman-type filter paper in a 50 ml tube over silica gel Dry tube[4]	Fridge Frozen	Months	Yes	$	$$	$$$
Plasma	Chemistry	Vitamins, minerals and metals	EDTA / heparinized tube, centrifuge (see text)	Frozen	Months	Unlikely	$	$$$	$$$
Serum	Biochemistry	Enzymes, electrolytes	Dry tube[4], centrifuge (see text)	Fridge	Few days	Possible	$	—	$$

	Endocrinology	Steroid hormones	Dry tube[4], centrifuge (see text)	Frozen	Months	Yes	$	$$$	$$
	Immunology	Globulins, antibodies	Dry tube[4], centrifuge (see text)	Fridge Frozen	Few days Months	Possible	$	$$$	$$$
	Toxicology	Pesticides	Dry tube[4], centrifuge (see text)	Fridge Frozen	Few days Months	Unlikely	$	$	$$$
Faeces	Parasitology	Intestinal parasites	(see text) Kit for nematode culture (see text)	Room temp.	Months	Easy	$	$	$
		Protozoa		Room temp.	Months to years	Easy	$	$$	$$
	Genetics	See Chapter 21	See Chapter 21	See Chapter 21	Few days to years	Possible	$$	$$	$$
	Viral/bacterial detection, toxicology	Non-invasive study of disease	RNAlater, frozen, dried or 10% glycerine (depending on what you want to detect)	Room temp. or frozen	Few days to years	Possible	$$	$$	
Urine	Endocrinology	Steroid hormones	See Chapter 20	See Chapter 20	Months	Possible	$	$$$	$$
	Urinalysis	Renal function	Dry tube, centrifuge, coloration kit, dipstix™	Room temp.	Day	Easy	$	$	$
	Pathogens	Virus, bacteria or parasite isolation	Sterile tube	Fridge	Few days	Possible	$	$$	$$$
Tissues	DNA/RNA extraction	General PCR screening for genetics.	Max. 0.5 × 0.5 cm² in RNAlater or formalin	Room temp	Weeks	Possible	$	$$	$$$
			Sterile tube	Frozen	Months				

Table 8.1 (cont.)

Sample	Area of investigation	Examples	Equipment and methods	Storage[1]	Storage time	In situ analysis	Costs[2] Sampling	Storage	Analysis
Pathogens	Cultivation, isolation and identification of viruses and bacteria	Swabs (e.g. rectal, oropharyngeal, lesion) come with a variety of transport media (often bacterial or viral specific) that prolong sample life. Or use RNAlater (−70 °C) or formalin	Room temp. Frozen	Weeks Months	no	$$	$$	$$
	Bacterial culture	Sterile tube (max. 0.3 × 0.3 cm)	Room temp. Frozen	Weeks Months				
Cytology	Alterations in cellular structure, presence of pathogens	Max. 0.5 cm × 0.5 cm in 10% formalin (10 × sample volume)	Room temp.	Years	Possible	$	$	$
Histology	Changes in tissues/cells and pathogens	Max. 0.5 cm × 0.5 cm in 10% formalin (10 × sample volume)	Room temp. (do not freeze)	Years	Unlikely	$	$	$$
Cytology	Female reproductive cycle	Reproductive status: dry swabs and stained smears of vaginal mucosa	Room temp.	Fixed smears – months	Easy	$	$	$

Secretions			Frozen				
	Fluids						
					Easy	$	$$
	Scent marking glands	Moisten dry sterile swabs with secretions and flash freeze in liquid nitrogen		Months			
Ectoparasites	Parasitology	Ectodermic agents	Place in plastic container with 70%–95% ethanol	Room temp.	Years	Possible	$ $
		Intradermic parasites (mites, etc.)	Scrape lesions with scalpel blade, place scrapings in lactophenol	Room temp.	Day	Easy	$ $

[1] Fridge, + 4 °C; frozen, − 20 °C or lower; room temperature, 15–25 °C

[2] Figures are indicative of the relative costs; from very low cost ($) to expensive ($$$)

[3] Kit type depends on what you are investigating, for example Gram stains for bacteria, Field's stain for various parasites, Diff Quik™ for blood smears

[4] Without anticoagulin

[5] Fraction containing leukocytes and platelets in a blood sample after centrifugation. In the centrifuge tube, this is the layer between the plasma (top) and the erythrocytes (bottom). Fresh chilled on dry ice is best

the tip of a needle or corner of a second slide for at least 30 seconds. After a few minutes at ambient temperature the smears are dry and can be sent to a laboratory or analysed *in situ* after staining or fixation (thin films only) in 90% methanol. Take a minimum of three to five smears per individual.

Plasma

Obtain plasma by centrifuging unclotted whole blood and separating the yellow/tan liquid material (plasma) from the clotted component (cell membranes and other blood composites). Divide this into aliquots in microtubes (0.5–2.5 ml) and store frozen (−20 °C or below). Manual centrifuges are available where electricity supply is an issue.

Serum

To collect serum, place whole blood in special serum separator tubes or dry sterile blood collection tubes immediately after venipuncture. Leave tubes undisturbed for at least one hour at ambient temperature to encourage clot formation and centrifuge at 2000 g for 5 minutes (or use a manual centrifuge). You can obtain serum without a centrifuge by letting the clot or blood cells settle for few hours, then aspirating the liquid (serum). To maximize serum quantity, allow the blood to clot with the collection tube inverted (rubber stopper down). After a few hours, turn the tube stopper up, carefully remove the stopper with the clot attached, leaving the serum in the tube (Munson, 2000). Divide the serum into several aliquots in small vials, refrigerate or freeze.

Faecal samples

You can collect faeces directly from the rectum when a primate is handled, or non-invasively during fieldwork. Storage in the field is easy and most analyses can be performed long after sample collection. Never handle faeces without gloves because of the risk of zoonosis transmission. Note colour, odour, size and consistency of faeces at the time of collection, along with major macroscopic elements, including presence or absence of worms. Preserve worms in ethanol solution for identification. Parasites are usually symbiotic with their hosts, so the presence of parasites in faeces does not necessarily mean that an animal is sick. To assess an individual's true parasitological status, repeat sample collection at regular intervals. In the absence of an outbreak, every 1–3 months is suitable for most parasites, with a single sample for protozoa and three consecutive daily samples for nematodes (to detect intermittent shedding).

Faecal parasite investigation

Flotation methods have been used routinely in the past to investigate faecal parasites, but the sensitivity of these methods is poor and a variety of different solutions are needed. The following formol–ether centrifugation technique is a simple and accurate method for investigating helminth and protozoal parasites and is a good screening test. You will need a 15 ml flat bottomed tube and lid; a 15 ml centrifuge tube; a 1 g plastic spoon; a Faecal Parasite Concentration (FPC) strainer; Trixon X-100; a cotton-tipped applicator (all commercially available in the Evergreen FPC kit, Fig. 8.4); see Useful Internet sites, a fresh faecal sample; 10% formol saline; diethyl ether (or ethyl acetate or petrol), a wooden stick; a plastic Pasteur pipette; gloves; microscope slides and coverslips; a centrifuge (hand-powered if you have no access to electricity) and a microscope.

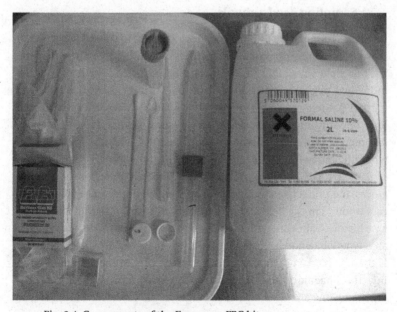

Fig. 8.4 Components of the Evergreen FPC kit.

Wearing latex (or similar) gloves, add 10 ml of formol saline 10% to the flat-bottomed tube and add one level spoon of fresh faeces. Mix thoroughly with the wooden stick until there are no visible lumps of faecal material. Put the lid on and shake vigorously. You can now store the sample if necessary. Ensure the FPC strainer is securely attached to the 15 ml centrifuge tube; it should be central on the vent-straw. Remove the white lid and attach the FPC strainer tightly to the flat-bottomed tube. Shake vigorously and shake the specimen through the strainer, pointing the conical end of the centrifuge tube downward. If necessary, tap the flat-bottomed tube to encourage the sample through the strainer. Unscrew the FPC strainer from the flat-bottomed tube. Add about 0.5 ml ether and 1–2 drops of Triton X-100 to the filtered sample in centrifuge tube, replace the lid and shake vigorously. Centrifuge for 1–2 min at 1500–3000 rpm, or for 2 min at about 90 rpm if using a manual centrifuge. If no centrifuge is available, leave the sample to settle for a minimum of 1 hour. Dispose of the fatty plug and supernatant and remove any debris from the tube using the cotton-tipped applicator. Add 10% formol saline to the sediment volume for volume and mix thoroughly. Add one drop of the sediment/formol saline mix to a microscope slide, add a coverslip and examine under a microscope. Survey the slide at × 10, then × 40 for further investigation. Consult Cheeseborough (2005) for diagnosis.

Sampling dead animals

Dead primates may be found in the field, brought to you for a number of reasons, or result from a failed capture procedure. Post-mortem examinations yield valuable data about cause of death and the health of the individual prior to death. Very little information is available on the natural pathology of wild primates and field studies offer one of the only opportunities for complete pathological examinations. Necropsy techniques are vital to obtaining useful samples, and are described elsewhere (Rabinowitz, 1993; Munson, 2000; Terio & Kinsel, 2008; Unwin *et al.*, 2009). The person performing the necropsy should wear a mask, gloves and protective clothing. This is especially important when a carcass is found in the field and no detailed information is available on the case history. Take care with investigating evidence of sudden death in an area where haemorrhagic fever or anthrax have recently been reported. If there is bleeding from any orifice of the carcass, or you suspect an infectious haemorrhagic disease, do not approach or touch the carcass. Ensure no-one else does either, photograph the carcass, and report it to the relevant authorities as soon as possible.

REFERENCES

Beck, B., Walkup, K., Rodrigues, M. *et al.* (2007). *Best Practice Guidelines for the Reintroduction of Great Apes.* Gland, Switzerland: SSC Primate Specialist Group of the World Conservation Union.

Charles-Dominique, P. (1977). *Ecology and Behaviour of Nocturnal Prosimians.* London: Duckworth.

Cheeseborough, M. (2005). *District Laboratory Practice in Tropical Countries.* (2nd edn.) Parts 1 and 2. *Tropical Health Technology.* Cambridge: Cambridge University Press.

Daszak, P., Cunningham, A. A. & Hyatt, A. D. (2000). Emerging infectious diseases of wildlife: threats to biodiversity and human health. *Science* 287, 443–9.

Deem, S. L., Karesh, W. B. & Weisman, W. (2001). Putting theory into practice: wildlife health in review. *Conserv. Biol.* 15, 1224–33.

Glander, K. E., Fedigan, L. M., Fedigan, L. & Chapman, C. (1991). Field methods for capture and measurements of three monkey species in Costa Rica. *Folia Primatol.* 57, 70–82.

Horne, W. A. (2001). Primate anaesthesia. *Vet. Clin. North Am. Exot. Anim. Pract.* 4, 239–66.

Jensen, S., Mundry, R., Nunn, C. L., Boesch, C. & Leendertz, F. (2009). Non-invasive body temperature measurement of wild chimpanzees using faecal temperature decline. *J. Wildl. Dis.* 45, 542–6.

Jones, W. T. & Bush B. B. (1988). Darting and marking techniques for an arboreal forest monkey, *Cercopithecus ascanius. Am. J. Primatol.* 14, 83–9.

Karesh, W. B., Wallace, R. B., Painter, R. E. *et al.* (1998). Immobilization and health assessment of free-ranging black spider monkeys (*Ateles paniscus chamek*). *Am. J. Primatol.* 44, 107–23.

Kondgen, S., Kuhl, H., Goran, P. K. *et al.* (2008). Pandemic human viruses cause decline of endangered great apes. *Curr. Biol.* 18, 1–5.

Leendertz, F., Boesch, C., Ellerbrok, H. *et al.* (2004). Non-invasive testing reveals a high prevalence of simian T-lymphocytic virus type 1 antibodies in wild adult chimpanzees of the Tai National Park, Cote d'Ivoire. *J. Gen. Virol.* 85, 3305–12.

Leendertz, F., Pauli, G., Matz-Rensing, K. *et al.* (2006). Pathogens as drivers of population declines: the importance of systematic monitoring in great apes and other threatened mammals. *Biol. Conserv.* 131, 325–37.

Lewis, J. (2004). Field use of isoflurane and air anaesthetic equipment in wildlife. *J. Zoo Wildl. Med.* 35, 303–11.

Muir, W. W., Hubbell, J. A. E., Skarda, R. T. & Berdnaski, R. M. (1995). *Handbook of Veterinary Anesthesia.* St Louis, MO: Mosby.

Munson, L. (2000). Necropsy procedures for wild animals. In *Conservation Research in the African Rain Forests: A Technical Handbook,* ed. L. White & A. Edwards, pp. 203–24. New York: Wildlife Conservation Society.

Osofsky, S. A. & Hirsch K. J. (2000). Chemical restraint of endangered mammals for conservation purposes: a practical primer. *Oryx* 34, 27–33.

Rabinowitz, A. (1993). *Wildlife Field Research and Conservation Training Manual.* New York: Wildlife Conservation Society.

Sapolsky, R. M. & Share L. J. (1998). Darting terrestrial primates in the wild: a primer. *Am. J. Primatol.* 44, 155–67.

Schobert, E. (1987). Telazol use in wild and exotic animals. *Vet. Med.* 82, 1080–8.

Sleeman, J. M., Cameron, K., Mudakikwa, A. B., *et al.* (2000). Field anaesthesia of free-living mountain gorillas (*Gorilla gorilla beringei*) from the Virunga Volcano Region, Central Africa. *J. Zoo Wildl. Med.* 31, 9–14.

Takako, M., Ryohei, N., Manabu, M., Nobuo, S. & Kiyoaki, M. (2001). Chemical restraint by medetomidine and medetomidine–midazolam and its reversal by atipamezole in Japanese macaques (*Macaca fuscata*). *Vet. Anaesth. Analg.* 28, 168.

Terio, K. & Kinsel M. (2008). *Primate Necropsy Manual.* Lincoln Park Zoo. pdf available from lpzoo.org.

Unwin, S., Boardman, W., Bailey, J. W. *et al.* (2009). *Pan African Sanctuary Alliance Veterinary Healthcare Manual.* (2nd edn.) Available at www.pasaprimates.org.

Vidal, C., Paredes, J. & Ancrenaz, M. (1998). Anesthesia en chimpancés: primer estudio comparativo de varias técnicas en trabajos de campo. *Consulta* 6, 1670–3.

West, G., Heard, D. & Caulkett, N. (2007). *Zoo Animal and Wildlife Immobilization and Anaesthesia.* (1st edn.) Oxford: Blackwell Publishing.

USEFUL INTERNET SITES

www.wildnetafrica.com – for chemical immobilization.

www.pasaprimates.org – for the *PASA Veterinary Healthcare Manual*, a peer-reviewed field manual. Chapters include biological sampling, gastrointestinal parasitology, field anaesthesia techniques and emergency medicine.

www.wcs.org – for post mortem protocols.

www.promedmail.org – for up-to-date information on disease outbreaks in wildlife.

www.mammalparasites.org – for a global parasite database on all mammals with an extensive section on findings in wild primates.

www.evergreensci.com/micro/index.htm – for the supplier of the Evergreen FPC kit.

9

Morphology, morphometrics and taxonomy

INTRODUCTION

Fieldworkers are often unaware of the value of making accurate records of their study animals' morphology. External morphology includes measurements, observations of glandular activity and detailed description of the pelage; internal morphology may include skull and postcranial measurements, recording of suture closure, epiphyseal fusion, dental eruption and wear, and observations on the gut. All these observations will yield information on taxonomy, age–sex class (Table 9.1), reproductive status, individual variation, growth, development, sexual dimorphism and so on. It may also be interesting to look at individuals' behaviour in the light of their external differences: does facial coloration correlate with behaviour in mandrills (*Mandrillus sphinx*), does flange development correlate with behaviour in orangutans (*Pongo* spp.), and so on?

Measurements may be made either on dead specimens, or on living animals when they are captured (Chapter 7) and, preferably, anaesthetized (Chapter 8). Valuable descriptive information, other than measurements, can be gathered from simple observations, or from photographs (Chapter 17). On occasion, measurements can even be made in this way. For example, Markham and Groves (1990) cite a personal communication from Herman Rijksen of how he weighed four wild orangutans (worth repeating here because it is such a wonderful example of lateral thinking) '. . . by "measuring" the bending arc of particular branches when supporting the full weight of the animal and hoisting up buckets of sand to match the same arc'.

The accurate identification of study animals has always been vital to understand them properly, especially today when remnant populations of once widespread species are in need of strict conservation,

Field and Laboratory Methods in Primatology: A Practical Guide, ed. Joanna M. Setchell and Deborah J. Curtis. Published by Cambridge University Press. © Cambridge University Press 2011.

Table 9.1 *Age classes*

Definable age classes will of course vary greatly among taxa; not only between major categories such as families, but even between quite closely related species. The terms in this table are often used.

Age class	Definition
Neonate	Infant showing signs of having been born very recently (a few days)
Infant	Unweaned; not independently moving (carried about, in nest, etc.)
Juvenile	Immature but independently moving
Immature	Any individual not evidently sexually mature
Subadult	Individual that is apparently sexually mature, but not physically mature
Adult	Both sexually and physically mature
Old adult	Adult showing apparent signs of age degeneration (like the first author)

scientific names are needed for them, and the question of their uniqueness arises. Is this the last remaining population of its species or subspecies? If the patterns of ecology and behaviour of the study population are different from those of others reputed to be of the same species, are we sure that the populations really are the same, or could there be some undetected taxonomic differences after all? It is increasingly important for fieldworkers to be aware of taxonomic niceties, and of morphological and morphometric matters in general, for a full appreciation of the biology of any study population.

TAXONOMIC DATA: SPECIES AND SUBSPECIES

Measurements and other observations will, in the first line, be relevant to taxonomy. This may matter crucially as far as interpreting behaviour is concerned: different species, for example, may differ from each other in behaviour and unless taxonomic differences are appreciated such behavioural differences might unwittingly be ascribed to different ecological settings or, worse, to chance. So a brief introduction to taxonomy is in order.

There is no such thing as 'the official' taxonomy; a taxonomic scheme is a hypothesis, a way of interpreting data, just like an ecological or a behavioural model. Most biologists who are not specialists in taxonomy probably think of species as populations, or groups of populations, that are reproductively isolated from others (the so-called

'Biological Species Concept'), but misunderstand what this means. It does not mean that species cannot interbreed with each other, nor that if they do interbreed their hybrids are sterile. Species of *Macaca* or of *Cercopithecus* that are widely sympatric may on occasion interbreed, and not merely F_1 hybrids but even backcrosses are to be seen mingling with pure-bred troops of one or other parental species. So species are not invariably reproductively isolated!

Then recall that there are innumerable examples of populations that are clearly closely related, but look different from each other, which are totally allopatric – separated from each other by rivers, or by unsuitable habitat, or on islands. How can one possibly judge whether or not they are reproductively isolated?

If this 'reproductive isolation' criterion is taken too seriously, it is obvious that the vast majority of organisms could never be classified as to species by any objective measure – even if we place asexual organisms on one side. This is a major reason why specialist taxonomists increasingly recommend the 'Phylogenetic Species Concept': species are what can be diagnosed. The meaning of diagnosability is simply that every single individual, within its age/sex class, can be allocated to its species. Any character, as long as it can be plausibly assumed to be genetic, is a candidate for the diagnosis: size, shape, colour, colour pattern, vocalization, or a particular nuclear DNA sequence. The implication is that a species has one or more fixed genetic differences from other species; it is genetically unique, not (or not necessarily) reproductively isolated.

At this point, we must specify that we exclude mitochondrial DNA sequences. These are inherited uniparentally, not biparentally, and are hence different from every other DNA sequence (Y-chromosome DNA of course excepted) and so also from all observable (phenotypic) features. The time to coalescence of the nuclear DNA sequences of two different populations is a function of their effective population sizes (N_e); the N_e of mitochondrial DNA (mtDNA) is effectively one-quarter that of a nuclear sequence, with a consequently much shorter time to coalescence (Nichols, 2001). Moreover, mtDNA is inherited, to all intents and purposes, only through the female line; it tells us nothing about what the males have been doing, nor (because there is no recombination) has it anything to say about population integration. MtDNA, consequently, offers us only incomplete information about populations; as species and subspecies are population concepts, it cannot, strictly speaking, be used in a taxonomic sense.

Subspecies are geographic populations of a species that differ as a whole, but not absolutely. Note that they are geographic: by definition, they cannot be sympatric. As is the case for species, mtDNA should not by itself be used to characterize subspecies.

PRESERVING SPECIMENS

If you have any doubt as to the taxonomic identity of your study animals, you should photograph them and try to compare the photos with the type specimen. This is the specimen, lodged in a museum somewhere, based on which the species was first described and named: it is the ultimate decider of a species/subspecies. A voucher specimen is even better than a photograph: we are absolutely not recommending that you should shoot a specimen for comparison, but all too often there are hunters around, and if a specimen is confiscated (N.B. *not* purchased) from a hunter the best disposal of it is in some museum where it can be studied, and from where it can be borrowed for comparison with other specimens (especially the type specimen), or at least photographed. Birds of prey also take a lot of primates and, usefully, bring up the bones in pellets underneath their nest perches, or pieces of prey may simply fall to the forest floor during dismemberment. Collect such specimens assiduously and present them to museums, where they will form permanent vouchers.

Bones

Bones are easy to preserve. If you find them in an already macerated state (i.e. separated from the soft tissues), label them clearly and place them in a box or phial, padded with cotton wool or some other soft material – if all else fails, then scrunched-up paper will do. If you find a dead specimen, or have to dispatch a live animal humanely for some reason, you need to separate the skin from the bone in some way.

If you don't need the skin, soak the specimen in warm water until the soft tissues float off. This might take several hours. Don't boil the specimen, as although many times quicker this method destroys *all* non-bony tissue, so that sutures spring apart and teeth fall out. Bio-active washing powders, which contain digestive enzymes, will clean off soft tissue from bones very well, but watch this as ultimately they have the same effect as boiling. Finally, in dry environments an ant-heap serves very well as a bone-cleaning laboratory, but the bones will be shifted about by the ants, so you need to hold them down in some way (wrapped in wire-netting, for example). How long this takes depends on the ambient conditions (in the Canberra summer it takes a few days!). Be warned that in too dry a climate bone will flake and ultimately falls apart, and bony splinters can penetrate your skin and set up infection.

Skins

If you need the skin, then turn the body on its back, make an incision up the ventral surface from groin to sternal notch, and transverse incisions between the two arms, to at least halfway down towards the elbows, and between the two legs to at least halfway down towards the knees. Insert your fingers carefully between skin and underlying fatty tissue, and separate them, peeling back the skin. The less fat that remains on the underside of the skin, the easier it will be to preserve, and the less unpleasant it will be to the touch. The phalanges can be preserved as part of the skeleton, or can stay inside the skin; the latter makes skinning a whole lot easier, but will work only with small primates, as there is too much soft tissue in the digits of larger species. The ear cartilages come away with the skin, as do those of the nose in the case of strepsirrhines. It is often convenient to leave most of the tail vertebrae in the skin, at least until you can get the specimen to a museum.

Skin your specimen, and preserve the skin, as soon as possible, because tropical heat and humidity cause the hair to come loose from the skin. The important thing is to keep it dry and make it unappealing to insects and mould. For best results, turn the skin inside out, and hold the hair surfaces apart with loose material such as scrunched-up tissue paper. Rub in salt, and leave it for some 24 h, pouring away any brine that is formed. Dry it either in the open (but not in full sunlight) or near a fire. Make sure it is dry and as fat-free as possible. Now rub in some preservative: in former days, an arsenic preparation was used, but borax is fine. Turn the skin back right-way out and (optionally) stuff it loosely with cotton-wool or tow (thick, dry, somewhat shredded rope). Store it in a dry place, preferably with naphthalene (mothballs).

Wet specimens

Wet specimens are sometimes useful, if special anatomical study is required. Store specimens in fresh formalin (formaldehyde solution, stock 37%–40% diluted with 10 parts water to 1 part formaldehyde). You will need twice the volume of fixative as specimen in your container. Inject the thorax with preservative and slit open the abdomen to allow the fixative to penetrate to internal organs. The concentration is important: too weak and the specimen rots, too strong and dehydration makes specimens hard and brittle. Do not keep the specimen in formalin too long (or at all, if DNA samples are required): it destroys the pelage colour and makes the bones brittle, and as for the human collector, it brings tears to the eyes; remember that it is POISONOUS! After a maximum of 30 days rinse your specimen in water and store it in 70% ethanol. Change the ethanol once after 30 days to eliminate the formalin. Other alcohols may be used temporarily if ethanol is not available.

Genetic and photographic type specimens

The *International Code of Zoological Nomenclature* (4th edn), which came into force on 1 January 2000, specifies that a type specimen must be 'an animal, or any part of an animal ...' (Article 72.5). If you think you may have found a new species, but can't collect a specimen, what do you do? When Thalmann and Geissmann (2000) met with an undescribed species of *Avahi* at Bemaraha they took good, crisp photos of it, but could not collect a specimen, both because the taxon is endangered (so killing one would have been, in effect, illegal as well as immoral) and because they quite simply could not bring themselves to kill one because of their moral scruples (which, incidentally, we share). Faced with the same quandary in the case of their new species of *Hapalemur*, Meier *et al.* (1987) captured animals alive and lodged them in a zoo; but this was not an option for *Avahi*, as they do not survive at all well in captivity. One option would be to keep the population and its predators under close observation, and keep an eye out for new and exciting bones under raptors' roosts, or individuals that are obviously suffering some terminal condition and are candidates for euthanasia for humane reasons.

Since the last edition of this book was published, DNA samples have been used as type specimens for various species, including primates. Even photographs are eligible: *Lophocebus* (now *Rungwecebus*) *kipunji* was described from a photograph (Jones *et al.*, 2005), the type specimen being not the photograph itself but the animal in the photo (despite the fact that it disappeared back into the vegetation a few moments after the photo was taken!). There were objections to this, but the Secretary of the Commission determined that it was a valid procedure. The next edition of the Code may take this up more specifically.

Labelling

Accurate and detailed labelling of specimens is extremely important. You can obtain standard printed labels from any Natural History Museum; these have spaces for geographic and habitat data, circumstances and date of collection, sex, and body measurements. Even if these printed labels are not available, it is essential to record the following data:

Date of collection

Collector

Method of collection (pick-up, euthanasia, confiscation etc.)

Locality: name (if there is one); distance from nearest settlement, river, forest block or other location easily found on a map or on the ground by another visitor; geographic coordinates; altitude (Chapter 4)

Habitat: forest type, other environmental details (Chapter 3)

Measurements: at the very least, head and body, tail, hindfoot, ear, mass (see below)

Other: details of phenotype, especially eye colour, pigmentation of exposed skin (ears, nose, palms and soles); whether pregnant.

Record the data in the field and copy it into a notebook or laptop computer as soon as possible. Make out a separate label for the skin, skull and any other sizable parts that are kept (pelvis, skinned body, etc.). Allocate the specimen a collector's number, and write this on **every** item as far as possible. Keep all the parts in the same specimen box, avoiding damage by wrapping fragile items (especially the skull) in cotton wool or other soft material; put small bones, like hand and foot bones and vertebrae, in special bags (one for each extremity, one for each spinal region).

SKULLS

Being the most complicated part of the skeleton, the skull (including the dentition) offers the most opportunities for diagnosis not only of taxonomic identity but also of sex, age and health status. Museum specimens commonly consist of the skin and the skull; preservation of the postcranial skeleton is unusual, of the soft tissues even more so.

Many traditional hunting and gathering cultures collect trophy skulls as a sign of their hunting success. The identification of these skulls to a species level can provide important information about species present in an area and can be used to confirm or contradict information acquired through interviewing local people.

Skulls are measured with callipers; these are a kind of metal ruler with two downward-pointing arms, a fixed arm and a sliding one. Some callipers have a digital read-out dial, which goes down to 0.1 mm or less; some have a vernier scale. Measurements should always be recorded in millimetres, or fractions thereof. Be only as accurate as the material warrants: a 300 mm gorilla (*Gorilla* spp.) skull is not worth measuring down to 0.1 mm! Callipers are small and light enough to take on any field trip, but in an emergency callipers used to measure machinery can be used, and are more easily available in local towns.

When it is not possible to remove specimens from the field, good photos are imperative (Chapter 17). Take photos of crania from the side (*norma lateralis*), front (*n. facialis*) and underneath (*n. basalis*), and photograph mandibles from the side, front and above (*n. dorsalis*), so that the teeth can be clearly seen. Photos are best taken with a plain backdrop and it is important that they are taken square on to avoid distortion. Place the callipers or a matchbox in the photo as a scale.

The skull consists of two moveable parts: **cranium** and **mandible**. The cranium is composed of many bones, joined together along **sutures**, which eventually fuse with age. Surrounding the brain is the **neurocranium**; hafted in front of it is the facial skeleton or **splanchnocranium**. These and some other standard anatomical terms are illustrated in Fig. 9.1.

In smaller strepsirrhines and in *Tarsius*, fusion of the two rami of the mandible is incomplete at the symphysis and the two halves can move separately. In larger strepsirrhines, and in all Simiiformes, the symphysis is firmly fused and the mandible is a single bone.

Measuring the skull

There is, unfortunately, no standard set of measurements for skulls. Fig. 9.2 illustrates some that are used in craniometric studies, and will enable you to make initial comparisons before turning the skull(s) over to a public collection.

Greatest length, total length, skull length (1–1)

From prosthion (the most anterior point on the bone between the upper central incisors) to opisthocranion (the most posterior point on the skull). Hold the fixed point of the callipers on the prosthion, and

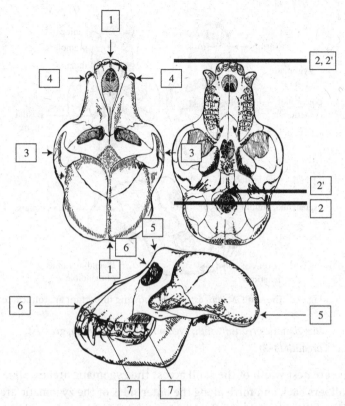

Fig. 9.1 Recommended cranial measurements. (See text for explanation.)

pull the sliding arm back until its point rests as far back on the cranium as possible.

Condylobasal length (2–2)

From prosthion to a tangent across the posterior borders of the occipital condyles. Lay the callipers sideways down; hold the fixed arm across the prosthion, and pull the sliding arm back until it rests against the posterior margins of both condyles.

Basal length (2'–2')

An alternative to condylobasal length; the two measurements can substitute for each other when different parts of the (often fragile) basicranium are missing. Place the fixed point on the prosthion, and move the sliding arm back until the point is on the basion, the anterior edge of the foramen magnum.

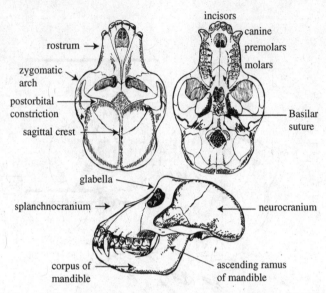

Fig. 9.2 The skull (*Macaca assamensis*) showing important anatomical terms.

Bizygomatic breadth/width, zygomatic breadth/width, greatest breadth (3–3)

The greatest width of the skull across the zygomatic arches. Slide the callipers back and forth along the outer sides of the zygomatic arches (perpendicular to the skull length, of course) until you have the greatest reading.

Bicanine breadth/width (4–4)

Greatest width of the muzzle, across the canine alveoli.

Cranial length, braincase length (5–5)

From *either* glabella *or* nasion to opisthocranion. The glabella is the midline swelling between the brow ridges; the nasion is the point where nasal bones and frontal bone(s) meet or, if you like, the point where internasal and frontonasal sutures meet. These two measurements may or may not be virtually the same.

Facial length, facial height, upper facial height (6–6)

From prosthion to nasion.

Maxillary toothrow length (7-7)

From the anterior end of the anterior upper premolar (P^3 in catarrhines and indrids, P^2 in most other primates) to the posterior end of the posterior upper molar (M^3 in nearly all primates, M^2 in callitrichines). Measure along the alveolar margin; this way, the measurement can be taken even if all the teeth have dropped out.

Others

Some authors take mandibular toothrow length; some include the canine in their toothrow lengths; some measure molar rows. Some authors take a cranial height measurement (usually basion to bregma); some favour a cranial width on the braincase itself, and/or on the external or internal margins of the orbits. However, by taking the above 7/8 measurements, you cannot go far wrong and, of course, if the specimen is preserved in a museum or other public institution, it is always available thereafter for more detailed studies.

Cranial capacity

Cranial capacity is a close approach to actual brain size. For this you will need a measuring cylinder or cup, a funnel, and some filler. Birdseed is ideal, as it packs well; sand is satisfactory, but may slip out through cracks such as incompletely fused sutures. Block up the optic foramina, preferably with Plasticine®; at a pinch, simple mud will probably do this effectively. Inspect the neurocranium to see whether there are any other holes. Hold the cranium upside down so that the foramen magnum is horizontal, and gradually pour in the filler. Every so often, shake the cranium, to make sure that the filler settles and packs in. When the filler has reached the rim of the foramen magnum, and cannot pack down any further, pour it out through the funnel (to make sure none is spilled) into the measuring cylinder, shaking this every so often to pack it down. Read off the volume.

Ageing and sexing the skull

The rate at which cranial sutures fuse varies both inter- and intraspecifically. The only suture for which fusion is within narrow age limits is the **basilar suture** (strictly, not a suture at all, but a synchondrosis: the

spheno-occipital synchondrosis), between the basioccipital and basisphenoid bones, some way in front of the foramen magnum, the hole through which the spinal cord passes. This fuses at maturity; afterwards, little or no further cranial growth takes place. It is therefore important to note whether it is widely open, in process of fusion, or fully fused.

The skull is impossible to sex without some prior knowledge of the way in which sexual dimorphism is expressed in that species. In most (not all) Simiiformes, males are larger than females and, because the facial skeleton grows with positive allometry, the jaws are relatively longer.

Cranial pathology

A more or less symmetrically developed pitting or channelling on the roofs of the orbits is called **cribra orbitalia**, and is due to anaemia. Anaemia (lowered haemoglobin levels) is a symptom of a whole range of conditions: dietary insufficiency, chronic disease, abnormal haemoglobins (commonly indicating a genetically determined response to malaria, in the human case at least). More severe anaemia results in thickening of the cranial vault and widescale pitting, especially on the parietals (**porotic hyperostosis**).

Thin, light, somewhat pitted skull bones are due to **osteoporosis**, indicating some type of dietary malabsorption.

A cancerous lesion (**osteosarcoma**) on the skull takes the form of a bony swelling, which has often burst to form a wide, saucer-like crater.

Healed fractures are detectable because part of the skull, such as the zygomatic arches or orbital margins, is markedly asymmetrical, and one side is irregularly shaped where the other is normal.

In species with strong muscular development, such as gorillas, damage to the muscle results in underdevelopment of the supporting bone on that side, making the entire skull asymmetrical.

INFORMATION FROM TEETH

Teeth consist of crowns (the part visible above the jawline) and roots (in alveoli, or sockets, in the jaw). A tooth is mainly formed of dentine, which is yellowish. In primates, the crowns are coated with enamel, a hard, white, shiny mineralized substance; roots are coated with cement, a greyish, rough-cast coating.

The teeth are **deciduous** in infants; these are shed, usually around the time of weaning, and replaced by **permanent** teeth. Record the eruption and wear status of the dentition in any skulls obtained: which teeth are present, which are in process of eruption, whether they are worn. Third molars erupt on the threshold of physical maturity and usually begin to show signs of wear shortly before the basilar suture begins to fuse. The sequence of eruption of both deciduous and permanent teeth varies interspecifically and may be slightly variable intraspecifically.

Tooth wear is initially in the form of faint scratch lines on the enamel, then as shiny facets; later, the enamel becomes worn through, revealing the softer, yellow-brown dentine beneath. Cusps on the cheek teeth (premolars and molars) wear at different rates; typically, the **lingual** (tongue-side) cusps on the upper (maxillary) cheek teeth are more worn than those on the **buccal** (cheek-side), and the reverse is true for the lower (mandibular) cheek teeth. Cusp shape and disposition is, within reason, a good indicator of diet. High, pointed cusps suggest insectivory; low, rounded cusps suggest frugivory. Taxa with a strongly folivorous diet commonly have high-crowned cheek teeth with the cusps to some degree joined by ridges (lophs); this condition is most strongly developed in the Old World monkeys (Cercopithecoidea), where the joining of the molar cusps in transverse pairs, accompanied by constriction of the crown between the lophs, is called bilophodonty. Different dietary items leave characteristic scratches on the crown, which can be analysed under a microscope (especially a scanning electron microscope).

Incisor teeth may occlude end-to-end (topbite), or the uppers in front of the lowers (overbite) or, rarely (but commonly in langurs of the genus *Presbytis*), the lowers in front of the uppers (underbite). If some incisors are missing, or only the cranium or only the mandible is present, the bite is indicated by the face of the remaining incisors that has wear marks.

In most Strepsirrhini, the lower incisors and canines are strongly procumbent (forming a tooth comb or scraper) and typically do not occlude at all with the uppers. In the true marmosets (genus *Callithrix*) the lower incisors and canines are also undifferentiated, compressed and slightly procumbent, although they do not form a dental comb in the same way; a closer approach to a toothcomb is seen in the sakis and uakaris (Pitheciinae), in which the incisors (but not the canines) of both jaws are procumbent and compressed.

Teeth must be measured with a dial or vernier calliper to tenths of a millimetre. The usual measurements are **mesiodistal length**,

buccolingual (labiolingual) breadth, and **crown height**, but others are possible; for example, in the case of lower molars with marked differentiation between trigonid and talonid (lemurs, for example), or a bilophodont construction (Old World monkeys), both mesial and distal breadths are often taken. Take the length and breadth measurements at the level of the alveolar line, i.e. just above the cemento-enamel junction. With age, teeth (especially cheek teeth) decrease in length due to interproximal wear (wearing against the teeth in front and behind); as the crown is worn away, the breadth may change too, first increasing (the crown usually flares somewhat below the occlusal surface), then decreasing. But these changes are minor, except with extreme wear. The only major change is the reduction in crown height; crown height measurement is of value mainly for documenting the degree of wear.

You can also measure teeth in live, anaesthetized animals (see, for example, Phillips-Conroy and Jolly (1986), who also made dental casts of their study animals).

Sexing the teeth

The canine teeth are reliable indicators of sex in all Catarrhini except gibbons (Hylobatidae) and humans (*Homo*). The male in all other catarrhines has longer canines than the female, and the anterior lower premolar is modified to act as a hone for it; this premolar is narrow, typically unicuspid, and with the enamel dipping down below the usual alveolar margin. This sexual dimorphism is gross in Cercopithecinae and many Colobinae, but in a few colobines the sexual difference is marginal. In many, but not all, Platyrrhini there is likewise sexual dimorphism in the canines. There is no such sex difference in tarsiers or strepsirrhines, and these are impossible to sex using the skull.

Dental pathology

Look for irregular, shallow, discoloured pits on the walls of the teeth (usually low down, near the neck of the tooth where it emerges from the alveolus), sometimes on the occlusal surface in the fissures between the cusps. These indicate **caries**.

Dental **abscesses** affect the surrounding bone and take the form of large crater-like holes in the alveolar bone overlying the infected tooth roots.

Empty alveoli (tooth sockets) generally mean that the tooth has fallen out after death (**postmortem**), but an alveolus that has been partly or completely filled in by bone indicates **antemortem** tooth loss. Teeth may be lost because of infection, or as a result of trauma.

Transverse lines across teeth, most obvious on incisors and canines, are **enamel hypoplasia** (inadequate formation of enamel). You can distinguish this condition from simple traumas on the teeth because the lines are symmetrical on the jaws, meaning that the lesion occurs at the same level on both antimeres (both lateral upper incisors, or whichever tooth is affected). They result from some acute stress, such as disease or poor nutrition. Because teeth form at regular rates, the measurement from the lesion to the cemento-enamel junction indicates the age of the animal when the hypoplasia-inducing stress occurred. Weaning, which often involves transferring to a poorer diet, is a likely cause of hypoplasia.

Look for broken teeth. Unlike bones, teeth do not regrow, so a breakage remains visible and, if the pulp cavity has been exposed, potentially a cause of infection. The jagged edges wear and become rounded. A badly decayed (carious) tooth may break, but the usual cause is either fighting (especially canines), or trying to bite too hard an object.

MEASURING THE BODY

Postcranial bones

The five bones commonly measured (lengths only, or shaft widths as well) are humerus, radius, femur, tibia and clavicle. Simply take the greatest length, from end to end, parallel with the shaft.

Humerus + radius expressed as a percent of femur + tibia is called the Intermembral Index. A very low index (below 70) tends to characterize vertical clingers; a very high one (above 100) indicates a brachiator, clamberer or knuckle-walker; the quadrupeds have indices in between, and the lower the index the more of a leaper the animal (as a rule of thumb).

Radius expressed as a percentage of humerus is the Brachial Index; tibia as a percent of femur is the Crural Index.

Growth of the body

During growth, the articular ends, and some other protuberances, of most postcranial bones remain separated from the shafts by cartilaginous plates. These are known as **epiphyses**, and they fuse to the shafts when growth is complete. As well as limb bones (including metapodials and phalanges as well as long bones), the clavicle, scapula and pelvis grow by means of epiphyses, and can therefore be used to indicate age.

Fractures

Note the presence of healed fractures in the skeleton. Jagged edges indicate that a bone has broken after death, or very shortly beforehand. When a breakage occurs during life, and the animal not only survives but lives for some time afterward, there are signs of healing: the broken ends grow over and join together again, but always leave some irregularity; in the case of a long bone or rib, the two halves of the bone override to some extent, instead of rejoining end-to-end. Illustrations of a variety of healed fractures and other pathologies are given by Schultz (1944) for white-handed gibbons (*Hylobates lar*); this classic study shows how unexpectedly common breakages and healing are in at least one species, and the grotesque injuries and deformities that some primates can live with.

EXTERNAL MEASUREMENTS

There are four standard linear measurements that are taken in all mammals:

> Head and body length (H+B, or HB). Tip of snout to anus (when head is stretched out as far as possible in line with back).
> Tail (T). From anus to tail tip, excluding hair.
> Hind foot (Hf). From heel to tip of longest toe, excluding nail. Record which toe is longest.
> Ear (E). Behind the conch, from the base of the ear to the tip.

These are taken straight, with a calliper if possible, not with a tape. They are best done in the flesh, as the skin stretches somewhat in skinning.

Additional measurements often taken are head length (an approximation to greatest skull length), trunk height (manubrium to pubis), hip breadth, hand length and breadth, foot breadth, and lengths of individual limb segments (thigh, lower leg, upper arm, forearm). Though not standard, they may be useful as approximations to bone lengths (see McArdle, 1981).

Weight (body mass) is nowadays, for better or worse, the most commonly employed standard size measurement. The animal should preferably be weighed at once; or, if the gut has been removed, the guts should be weighed separately, losing as little of their contents as possible.

From the mass and body length, the Body Mass Index can be calculated as (kg mass)/(m body length)^2. Skin-folds and limb circumferences can also be useful measures of body condition.

GENITAL ANATOMY

Many primates, especially strepsirrhines, show marked changes in testicular size according to breeding condition. Measure the length and width of the testes with callipers, and calculate testicular volume (TV) as a regular ellipsoid:

$$TV = [\pi \times (width)^2 \times length]/6 \tag{9.1}$$

A scrotal skin-fold is often subtracted from the length and width of the testes.

Females may undergo periodic opening and closing of the vulva, or exhibit sexual swellings, and some fairly standard scores exist to describe these (see, for example, Scott, 1994). Nipples become elongated with suckling, and can indicate whether a female has borne young. Note pregnancy.

GUT MORPHOLOGY

If a dead animal is to be preserved whole, always remove the alimentary tract (gut) first and preserve it separately, because this is a prime site for bacterial activity. Gut contents can be useful for dietary studies (Chapter 12). Gut size and morphology is subject to significant variations across the primates as a whole, and even between quite closely related taxa. In most primates the stomach is a simple sac, somewhat elongated transversely; the oesophagus enters it from above, and the duodenum (the first part of the small intestine) leaves it at one side; the two surfaces between oesophagus and duodenum are called the lesser and greater curvatures (self-explanatory terms). In the subfamily Colobinae (Catarrhini, Cercopithecidae), however, the stomach is sacculated, and divided into three or four compartments: the **saccus** (the blind, sac-like anterior compartment), the **tubus** (the middle section) and the **pars pylorica** (the posterior compartment); the fourth compartment, the **praesaccus**, if it occurs at all, is located at the junction of the saccus and the oesophagus. The saccus (and praesaccus) and anterior tubus are fermentation chambers, where cellulose and hemicellulose are fermented by symbiotic bacteria into short-chain fatty acids and waste gases. A groove, the **reticular groove**, the edges of which may be contracted to make it into a tube, runs along the dorsal side of the tubus, and conducts fluid ingesta that need no fermentation.

The other main gut variable is the **caecum**, which may be extremely long, and full of cellulolytic bacteria (e.g. *Lepilemur*), or very reduced but with a **vermiform appendix**, containing lymphatic tissue (all the Hominoidea). The **colon** may also be a significant site for fermentation, and varies from thick and sacculated (in Simiiformes) to slender but folded into an M-shaped loop, the **ansa coli** (all Lemuriformes except the Cheirogaleidae).

Traditionally, the lengths of each of the components of the alimentary tract are measured: oesophagus, stomach, small intestine, caecum and colon (= large intestine), rectum. The small intestine is complexly coiled, and the colon has ascending, transverse and descending portions (plus a median loop, the ansa coli, in strepsirrhines), and the mesenteries holding them in place must be cut through before they can be straightened out for measuring.

Chivers and Hladik (1980) attempted a more sophisticated method of measurement. First of all they flattened each segment and measured its breadth at points, from which an average breadth was calculated, and a surface area (breadth × length), except for the stomach, where they measured the length of the greater curvature as a measure of its circumference. From these areas, they calculated volumes using the formula for the volume of a sphere in the case of the stomach, and that of a cylinder in the case of the other components. This is probably the best that can be achieved in the field, but is still in the end a rather crude approximation to the complexities of villous formation (small intestine) and sacculation (large intestine).

PELAGE

The pelage in most primates is patterned; this may consist only of the ventral surface (plus, usually, the inner aspects of the limbs) being lighter in colour than the dorsal, or the entire body may be divided into different chromogenetic fields. Record the boundaries between any such fields, along with their approximate colour. Colour photographs vary deceptively in different light or under different exposures (Chapter 17), so include a standard colour strip in your photos.

Some or all of the dorsal hairs are **agouti** patterned: that is, with alternating light (phaeomelanic) and dark (eumelanic) bands. Record the number of band-pairs, along with the approximate tone of each band. Non-agouti hairs are called **saturated**.

Hair tracts on the cheeks, crown, nape, throat and other places are usually in the form of **whorls**. These should be recorded, and so should the often quite complicated arrangements of tracts on the top and sides of the head.

In some species, the sexes are different colours and differently patterned (most strikingly in some lemurs, genus *Eulemur*; sakis, genus *Pithecia*; and gibbons, Hylobatidae). Age changes in pelage are usual in catarrhines, especially from neonate coat to juvenile/adult coat.

GLANDS

An important feature of many primates, but one liable to be overlooked by casual observers, is skin glands. Their presence and location is not always obvious, but even if there is not a clearly demarcated glabrous area there is generally indication in the form of irregularity or sparse hair-covering of the skin, and sometimes the secretion itself is visible. Lemuriformes and Lorisiformes have glands on the perineum, scrotum, labia, wrist, inner elbow, shoulder, jaw angle or throat. Probably all platyrrhines have a gland on the sternum, and so do some catarrhines including mandrills, gibbons and orangutans. You can describe or score the activity of glands for individual animals, and collect samples of glandular exudates (Chapter 8).

REMOVAL OF SPECIMENS FROM THE FIELD

It is often impracticable, owing to a large number of samples, or unethical, to remove trophy skulls or skins from the field. It is therefore imperative to collect as much information as possible about the specimens while in the field. If there is a facility that can keep them, specimens should be kept in their country of origin so that other researchers can benefit from them. If all measurements and photos can be taken before leaving the research country there should be no need to remove the specimens from their country of origin. If you do need to export specimens from the research country there are a number of procedures to follow (for example, Convention on the International Trade in Endangered Species (CITES) permits, health/ agriculture department authorizations, and other permits may be required for import and export).

REFERENCES

Chivers, D. J. & Hladik, C. M. (1980). Morphology of the gastrointestinal tract in primates: comparisons with other mammals in relation to diet. *J. Morphol.* 166, 337–86.

Jones, T., Ehardt, C. L., Butynski, T. M. *et al.* (2005). The highland mangabey *Lophocebus kipunji*: a new species of African monkey. *Science* 308, 1161–4.

McArdle, J. E. (1981). *The Functional Morphology of the Hip and Thigh of the Lorisiformes.* 1 (Contributions to Primatology 17.) Basel: Karger.

Markham, R. & Groves, C. P. (1990). Weights of wild orang utans. *Am. J. Phys. Anthropol.* 81, 1–3.

Meier, B., Albignac, R., Peyriéras, A., Rumpler, Y. & Wright, P. (1987). A new species of *Hapalemur* (Primates) from South East Madagascar. *Folia Primatol.* 48, 211–15.

Nichols, R. (2001). Gene trees and species trees are not the same. *Trends Ecol. Evol.* 16, 358–64.

Phillips-Conroy, J. E. & Jolly, C. J. (1986). Changes in the structure of a baboon hybrid zone in the Awash National Park, Ethiopia. *Am. J. Phys. Anthropol.* 71, 337–50.

Schultz, A. H. (1944). Age changes and variability in gibbons. A morphological study on a population sample of a man-like ape. *Am. J. Phys. Anthropol.* 2, 1–129.

Scott, L. (1984). Reproductive behavior of adolescent female baboons (*Papio anubis*) in Kenya. In *Female Primates: Studies by Women Primatologists*, ed. M. F. Small, pp. 77–100. New York: Alan R. Liss, Inc.

Thalmann, U. & Geissmann, T. (2000). Distribution and geographic variation in the Western woolly lemur (*Avahi occidentalis*) with description of a new species (*A. unicolor*). *Int. J. Primatol.* 21, 915–41.

PAUL E. HONESS AND DAVID W. MACDONALD

10

Marking and radio-tracking primates

INTRODUCTION

Although the past two decades have seen a revolution in many aspects of field biology due to advances in radio-tracking and telemetry, including the development of satellite and GPS technology, these techniques have been less applied to primates than to other orders of mammals or to vertebrates in general (Casperd, 1992). A comprehensive account of animal tagging and radio-tagging would warrant a book in itself, but here we aim to draw the attention of primatologists to the advances in this family of techniques. We review a number of important studies of primates that have used them, and introduce the practicalities involved. If you are considering the use of these techniques then you should advance no further without exploring the wider, and vast, literature associated with radio-tracking other mammals. General reviews of radio-tracking are presented in Amlaner and Macdonald (1980), Kenward (2001), Millspaugh and Marzluff (2001) and *Telonics Quarterly* (see 'List of supplies and useful Internet sites').

The term 'radio-tracking' is correctly applied only to the use of radio transmitters and receivers to record location information. Traditionally, most field biologists have used VHF (Very High Frequency) or UHF (Ultra High Frequency) radio-tracking transmitters, but recently there has been a rapid growth in satellite tracking systems and associated technology (Chapter 4). 'Biotelemetry' is, strictly, the remote measurement of biological, particularly physiological, data (e.g. heart rate, blood pressure, body temperature, brain wave activity) (Amlaner, 1978; Chapter 19).

An almost invariable prerequisite to marking animals, by whatever means, is capture and restraint (Chapters 7 and 8). It is important to ensure that capture, handling and, ultimately, wearing the tag are minimally stressful (Halloren *et al.*, 1989), and appropriate veterinary

Field and Laboratory Methods in Primatology: A Practical Guide, ed. Joanna M. Setchell and Deborah J. Curtis. Published by Cambridge University Press. © Cambridge University Press 2011.

guidance is essential (Karesh *et al.*, 1998; Chapter 8). For both ethical and scientific reasons, the impact of the research on the animal, and of the number of subjects involved, must be minimized as a priority, while still achieving results with adequate statistical power.

MARKING

Individual characteristics may allow the accurate identification and recognition of primates (Ingram, 1978) (Chapter 2), but you often require artificial marking for the unambiguous recognition of individuals. Circumstances under which this may be useful include those where the animals under study:

- do not have readily recognizable naturally occurring individual characteristics (e.g. scars, missing digits);
- live in large social groups;
- live in social/foraging groups of changing composition such as fission/fusion societies (e.g. spider monkeys, *Ateles geoffroyi*; Campbell & Sussman, 1994);
- are nocturnal and/or live in habitat that affords poor visibility, but where individuals are frequently encountered;
- are the subjects of long-term studies, where different researchers may be involved;
- are the subjects of capture–mark–recapture studies.

The following methods of marking have been applied to primates. Consider the need to identify individuals from whichever side they are viewed (Rasmussen, 1991). You may need to use a combination of methods to achieve identification of all animals (see, for example, Oluput, 2000; Di Fiore *et al.*, 2009).

Fur shaving

Shaving patterns of rings on limbs suffers from limited longevity owing to hair regrowth (e.g. within 6 months; Oluput, 2000), as well as risks associated with exposing skin to sunburn, insect bites and abrasion (Rasmussen, 1991).

Fur dying

Application of picric acid to white hair turns it yellow. Halloren *et al.* (1989) report that there were no adverse effects on body mass or survivorship at 12 months of age in marked captive callitrichid infants compared with unmarked animals.

Permanent black dye (Nyanzol-D, J.Belmar Inc.) lasts up to one year in adults but less than four months in young monkeys (species not specified; Rasmussen, 1991). Megna (2001) recommends the following preparation: 'Mix 12 cc water, 12 cc over-the-counter hydrogen peroxide, 48 cc rubbing alcohol and 1 heaped teaspoon of the dye. Stir well, allow to sit a while, stir again and apply with a sponge-type paint/craft brush.' This dye is reported to work well on skin and fur, and has been used on adults and juveniles, but not infants (Drea & Wallen, 1999).

Savage *et al.* (1993) used a commercially available hair dye, Redken DecoColors, in different combinations of red, yellow, orange, purple and brown to mark the naturally white parts of the pelage of cotton-top tamarins (*Saguinus oedipus oedipus*).

Freeze branding

Clipping the fur away and spraying the skin with freon through a template of a unique code, for 12 s (Jones & Bush, 1988a) 'bleaches' the hair follicles, which subsequently produce white hair even after moulting. This is really only valuable for marking dark-haired species. When done properly, freeze branding will not cause blistering or permanent damage to the skin, but consider anaesthesia or sedation for restraint and to allow accurate branding. You may need large brand marks to ensure visibility and distinctness, so take care not to disrupt natural hair colours/patterns that may form part of the species' signalling system. Fernandez-Duque and Rotundo (2003) report the loss of distal parts of the tail in some owl monkeys (*Aotus azarai azarai*) where fur shaving marks on the tail were reinforced with freeze branding.

Tattooing

You can tattoo primates as small as callitrichids (Savage *et al.*, 1993). Tattooing should be conducted under anaesthetic. Disinfect the tattoo site and needles (e.g. with sealed swabs). This method is of little value for dark-skinned species and you need to consider which part of the body is tattooed, in respect of both the longevity and the visibility of the mark. Ear tattoos tend to last longer, but are less visible from a distance; those on the chest or inner thigh fade more rapidly and may become obscured by regrowth of hair (Rasmussen, 1991). Tattoo guns that run on batteries are useful in the field, but often constrained by poor needle speed.

Ear notches or punches

If you take ear notches or punches for DNA samples, you can use the pattern or position of these marks to identify individuals, although they can be confused with natural marks (Rasmussen, 1991). Sterilize equipment between individuals. Where other sources of DNA may be available (e.g. from hair or faeces, Chapter 21) it may be ethically questionable to take ear notches or punches.

Toe clipping

Formerly, studies of small mammals often removed the terminal joint of a digit as an individual mark. Increasingly, such mutilation is considered unethical, especially in primates, which depend on fine manipulative skills.

Ear tags

Coloured ear tags, such as those for livestock, can provide an easily visible marking system. Taylor *et al.* (1988) note that rigid, plastic tags can be difficult to see, often being obscured by fur. They trialled two types of tag on captive anubis baboons (*Papio anubis*) in which the fitting, under anaesthetic, of combined Jumbo Rototag (back) and Allflex tags (front) (NASCO) resulted in high rates of grooming out and infection. No such problems occurred with Duflex sheep/goat tags (Fearing Manufacturing Co. Inc.), which incorporate a slow-release, antibacterial coating (chlorhexidine).

Microchips

Microchips (passive integrated transponder, or PIT, tags) encased in glass, frequently with a bio-compatible coating, are injected subcutaneously by using a syringe or injector 'pistol'. In primates, they are typically injected into the lateral thigh area; avoid thin-skinned areas such as the hands and feet (Rasmussen, 1991). These microchips (approximately $12\,\text{mm} \times 2\,\text{mm}$ and about $0.06\,\text{g}$) function as radio-frequency transmitters and contain no batteries, but are activated and read by using a special microchip reader held less than $20\,\text{cm}$

away (depending on the reader). The unique microchip code is detected by the reader and the chip will last for around one million activations. They can also contain body temperature sensors (Chapter 19). A practical difficulty is the need to get the reader sufficiently close to the chip, especially as the chip may migrate short distances under the skin (Wolfensohn, 1993). However, this system works well in combination with other more transient marks.

Collars or belts

Brightly coloured collars or belts, secured with rivets or cable ties (be careful to ensure no sharp ends or edges protrude) can be efficient markers. Collars made of butyl material are commercially available (e.g. Telonics Inc.). Plastic disks fixed directly to collars can be fragile (Jones & Bush, 1988a), whereas anodized aluminium discs (e.g. pet tags) are more durable. Combinations of coloured beads on ball-chain collars are a good option for smaller primates (Fernandez-Duque & Rotundo, 2003). Take care with loose-hanging metal tags that may act as bells and could compromise the ability of tagged animals to avoid predators, and may impair behaviour such as in social interaction and foraging or hunting.

RADIO-TRACKING

Radio-tracking can facilitate the habituation of study animals by increasing contact time though the rapid location of collared individuals (Campbell & Sussman, 1994). It can also prove a useful tool to measure decrease in flight distance with habituation (Rasmussen, 1998; Chapter 2). It may be that more rapid habituation through radio-tracking serves only to counter the timidity caused by capture (Campbell & Sussman, 1994; Gursky, 1998; Jones & Bush, 1988a; Chapter 7), but all these effects merit note as potentially confounding variables, and the link between radio-tracking and enhanced habituation is generally a bonus to most field studies (Karesh et al., 1998). Radio-tracking can also make it easier and quicker to obtain accurate home range information and lead you to areas of the animals' ranges that you would not otherwise have discovered (Campbell & Sussman, 1994; Fedigan et al., 1988).

In primates, radio-tracking has primarily been used to investigate aspects of ecology such as seasonal changes in home range (Li et al., 1999), substrate use (Harcourt & Nash, 1986), polyspecific associations (Buchanan-Smith, 1990) and activity patterns (Donati et al., 2009), but it has also proved an extremely valuable tool for studying mating systems and social organization (Bearder & Martin, 1980; Crofoot *et al.*, 2008). Combining radio-tracking with molecular techniques has also proved valuable in the study of dispersal patterns (Di Fiore *et al.*, 2009). It is a particularly useful tool for studying primates that are nocturnal or live in habitats that afford poor visibility. More specialized studies have examined locomotion (Sellers & Crompton, 2004), predation (Wiens & Zitzmann, 1999), urine marking or washing (Charles-Dominique, 1977; Harcourt, 1981), and calling frequency (Gautier & Gautier-Hion, 1982), and monitored translocated or reintroduced animals (Medici *et al.*, 2003; Peignot *et al.*, 2008) and solitary males (Jones & Bush, 1988b). You can also use collar-mounted temperature sensors to monitor skin temperature (Chapter 19).

The basic equipment needs for a radio-tracking study include: a selection of transmitters (tuned to transmit on different frequencies) with batteries, transmitter mounts, a receiver (plus batteries), a set of earphones, a coaxial cable (with suitable antenna and receiver connectors), and an antenna. It is prudent to have duplicates of much of this list, as back-ups. When ordering transmitters (see 'List of suppliers and useful Internet sites') ensure that they are set to transmit at different frequencies from those of any other researchers working within transmitter or dispersal range.

Transmitters

Transmitters are now available that weigh as little as 0.9 g (e.g. Sirtrack Single Stage Transmitter), making radio-tracking of even the smallest primate species possible. Standard VHF/UHF transmitters emit a pulse on a tuned radio frequency, which can be detected by using an antenna and receiver and used to locate the source of the signal, generally through triangulation. An important distinction between VHF and UHF transmitters is that higher frequencies are associated with reduced transmission range and high attenuation by environmental features such as vegetation. Another important consideration when selecting transmitter frequency is the associated constraint on the size of both transmitting and receiving antennae. Kenward (2001) discusses the relationship between frequency, wavelength and receiving and transmitting antennae in some detail. In summary, a higher frequency (e.g. up to 230 MHz) will allow the use of shorter, more convenient antennae, but this is compromised by the reduced range of high-frequency signals, which are more susceptible to absorption by vegetation and attenuate more rapidly. Lower frequencies (e.g. 142–150 MHz) therefore have an advantage in terms of signal range in areas where vegetation may be damp and dense, such as in rain forest. However, increasing the frequency of the transmitter to 216–230 MHz would require a transmitting antenna of only about two-thirds of the length of that for a lower-frequency transmitter for the same efficiency (Kenward, 2001).

Transmitter mounting

You need to attach external transmitters, with their enclosed batteries, to the study animal. Body size and anatomy will determine the best attachment method, for example on a neck collar, harness or waistband. Obvious welfare priorities necessitate careful attention to potential growth and seasonal mass changes. Devices designed to drop off on a given time-scale, or to expand, offer some solutions.

Collars are the most frequently used means for attaching transmitters. Nylon webbing is often the most practical and cost-effective material, having a good strength to weight ratio, and being non-absorbent and relatively cheap. Other options for collar materials include butyl, urethane and cable ties (see, for example, Biebouw, 2009). Test different materials in captivity before attachment to wild animals. For example, Kierulff *et al.* (2005) tested neoprene, nylon tubing and ball-chain collars on capuchins in zoos before fieldwork, finding the ball-chain option the only one that did not cause skin irritation or lesions. The transmitter antenna can either be a tuned loop of wire or a brass strip, which can form the basis of the collar, or be of whip form protruding from the assembly, wrapped around inside the transmitter casing or attached to the perimeter of the collar, belt or harness. Tuned loops may be more efficient and are certainly more useful for species that might significantly damage a whip antenna, but they suffer from size (diameter) constraints of what is easily tuneable and can become detuned during use or stop transmitting altogether if the loop is broken (e.g. by a predator) (Kenward, 2001). Collars can be fastened with pop rivets or nylon locking nuts through metal backing plates or a similarly secure system. Take care that no protuberances are left on the inside of the collar that may harm the animal. Fit collars loosely enough not to affect feeding or comfort but tight enough not to come off; a common guide is to be able to place two fingers under the fixed collar. However, collar attachment may be a serious difficulty for animals without a defined neck, such as the slow loris (*Nycticebus coucang*) (Barrett, 1985). Some collars are designed to expand with growth or seasonal mass change, using degradable foam on the inside or pleats held with degradable or forcible thread.

Transmitter retrieval

For both scientific and ethical reasons, it is essential to minimize the effects of radio-tracking on the subject and even minimal effects should be measured carefully to reduce the risk of confounded results. An obvious question is whether it is possible to retrieve transmitters from study animals, either at the end of the study or at the end of the transmitter's life. Transmitter retrieval is not often reported in the literature (but see Gursky, 1998; Biebouw, 2009). Do a cost–benefit analysis of the impact of leaving the transmitter in place versus the risks associated with retrieval.

Various devices can be built into a collar or harness with the aim of causing it to drop off after a timely interval. However, there are trade-offs, such as the extra bulk of these detachment devices and the risk of premature detachment. Most manufacturers will refurbish collars and transmitters (e.g. replace the batteries), which may represent a considerable saving over the purchase of new equipment.

Attaching a transmitter weighing 3% of an animal's body mass is the equivalent of an 80 kg person carrying an additional 2.4 kg, so take care to detect, minimize and monitor any adverse affects of transmitters. Balance any reduction in the mass of the assembly against reduced transmitter life and/or range due to minimizing the primary flexible component: the battery. Gursky (1998) found no significant difference in the behaviour of spectral tarsiers fitted with collars of 5% or 7% of their body mass. Juarez *et al.* (2008) found that radio-collars did not negatively affect long-term reproductive abilities compared with control animals in owl monkeys (*Aotus azarai*). However, Hilpert and Jones (2005) found that a radio-collared female mantled howler monkey (*Alouatta palliata*) was more often alone, interacted less with males than females with identification collars, and suggest her death/disappearance may have been related to being radio-collared.

Transmitter reliability

Manufacturers rarely guarantee the life of the transmitter's battery without additional cost, with low temperatures, in particular, having an adverse effect on battery life. Transmitters are frequently shipped and stored with a detachable magnet fixed to the outside to prevent the closure of a reed switch that would otherwise complete the power circuit and activate the transmitter. On receipt of the transmitters, and periodically during storage, check that the magnets are preventing transmission by testing whether you can detect the transmitter using tracking equipment with the magnet in place. If the magnets or reed switches fail, the batteries may be exhausted when you come to use the transmitter. Also check and record the frequency of an activated transmitter and do not simply rely on the label.

Receivers

Receivers are generally the most expensive element of a radio-tracking system. There are two main types of receiver: standard models (approx. US$650–850), and more specialized, programmable ones (over US$2000). Although most researchers chose portable options, you can also locate receivers at static monitoring points with better reception ranges (see, for example, Oluput, 2000; Crofoot et al., 2008).

Standard receivers are pre-tuned to a specific frequency range. Transmitters must match the frequency range of your receiver. Three of the most commonly used receivers are compared in Table 10.1. Most receivers have a monitoring speaker built in and can also be used with headphones. Noise reduction devices are also available (e.g. Telonics

Table 10.1 *Comparison of the mass, size and power of three commonly used receivers*

Receiver	Size (mm × mm × mm)	Weight (including batteries) (g)	Power
Telonics TR-4	170 × 90 × 45	425 (including 2 batteries)	1 or 2 × 9 V batteries[a]
AVM LA12-Q	150 × 150 × 60	750	6 × NickelCadmium (Ni-Cad) 'AA' cells[b]
Lotek Biotracker / Biotrack Sika	150 × 85 × 55	800	4 × 'A' NiMH rechargeable cells or 4 × 'AA' primary cells[c]

[a] 12 V DC, 220 V DC adapters available.
[b] 110 and 220 V DC adapters available.
[c] 10–15 V DC and international adapters supplied.

TNR-3000; 190 g, 9V AC/DC) that are placed in line between the receiver and earphones. These are designed to improve signal quality against external sources of interference (e.g. vehicle engines, power lines and atmospheric interference). High humidity can seriously damage receivers and batteries, so dry receivers when not in use, remove the batteries and store both with silica gel.

Receiving antennas

Although more complex antenna systems are available (e.g. aircraft mounted and static systems; Kenward, 2001), most field studies require portable antennae. The most basic type of antenna, the omni-directional, is typically of whip form and can be attached to the roof of a vehicle by a magnet. It is particularly useful for detecting the general proximity of a transmitter, without providing any information about the bearing from the observer.

Directional, hand-held antennae (e.g. multi-element Yagi, 2-element Adcock or Telonics 'H' and loop) are the most practical for studying primates in forest or woodland habitats. A common variant is the multi-element Yagi gain-type antenna, which consists of a main driven element (connected via a coaxial cable to the receiver) with a number of parasitic directing elements in front of it and a reflector element behind, all arranged in a specific mathematical relationship. Although the spacing relationship between the elements is defined, the number of elements can vary; increasing the number of elements increases the directionality of the antenna (Burger, 1991). Collapsible versions of the Yagi are available, and are a good option for use in dense vegetation. You can use a pair of Yagis, typically of four or more elements, in tandem (a Null-Peak System). This is significantly more accurate than a single hand-held antenna, but carries the disadvantage of bulk. Macdonald and Amlaner (1980) outline some of the problems associated with using directional antennae and include some tips for their solution, as does Kenward (2001). Kenward (2001) suggests a procedure for determining the bearing of the transmitter from the observer using a Yagi antenna:

1. Find the strongest signal that is not either a 'back bearing' or a reflection.
2. Sweep the antenna to one side until there is no more signal reception.

3.   Swing the antenna back towards the strongest signal until it is just detectable and mark this bearing by using a landmark or compass.
4.   Repeat 2 and 3 on the other side of the strongest signal and again mark the bearing of the die-off of the signal.
5.   The bearing of the transmitter for triangulation is the mid-point between those marked in 3 and 4. If you can, confirm the exact location derived through triangulation by making visual contact with the subject. You may be able to locate an animal by simply following the bearing of the transmitter signal, correcting as you go, until you find it. However, if the study animal is nervous, poorly habituated or in habitat where stealth is difficult, it may be continually displaced ahead of you. In such cases a triangulated fix may provide the only genuine measure of the animal's movements.

Environmental effects

Radio signals can be affected by the environment, including reduction of signal intensity through absorption by vegetation. Fedigan *et al.* (1988) found that changes in vegetation and climate between the dry and wet season in Santa Rosa National Park (Costa Rica) resulted in a decrease in reception range from 500 to 150 m, and suspended radio-tracking during the wet season. You can also pick up false bearings due to the reflection of the signal by natural topographic features (e.g. cliffs, hills, gullies, rocks) as well as by woodland, individual trees and artificial features such as buildings. However, reflected signals can prove useful where the direct signal may be totally blocked. Diffraction of the signal is a particular problem in woodland or forest, where it can be difficult to get an accurate bearing owing to interference effects around tree trunks (Kenward, 2001). Although holding an antenna so that the elements are vertical in woodland or forest may improve the accuracy, it is still preferable to hold it horizontally owing to the amount of vertical reflection and diffraction produced by trees (Kenward, 2001).

SATELLITE TRACKING

The obvious feature distinguishing satellite tracking from traditional radio-tracking is that it does not require the continuous presence of the researcher. This facilitates the tracking of animals that move considerable distances or that use seasonally inaccessible areas. Two types of device are commonly used for satellite tracking, the GPS (Global Positioning System) and ARGOS (Advanced Research and Global Observation Satellite)/Globalstar (via North Star Science and Technology) systems. The GPS uses a ground-based receiver, which downloads information from a specialized constellation of low-orbit satellites. The receiver requires data from at least three satellites, simultaneously, to calculate its position. The ARGOS system works in reverse to GPS, in that a ground-based transmitter sends information to another specialized satellite constellation. Only one satellite is required to calculate a fix, which is in turn transmitted to a ground-based station for retrieval (Chapter 4).

GPS receivers produce positional accuracy of 5–15 m, which you can improve further with differential correction (Differential GPS (DGPS); Chapter 4). DGPS typically produces sub-metre accuracy and is available built into some receivers along with other options (e.g. temperature and mortality sensors). ARGOS generates variable accuracy, ranging between hundreds of metres and kilometres, making it impracticable for animals with small ranges, and is thus not entered into any further here. However, this may improve, so look into this when planning a study.

GPS receivers store data onboard, which you have to access directly by retrieving the collar (facilitated by an automatic drop-off mechanism, or recapture), via wireless ground to ground download (a VHF link or the mobile phone (Global System for Mobile Communication, or GSM) network), or via satellite uplink. Direct connection risks the loss of data if you are unable to relocate the receiver. You can solve this by incorporating a drop-off mechanism and a standard VHF transmitter to recover the collar. VHF links work over a range of 0.1–2 km, ground to ground, and up to 5 km from the air, depending on the topography and vegetation obscuring the line of sight from transmitter to receiver. Downloading data via a standard radio-link can take a long time (e.g. 30 minutes for two months' worth of data) but you can shorten this with data compression. Achieving this within the specified distance, and at a particular time, may be excessively challenging in the field and in almost all cases the data retrieval system (extra batteries, VHF transmitters, drop-off mechanism, etc.) adds to the overall weight of the collar.

Downloading data via GSM does not require ground-based personnel, but only works where there is mobile phone coverage. Satellite uplinks, like the ARGOS system, also require no ground-based reception or personnel, but are limited by the number of fixes that can be sent at a time. GPS can store up to one fix per second, whereas on average ARGOS can resolve just one a day.

Although GPS receiver-based animal tags, and ARGOS transmitters, can now weigh as little as 30 g (plus the collar/belt/harness assembly), bringing this technology within the range of primates weighing as little as about 1.5 kg, GPS reception and ARGOS transmission can be seriously attenuated by thick vegetation. This can mean that in tropical forest neither method is suitable, owing to poor fix acquisition rate. Sprague *et al.* (2004) found that positioning attempts on a GPS collared Japanese macaque (*Macaca fuscata*) within a mixed urban/rural habitat were only 20% successful and almost all the failed attempts were in forest habitat. This explains why take-up of this technology by primate researchers has been limited. However, Markham and Altmann (2008) demonstrated the value of automated GPS tracking for large primates in open habitats, using a store-on-board collar with temperature logger, activity sensor and VHS transmitter (ATS Systems GPS G2110 collar: 350 g).

BIOTELEMETRY

You can achieve biotelemetry via implantable or external transmitters, the latter sometimes fitted to a jacket or harness. External devices include those with tilt switches to monitor activity and posture (Kenward, 2001; Markham & Altmann, 2008). Thus far, implantable devices have had limited use in wild primates (Chapter 19). Restrictions on implant size constrain battery life, further diminished by frequent sampling (Rasmussen, 1991). Parameters such as body temperature vary gradually and need less frequent sampling than, for example, brainwave activity or heart rate. Sampling multiple parameters requires a multi channel device, greater signal band width, and thus more battery power. The choice of transmission frequency is also important; higher frequencies attenuate more rapidly through tissue, which reduces reception range (Rasmussen, 1991).

REGULATIONS

Radio-tracking studies typically fall under both general (access and research) and specific regulations (governing radio-frequency use and invasive procedures). Flaunting regulations may affect not only your ability to undertake research, but also that of other researchers. Whereas there has been a strategy to harmonize frequency band allocation across the European Union beyond 2008, there is little consistency elsewhere, though many countries allocate 150 kHz to 2 MHz to wildlife radio-tracking (Kenward, 2001). Although VHF and UHF transmitters are very low energy, have limited range and are consequently unlikely to interfere with, or suffer interference from, other radio-transmitter users, it is important to consult both the authorities and others who may have used radio-tags in your study country.

It is unacceptable to attempt any invasive procedure, such as the implantation of biotelemetry devices, in a 'trial and error' manner. Training and supervision are morally, and generally legally, essential.

ACKNOWLEDGEMENTS

Many thanks to Stephen Ellwood for his guidance on the section in this chapter concerning satellite tracking.

REFERENCES

Amlaner, C.J. Jr (1978). Biotelemetry from free-ranging animals. In *Animal Marking: Recognition Marking of Animals in Research*, ed. B. Stonehouse, pp. 205–28. London: Macmillan Press.

Amlaner, C.J. Jr, & Macdonald, D.W. (ed) (1980). *A Handbook on Biotelemetry and Radio Tracking*. Oxford: Pergamon Press.

Barrett, E. (1985). The ecology of some nocturnal arboreal mammals in the rain forest of Peninsular Malaysia. PhD thesis, University of Cambridge.

Bearder, S. & Martin, R. (1980). The social organization of a nocturnal primate revealed by radio tracking. In *A Handbook on Biotelemetry and Radio Tracking*, ed. C. Amlaner & D. Macdonald, pp. 633–48. Oxford: Pergamon Press.

Biebouw, K. (2009). Home range size and use in *Allocebus trichotis* in Analamazaotra Special Reserve, central eastern Madagascar. *Int. J. Primatol.* 30, 367–86.

Buchanan-Smith, H. (1990). Polyspecific association of two tamarin species, *Saguinus labiatus* and *Saguinus fuscicollis*, in Bolivia. *Am. J. Primatol.* 22, 205–14.

Burger, W. (1991). Receiving antenna 'accuracy'. *Telonics Q.* 4, 2–3.

Campbell, A. & Sussman, R. (1994). The value of radio tracking in the study of neotropical rainforest monkeys. *Am. J. Primatol.* 32, 291–301.

Casperd, J. (1992). Primate radiotracking and biotelemetry. *Primate Eye* 47, 18–23.

Charles-Dominique, P. (1977). Urine marking and territoriality in *Galago alleni* (Waterhouse, 1937 - Lorisoidea, Primates): a field study by radio-telemetry. *Z. Tierpsychol.* 43, 113–38.

Crofoot, M.C., Gilby, I.C., Wikelski, M.C. & Kays, R.W. (2008). Interaction location outweighs the competitive advantage of numerical superiority in *Cebus capucinus* intergroup contests. *Proc. Nat. Acad. Sci. USA* 105, 577–81.

Di Fiore, A., Link, A., Schmitt, C.A. & Spehar, S.N. (2009). Dispersal patterns in sympatric woolly and spider monkeys: integrating molecular and observational data. *Behaviour* 146, 437–70.

Donati, G., Baldi, N., Morelli, V. Ganzhorn, J. U. (2009). Proximate and ultimate determinants of cathemeral activity in brown lemurs. *Anim. Behav.* 77, 317–25.

Drea, C. & Wallen, K. (1999). Low status monkeys 'play dumb' when learning in mixed social groups. *Proc. Natl. Acad. Sci. USA* 96, 12965–9.

Fedigan, L., Fedigan, L. M., Chapman, C. & Glander, K. (1988). Spider monkey home ranges: a comparison of radio telemetry and direct observation. *Am. J. Primatol.* 16, 19–29.

Fernandez-Duque, E. & Rotundo, M. (2003). Field methods for capturing and marking azari night monkeys. *Int. J. Primatol.* 24, 1113–20.

Gautier, J. & Gautier-Hion, A. (1982). Vocal communication within a group of monkeys: analysis by biotelemetry. In *Primate Communication*, ed. C. Snowdon, C. Brown & M. Petersen, pp. 5–29. Cambridge: Cambridge University Press.

Gursky, S. (1998). Effect of radio transmitter weight on a small nocturnal primate. *Am J. Primatol.* 46, 145–55.

Halloren, E., Price, E. & McGrew, W. (1989). Technique for non-invasive marking of primate infants. *Lab. Primate Newsl.* 28, 13–5.

Harcourt, C. (1981). An examination of the function of urine washing in *Galago senegalensis. Z. Tierpsychol.* 55, 119–28.

Harcourt, C. & Nash, L. (1986). Species differences in substrate use and diet between sympatric galagos in two Kenyan coastal forests. *Primates* 27, 41–52.

Hilpert, A. L. & Jones, C. B. (2005). Possible costs of radio-tracking a young adult female mantled howler monkey (*Alouatta palliata*) in deciduous habitat of Costa Rican Tropical Dry Forest. *J. Appl. Anim. Welf. Sci.* 8, 227–32.

Ingram, J. C. (1978). Primate markings. In *Animal Marking: Recognition Marking of Animals in Research*, ed. B. Stonehouse, pp. 169–74. London: The Macmillan Press.

Jones, W. & Bush, B. (1988a). Darting and marking techniques for an arboreal forest monkey, *Cercopithecus ascanius. Am. J. Primatol.* 14, 83–9.

Jones, W. & Bush, B. (1988b). Movement and reproductive behaviour of solitary male redtail guenons (*Cercopithecus ascanius*). *Am. J. Primatol.* 14, 203–22.

Juarez, C. P., Berg, W. J. & Fernandez-Duque, E. (2008). An evaluation of the potential long-term effects of radio-collars on the reproduction and demography of owl monkeys (*Aotus azarai*) in Formosa, Argentina. *Am. J. Primatol.* 70 (Suppl. 1), 49.

Karesh, W., Wallace, R., Painter, L., Rumiz, D., Braselton, W., Dierenfeld, E. & Puche, H. (1998). Immobilisation and health assessment of free-ranging black spider monkeys (*Ateles paniscus chamek*). *Am. J. Primatol.* 44, 107–23.

Kenward, R. (2001). *A Manual for Wildlife Radio Tracking.* London: Academic Press.

Kierulff, M. C. M., Canale, G. & Gouveia, P. S. (2005). Monitoring the yellow-breasted capuchin monkey (*Cebus xanthosternos*) with radiotelemetry: choosing the best radio-collar. *Neotropical Primates* 13, 32–3.

Li, B., Chen, C., Ji, W. & Ren, B. (1999). Seasonal home range changes of the Sichuan snub-nosed monkey (*Rhinopithecus roxellana*) in the Qinling Mountains of China. *Folia Primatol.* 71, 375–86.

Macdonald, D. & Amlaner, C. Jr (1980). A practical guide to radio tracking. In *A Handbook on Biotelemetry and Radio Tracking*, ed. C. Amlaner, Jr & D. Macdonald, pp. 143–59. Oxford: Pergamon Press.

Markham, A. C. & Altmann, J. (2008). Remote monitoring of primates using automated GPS technology in open habitats. *Am. J. Primatol.* 70, 495–9.

Medici, E. P., Valladares-Padua, C. B., Rylands, A. B. & Martins, C. S. (2003). Translocation as a metapopulation management tool for the black lion tamarin, *Leontopithecus chyrsopygus. Prim. Cons.* 19, 23–31.

Megna, N. (2001). Marking monkeys – Nyanzol-D (discussion). *Lab. Primate Newsl.* 40, 5.

Millspaugh, J. & Marzluff, J. (eds.) (2001). *Radio Tracking and Animal Populations.* San Diego, CA: Academic Press.

Oluput, W. (2000). Darting, individual recognition, and radio-tracking techniques in grey-cheeked mangabeys *Lophocebus albigena* of Kibale National Park, Uganda. *African Primates* 4, 40–50.

Peignot, P., Charpentier, M. J. E., Bout, N. *et al.* (2008). Learning from the first release project of captive mandrills *Mandrillus sphinx* in Gabon. *Oryx* 42, 122–31.

Rasmussen, D. (1998). Changes in the range use of Geoffroy's tamarins (*Saguinus geoffroyi*) associated with habituation to observers. *Folia Primatol.* 69, 153–9.

Rasmussen, K. (1991). Identification, capture and biotelemetry of socially living monkeys. *Lab. Anim. Sci.* 41, 350–4.

Savage, A., Giraldo, L., Blumer, E. *et al.* (1993). Field techniques for monitoring cotton-top tamarins (*Saguinus oedipus oedipus*) in Columbia. *Am. J. Primatol.* 31, 189–96.

Sellers, W. & Crompton, R. H. (2004). Automatic monitoring of primate locomotor behaviour using accelerometers. *Folia Primatol.* 75, 279–93.

Sprague, D. S. Kabaya, H. & Hagihara, K. (2004). Field testing a global positioning system (GPS) collar on a Japanese monkey: reliability of automatic positioning in a Japanese forest. *Primates* 45, 151–4.

Taylor, L., Easley, S. & Coelho, Jr. A. (1988). Ear tags for long-term identification of baboons. *Lab. Primate Newsl.* 27, 8–10.

Wiens, F. & Zitzmann, A. (1999). Predation on a wild slow loris (*Nycticebus coucang*) by a reticulated python (*Python reticulatus*). *Folia Primatol.* 70, 362–4.

Wolfensohn, S. (1993). The use of microchip implants in identification of two species of macaque. *Anim. Welfare* 2, 353–9.

LIST OF SUPPLIERS AND USEFUL INTERNET SITES

www.biotelem.org/manufact.htm – for a directory of biotelemetry equipment manufacturers.

Some manufacturers

www.avidplc.com – for microchips and readers.

www.atstrack.com – for radio-telemetry and GPS systems, data collection computers, antennas, etc.

www.avminstruments.com – for transmitters, receivers, antennas, etc.

www.biomark.com – for microchips and readers.

www.biotrack.co.uk – for radio-transmitters and receivers.

www.destronfearing.com – for eartags, microchips and readers.

www.lotek.com – for radio-telemetry and GPS systems, archival tags, etc.

www.enasco.com – for an online catalogue including outside USA (eartags, applicators, etc.)

www.navsys.com – for GPS systems.

www.northstarst.com – for GPS and Globalstar satellite systems, and radio-transmitters.

www.followit.se – for radio-telemetry and GPS systems, etc.

www.sirtrack.com – for radio-telemetry and GPS systems, etc.

www.telonics.com – for radio-telemetry and GPS systems, accessories, etc., and *Telonics Quarterly* on-line.

KLAUS ZUBERBÜHLER AND ROMAN M. WITTIG

11

Field experiments with non-human primates: a tutorial

INTRODUCTION

Field experiments are a powerful way of investigating the mechanisms of primate behaviour, their adaptive functions, and the cognitive forces responsible for them. They have several major advantages over observational data. First, they can reduce ambiguity about the causal relationship between stimuli and behavioural responses; second, they allow systematic investigations of even rare events; and third, they can test hypotheses more directly by systematically controlling for confounding variables. Although observational data can sometimes achieve the same results, they typically require more complicated statistical procedures and more observation effort. Field experiments can contribute to a range of scientific disciplines, but they have been used most extensively by psychologists interested in the primate mind, behavioural ecologists working on anti-predator behaviour, and anthropologists dealing with proto-human behaviour and primate culture.

In this chapter, we begin with a brief overview of the main observational methods used in primate fieldwork. We then discuss a number of commonly used experimental designs and stimuli, as well as some further techniques with considerable potential for work with wild primates and conclude by drawing attention to common problems and pitfalls.

OBSERVATIONAL METHODS

Field experiments require a profound understanding of the causes and consequences of the behaviours under study, and they should always be the final step in a research programme based on lengthy, detailed

Field and Laboratory Methods in Primatology: A Practical Guide, ed. Joanna M. Setchell and Deborah J. Curtis. Published by Cambridge University Press. © Cambridge University Press 2011.

and careful behavioural observations. Without such knowledge, there is a considerable danger of data misinterpretation, inadequate experimental design and meaningless findings. Four main observational data-gathering techniques are particularly useful for studying primates: focal animal sampling, scan sampling, all occurrence sampling, and ad libitum sampling. We provide an outline of the main features of each method here; detailed descriptions can be found in Altmann (1974) or Martin and Bateson (1993). One useful distinction is between behavioural events (e.g. alarm calling) and basic behavioural states (e.g. resting). Behavioural elements are usually compiled as a comprehensive list that includes all possible states and events that can be observed in the study species: an ethogram. The coding of behavioural elements can be direct, i.e. by scoring ongoing behaviour, or indirect, e.g. by coding from video material or other stored raw data. Some states are not part of the ethogram, for instance if they are properties of a relationship, such as proximity to other group members or temporal associations with specific partners.

For data analysis, raw observational data are typically converted into frequencies (percentages), rates (e.g. units per time) or indices (usually a relationship between different variables). Examples include the 'Shannon Wiener Diversity Index', used to describe the diversity of an individual's grooming partners (see, for example, Crockford *et al.*, 2008), or the 'Relationship Benefit Index', which describes how beneficial the relationship of one dyad is compared with other dyads (Wittig & Boesch, 2003).

Focal animal sampling

Focal animal sampling is perhaps the most frequently used method in studies of primate behaviour. The idea is to follow a particular individual over a pre-determined time period, which can be anything from 10 min to a full day, or even consecutive days, and to collect data on all behaviours relevant to the study. The main advantage of this method is that it produces raw data that can easily be converted into frequencies, rates or indices, allowing systematic comparisons within and between individuals. It is important to balance the final dataset such that all focal animals contribute roughly the same amount of observation time.

Scan sampling

Scan sampling involves regular scans during which you record the behaviour of any individual visible. This provides raw data that can be converted into frequency information for the different behaviours sampled. It is particularly useful for producing insights into the nature of social relationships, as assessed by grooming behaviour, proximity or association patterns.

All occurrence sampling

All occurrence sampling focusses on specific behaviours that are perhaps rare but nevertheless of particular interest (e.g. hunting behaviour in chimpanzees, *Pan troglodytes*) and involves monitoring all relevant individuals simultaneously. This is usually only feasible if the group is small and cohesive. Frequencies and indices can be calculated without problems, but rates can be more difficult because it is often impossible to record the duration of the behaviours for several individuals simultaneously.

Ad libitum sampling

Ad libitum sampling has the fewest stipulations. Here, you follow the same individual or group and record anything that appears to be noteworthy. This method is not suitable for calculating derived measures, such as frequencies, rates or indices, since data collection is not standardized. Ad libitum sampling generates a collection of events, which can be very helpful in providing a basis for more systematic follow-up research or for controlled experimental procedures. If conducted over an adequate time period, ad libitum data can provide valuable information on the presence or absence of certain behaviours, which is important in cross-population comparisons. In some cases, especially if the behaviour is conspicuous, the group cohesive, and observation conditions excellent, ad libitum data can be treated as all occurrence data.

Dealing with incomplete samples

One problem with focal animal, scan, and all occurrence sampling is that they all depend on perfect visibility, a condition that is rarely met in the wild, as you often lose sight of the animal or group. There are two main options for dealing with this. One is simply to abandon data collection and discard the entire sample. This is a reasonable strategy if the predetermined sampling period is short (say 10 minutes), if samples are not difficult to obtain, or if losing sight is very rare. However, some studies depend on dawn-to-dusk follows and it is not feasible to discard an entire day's data. A sensible approach here is to add a specific element to the ethogram that accounts for such lapses in data collection systematically (e.g. 'bad observation period' or 'out of sight'), and consider this in the final weighting and analysis of the data. This is especially suitable for studies that depend on long observation periods or if losing sight temporarily of the animal is simply unavoidable, for instance due to obstructive habitat or arboreal lifestyle.

EXPERIMENTAL DESIGNS

General points

The basic logic of experimentation is to keep everything equal, except for the variable of interest. It is imperative to consider basic statistical constraints from the start, e.g. whether subjects will be exposed to all experimental conditions (matched design) or whether they will be randomly assigned to one of them (independent design). In field conditions, matched designs are usually chosen, mainly because access to groups and individuals is so limited. When carrying out an experimental programme, you need to alter trial types randomly or in a systematic and predetermined way, to prevent biases in the data and minimize habituation effects (see below). Once you have selected the final design and begun data collection, it is not usually acceptable to discard trials on an ad hoc basis, as this is likely to introduce observer bias. Instead, you should include all trials in the final dataset, provided they meet the experimental conditions and there are no equipment faults. Using stimuli unknown to the subject (artificial sounds) is often of limited value, unless the study requires this specifically (see below). Although broadcasting unfamiliar sounds may generate measurable responses, it is usually difficult to interpret such results. Finally, at the analysis stage, it is crucial to ask someone blind to the hypotheses to check the coding of video material, and you should always report inter-observer reliability data.

Simple stimulus presentation

The aim of most field experiments is to simulate a natural event that has not actually taken place, such as the sudden appearance of a predator, the occurrence of a specific social event, or the presence of another conspecific. It is important to keep in mind that any behavioural response to such simple manipulations remains meaningless unless it can be compared to a response produced to an adequate control condition. This is because each stimulus presentation also represents a sudden and novel event, which makes it theoretically impossible to decide whether subjects have responded to the particulars of the simulated event or more broadly to the novelty factor generated by the experimental manipulation. A meaningful field experiment thus contains at least two types of stimulus, a test and a control condition. For instance, observing a group of monkeys responding to a leopard model generates non-interpretable data unless their response can be compared to their response to an equally conspicuous object, presented in the same way. A more elegant design is to ensure that two or more test conditions act as each other's controls. For example, if you test a study population with both a leopard and an eagle model, then the two manipulations automatically control for novelty, so that a specific control is obsolete.

The violation-of-expectation paradigm

More sophisticated designs that go beyond simple stimulus presentations are sometimes used to address specific theoretical questions. 'Violation-of-expectation' paradigms are based on stimulus presentations that, in one way or another, violate a natural pattern of behaviour. They test whether individuals perceive this manipulation as relevant, which reveals something about the mental representations subjects are able to maintain of their world. Operationally, responses to violations are recognized if individuals respond more strongly to unnatural patterns compared with natural ones (see, for example, Bergman et al., 2003; Crockford et al., 2007). A recent example is an experiment with wild chimpanzees, which compared responses to the sudden arrival of neighbours with the sudden arrival of unfamiliar strangers, something that subjects were able to discriminate between by listening to their vocalizations alone, suggesting that chimpanzees are aware of the different social consequences associated with encountering such individuals (Herbinger et al., 2009).

The habituation–dishabituation paradigm

When studying the response of primates to their vocalizations (or any other stimuli) it is often not clear whether subjects really process the potential semantic features of these stimuli (i.e. their meaning), rather than some surface physical properties, a point often made by behaviourists critical of animal cognition studies (e.g. Rendall *et al.*, 2009). One way of addressing the two alternatives is to expose individuals to more complicated designs, such as habituation–dishabituation or prime–probe experiments. In habituation–dishabituation experiments individuals are exposed to repeated presentations of the same stimulus type, usually until they cease to respond. After this has happened, you present the test stimulus. If subjects show increased response strength again, you can conclude that the subjects discriminate between the two stimuli. If not, subjects are likely to treat the two stimuli the same. In one such study, free-ranging vervet monkeys (*Chlorocebus pygerythrus*, which were referred to as *Cercopithecus aethiops*) at the time were exposed to a series of inter-group calls given by a familiar group member in the absence of any neighbouring group. Subjects soon stopped responding to the individual's calls and subsequently also ignored playback of an acoustically different inter-group call given by the same individual, an effect not observed in a control condition where the two calls of the individual had different referents (Cheney & Seyfarth, 1988).

Prime–probe experiments represent a related technique where individuals are primed with only one exemplar (the prime stimulus) followed by another exemplar (the probe stimulus). In one such experiment, Diana monkeys (*Cercopithecus diana*) were first primed with conspecific alarms to a predator, for instance to a crowned eagle, which triggered the corresponding alarm calls and anti-predator responses reliably. In the probe condition, monkeys then heard the corresponding or non-corresponding predator vocalizations, i.e. eagle shrieks or leopard growls. If primed with eagle alarm calls, monkeys did not respond to eagle shrieks (normally a very powerful stimulus to elicit anti-predator responses and alarm calls), but they responded normally to leopard growls (Fig 11.1). This study thus demonstrated that monkeys processed the semantic features associated with their own alarm calls, not just the acoustic features of the different vocalizations (Zuberbühler *et al.*, 1999).

EXPERIMENTAL STIMULI

Sound presentations

The classic field experiment is undoubtedly the playback study. It has a relatively long history in science, although it was not used systematically with wild primates until the 1970s (see, for example,

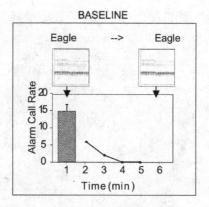

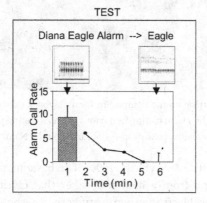

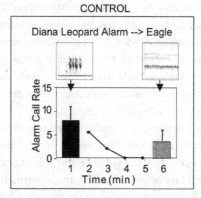

Fig. 11.1. A prime–probe experiment designed to investigate whether non-human primates are capable of forming mental representations when hearing conspecific alarm calls (from Zuberbühler *et al.*, 1999; Copyright © [1999] by the American Psychological Association. Adapted with permission. The use of APA information does not imply endorsement by APA.).

Fig. 11.2. Setting up the equipment for a playback experiment to
investigate the alarm calling behaviour of free-ranging putty-nosed
monkeys, *Cercopithecus nictitans* (photo by Aylin McNamara).

Waser 1975; Cheney & Seyfarth, 1990). The basic manipulation is to
broadcast a sound from a hidden speaker that carries a particular
relevance to the subjects in order to study their responses. Most playback experiments require at least two experimenters (Fig. 11.2): one to
operate the speaker and one to focus on the behavioural response. The
two observers will usually need to be in constant contact to determine
the right moment to broadcast the sound (e.g. via radio or mobile
phone). It is important to hide the speaker well in vegetation so
that individuals cannot form an association between the sound and
playback equipment. It is useful to conceal it in a backpack, with a
custom-made aperture to ensure unobstructed sound radiation (see, for
example, Slocombe *et al.*, 2009). This allows you to choose the perfect
spot and moment to conceal the speaker and you can carry the speaker
away when individuals approach after an experiment, so that they
don't discover it. Responses are generally recorded on film or scored
directly. Analyses based on vocal responses typically require specialist
recording equipment (Chapter 16). It is good practice to start data recording several minutes before the experimental stimulus is presented,
and to do so continuously until well after the experimental intervention.
For some questions, the focus is on the potential long-term influences

of the experimental event, and then it may even be necessary to conduct continuous focal animal sampling for hours or even days after the event (see, for example, Herbinger *et al.*, 2009).

It is usually a good idea to run a few practice trials before starting data collection, to become accustomed to the equipment and the requirements of the experimental setup. A particular problem with playback experiments is unwanted noises produced by the speaker, for instance when switching the power on or off. Practice trials help to identify such problems beforehand. Another recommendation is to run pilot trials with individuals that will not participate in the actual experiment, to obtain a better sense of the expected range of behavioural responses, and to facilitate subsequent data collection.

Playback experiments are useful to study the meaning of vocalizations, as in the seminal study on vervet monkeys in Amboseli National Park, Kenya (Seyfarth *et al.*, 1980) and to simulate social interactions (see, for example, Cheney & Seyfarth, 2007). An interesting technical variant is to simulate interactions with two different speakers, broadcasting from opposite directions, to simulate two or more individuals. This increases the organizational burden considerably but generates a range of interesting research possibilities, as the subject now needs to integrate two stimuli and the potential social implications of the interaction (see, for example, Crockford *et al.*, 2007).

Object presentations

A relatively simple and straightforward way of conducting field experiments with free-ranging primates is to present some type of object that carries a particular relevance for the animals. This method has a long tradition in animal behaviour research, an early example being Kortlandt's legendary stuffed leopard experiments with chimpanzees (Kortlandt & Kooij, 1963). You can use object presentations to address a range of questions in behavioural ecology, communication research, spatial cognition and culture studies. Although the details of presentation vary, a general requirement is that the animals should not witness the positioning of the object, mainly to avoid associations between the object and the observer. This is not just for ethical reasons but also due to the fact that many wild primates ignore artefacts that they associate with humans. It will also usually be important to position the object in a natural location, so the subjects detect it spontaneously during their natural travels (Fig. 11.3). More general requirements that also apply to other experiments are discussed in a separate section (below).

Fig. 11.3. Using a catapult to launch a fishing line over a high branch. A rope is then placed over the branch, with which a crowned eagle model can be positioned in the upper canopy within the anticipated travelling path of a monkey group.

A pioneering field experiment involved the presentation of conspecifics, via manipulation of social structure and transfer experiments in hamadryas baboons (*Papio hamadryas*) (Kummer, 1968). However, our understanding of primate social behaviour has now considerably improved, and transfer experiments with wild primates are generally no longer considered to be ethical (H. Kummer, personal communication). More recent experiments with object presentations focus on the presentation of predator models (see, for example, Bshary & Noë, 1997; Arnold *et al.*, 2008) or food. A primate home range can contain thousands of potential food trees, making systematic study a daunting task. Experimental presentation of food is therefore a particularly promising technique to study foraging decisions. One technique, developed by Janson (1998), is to provide natural foods on feeding platforms distributed through the home range of the study group. Once the primates have learned to exploit this new food source, you can systematically control their location, productivity and temporal availability.

Experimental food presentations are also useful in the study of primate culture, which first became a focus of interest after the spread of sweet potato washing behaviour in a group of Japanese macaques (*Macaca fuscata*) at Koshima (Kawai, 1965). However, this topic is still hotly debated (Laland & Janik, 2006), partly because the necessary field experiments have not yet been conducted. A first attempt has recently been made by Gruber *et al.* (2009), who compared the performance of chimpanzees of the same subspecies (*Pan troglodytes schweinfurthii*) from two different Ugandan communities in extracting honey from cavities drilled into horizontal logs. Individuals from the two sites solved the problem very differently but based on their previous tool use behaviour.

In laboratory studies, food is routinely used in behavioural experiments, typically as the unconditioned stimulus in associative learning or operant conditioning paradigms. Such studies are rare in the wild, but there is no principle reason against it. For example, an experimental study of wild marmosets (*Callithrix jaccus*) highlighted the potential role of individual learning (or operant conditioning) in studies of animal culture (Pesendorfer *et al.*, 2009), a hypothesis that will need to be addressed with proper field experiments of the kind already done with laboratory animals (Whiten *et al.*, 2005). For example, it would be useful to demonstrate a new foraging technique or food choice to one member of a primate group to investigate whether this will spread throughout the community via social learning.

On a cautionary note, food provisioning carries some potentially very serious health hazards for wild primates. For example, food handled by humans is one of the potential vehicles for disease transmission from humans to chimpanzees (Boesch, 2008). Furthermore, if provisioning is continued for long periods, individuals may start to rely on it, which can have a range of adverse effects. Nonetheless, if conducted with appropriate care and caution, food presentation experiments can be useful tools to investigate a range of aspects of primate behaviour.

Finally, object presentations are particularly useful for studies on tool use behaviour of wild animals. For example, Carvalho *et al.* (2009) provided chimpanzees with nuts and about 50 individually marked stones of various sizes and shapes in an 'outdoor laboratory' to investigate whether subjects choose stones randomly, or whether they possess a pre-existing notion of their functional suitability as hammers or anvils, or their suitability as a composite tool.

Other possibilities

Experiments with olfactory cues have been conducted successfully with humans and non-primate mammals, typically in the context of sexual behaviour and reproduction (Müller & Manser, 2007). Olfactory stimuli have been used successfully with captive primates, particularly ring-tailed lemurs (*Lemur catta*) (see, for example, Scordato & Drea 2007), and they show considerable promise in addressing questions on the role of olfaction in primate communication in natural populations, including territorial behaviour, pheromones and MHC-related mating patterns (Wedekind & Furi, 1997). So far, primate fieldworkers have not made much of these possibilities, possibly because they rarely have the option of interacting closely enough with their subjects to collect the necessary stimuli. An interesting exception is urine, which appears to function as an olfactory signal in some species (Campos *et al.*, 2007). In some cases, it is possible to catch and handle individuals without detrimental effects to them (Chapters 7 and 8). Here, it may thus be possible to systematically collect samples from the scent glands, which would open a wide range of possibilities. For instance, you could deposit scent samples strategically in a subject's home range to study their role as markers of individual identity, age/sex class membership, fertility and social status or in territorial defence. Useful response measures are the amount of counter scent-marking involving different glands and the duration of sniffing (Ramsay & Giller, 1996), but other effects may also be observed. There is no reason to restrict olfactory experiments to simple stimulus presentations; more complicated designs such as habituation–dishabituation experiments or violation of expectation paradigms (see above) are equally conceivable.

Another technique hardly exploited for work with wild primates is the use of images and video material, a methodology where progress has been made with captive primates (see, for example, Köhler, 1921; Price *et al.*, 2009). For instance, it may be possible to expose wild primates to video footage of a conspecific showing some new behaviour, such as manufacturing a new tool or consuming an untested food, to investigate whether subjects can socially learn from such experiences. Obviously, the main problem with this technique is finding appropriate ways of presenting the stimuli. Some primates forage in areas used by humans and consequently are sufficiently accustomed to human artefacts. In such populations, it may be possible to set up a mini field 'movie theatre', e.g. an indestructible box with a peephole, containing a DVD player.

Pseudo-replication

As attractive as they are, field experiments are prone to a number of problems, which can make a study worthless if left unaddressed. A widespread problem is pseudo-replication, which can take various forms. For example, it is not usually acceptable to test a particular subject several times with the same experimental condition. If you do, then do not treat data points as independent, as this will artificially inflate the sample size, and lead to problems in the statistical analyses. For most experimental designs, you should not use a particular stimulus more than once. This is sometimes difficult to achieve, for instance when working with elaborate predator models, but it is relatively easy to address in playback studies. For instance, if you are interested in whether monkeys can discriminate between two acoustically distinct grunt types, A and B, then use different exemplars of grunts for each trial. Otherwise the results may simply report the monkeys' response to a particular recording, rather than a grunt type as such. However, other designs require you to broadcast the same call in more than one condition, particularly if the social context is the independent variable and varies between conditions, or if the intensity of the stimulus is predicted to change with the intensity of the signal (see, for example, Wittig et al. 2007a). Related to the pseudo-replication, it is a particularly bad idea to use repetitions of the same call element when constructing a playback stimulus. Subjects are likely to perceive such sequences as unnatural, perhaps similar to humans listening to a skipping compact disc. Finally, individuals should be prevented from hearing their own calls, as they are likely to interpret this as evidence for the presence of an unfamiliar intruder.

Habituation

Habituation is particularly problematic when experimenting with non-human primates. A common manifestation of habituation is that subjects cease to respond to a previously powerful stimulus or that the response amplitude diminishes over trials. Most likely, this is because individuals have learned that the stimuli are not linked with any consequences, in contrast to the corresponding naturally occurring events. The only safe way to avoid habituation is by exposing each individual only once to each stimulus, although this is often difficult to implement when working with wild primates. If you need to test subjects more than once, then apply the same rule to all individuals, to rule out some subjects being more affected by habituation than others. For paradigms that require you to use the same call in more than one condition, a possible solution is to alternate the order of the conditions. Otherwise a reduced reaction to the second presentation may not be due to the condition, but to habituation. Another way of preventing habituation is to avoid presenting experimental stimuli more often than the subject would encounter them naturally. For example, if leopard attacks occur on average once per month, then model presentations of leopards should not be more frequent than that. Ideally, trials should take place after a one-month time period when no leopard attack has been observed.

Authenticity

A key issue with field experiments is whether the models are of sufficient quality so that they are perceived as accurate representations of the real world object or event. With playback experiments, it is crucial that the master recordings are of very high quality. Poor recordings or recordings with strong background noise may confuse the subject and generate uncharacteristic responses. In some cases, it is possible to cleanse unwanted background noise from master recordings using specialist software (see 'Useful Internet sites'). Professional recording equipment is essential for obtaining high-quality recordings for a successful playback experiment (Chapter 16).

If you use conspecific calls as playback stimuli, then the identity of the caller is usually important (see, for example, Wittig *et al.*, 2007b). If calls are produced in sequences, it can be vital to keep the temporal structure of the sequence intact, as inter-call intervals or call order can carry meaningful information (Ouattara *et al.*, 2009b). You should also have some idea of the natural amplitude of the sounds, and make sure the stimuli are broadcast within the same range. A sound pressure meter can be helpful when calibrating stimulus amplitude. It may also be important to choose a playback location that matches the one where the original sound was recorded. Vocal interactions often follow a certain sequence structure and individual variation is also apparent. When creating playback stimuli, it is advisable to use a natural call sequence as a master structure and copy the individual calls into the structure. This produces stimuli that appear natural in terms of their dialogue structure and are of comparable context and intensity.

Authenticity can also be a considerable problem when conducting experiments with objects, especially replicas of live animals (see, for example, Ouattara *et al.*, 2009a). There is no completely satisfactory way of addressing this problem, apart from ensuring that responses to the model and the corresponding real object are the same or very similar.

There may also be population differences within a species' ability to recognize predator models. For instance, Bshary (2001) found that Diana monkeys in forest fragments where hunting was common did not respond to imitations of crowned eagle shrieks, whereas groups in protected forest exhibited strong responses to real and imitated eagle shrieks. This suggests that discrimination abilities can be acquired and lost within the lifespan of an individual. Authenticity is less of a problem if primates respond strongly to a key feature of the experimental object. A well-known example is the spotted coat of the leopard, which can trigger strong responses in African monkeys, no matter how it is presented to them (see, for example, Schel *et al.*, 2009).

When using predator models, you need to ensure that the model is positioned in a naturalistic location, so that the monkeys will detect it as part of their natural undisturbed behaviour. This usually requires considerable knowledge of the monkeys' daily travel routines. Even if you have a pretty good understanding of the possible travel routes, it is often striking how easily monkeys can overlook predators positioned within their travel path, which can make model predator presentations an extremely frustrating way of collecting data.

SUMMARY

Field experiments can be a powerful way of investigating the behaviour of non-human primates in their natural habitat. Various techniques are available, but all rely on robust knowledge of the species' natural behaviour. Successful experimentation in the wild adheres to the same principles that apply in the laboratory, although the various field experimental techniques have a number of idiosyncratic problems and weaknesses. Among the different techniques, object and sound presentations have been used most successfully with wild primates, but there is potential for other techniques as well, including odour and video cues.

ACKNOWLEDGEMENTS

Preparation of this chapter was made possible with the support of the Leverhulme Trust and the Wissenschaftskolleg zu Berlin. We thank the editors for their helpful comments.

REFERENCES

Altmann, J. (1974). Observational study of behavior: sampling methods. *Behaviour* 49, 227-67.

Arnold, K., Pohlner, Y. & Zuberbühler, K. (2008). A forest monkey's alarm call series to predator models. *Behav. Ecol. Sociobiol.* 62, 549-9.

Bergman, T. J., Beehner, J. C., Cheney, D. L. & Seyfarth, R. M. (2003). Hierarchical classification by rank and kinship in baboons. *Science* 302, 1234-6.

Boesch, C. (2008). Why do chimpanzees die in the forest? The challenges of understanding and controlling for wild ape health. *Am. J. Primatol.* 70, 722-6.

Bshary, R. (2001). Diana monkeys, *Cercopithecus diana*, adjust their anti-predator response behaviour to human hunting strategies. *Behav. Ecol. Sociobiol.* 50, 251-6.

Bshary, R. & Noë, R. (1997). Red colobus and Diana monkeys provide mutual protection against predators. *Anim. Behav.* 54, 1461-74.

Campos, F., Manson, J. H. & Perry, S. (2007). Urine washing and sniffing in wild white-faced capuchins (*Cebus capucinus*): testing functional hypotheses. *Int. J. Primatol.* 28, 55-72.

Carvalho, S., Biro, D., McGrew, W. C. & Matsuzawa, T. (2009). Tool-composite reuse in wild chimpanzees (*Pan troglodytes*): archaeologically invisible steps in the technological evolution of early hominins? *Anim. Cogn.* 12, S103-S114.

Cheney, D. L. & Seyfarth, R. M. (1988). Assessment of meaning and the detection of unreliable signals by vervet monkeys. *Anim. Behav.* 36, 477-86.

Cheney, D. L. & Seyfarth, R. M. (1990). *How Monkeys See the World*. Chicago, IL: University of Chicago Press.

Cheney, D. L. & Seyfarth, R. M. (2007). *Baboon Metaphysics*. Chicago, IL: University of Chicago Press.

Crockford, C., Wittig, R. M., Seyfarth, R. M. & Cheney, D. L. (2007). Baboons eavesdrop to deduce mating opportunities. *Anim. Behav.* 73, 885-90.

Crockford, C., Wittig, R. M., Whitten, P. L., Seyfarth, R. M. & Cheney, D. L. (2008). Social stressors and coping mechanisms in wild female baboons (*Papio hamadryas ursinus*). *Horm. Behav.* 53, 254-65.

Gruber, T., Strimling, P., Muller, M., Wrangham, R. W. & Zuberbuhler, K. (2009). Wild chimpanzees rely on cultural knowledge to solve an experimental honey acquisition task. *Curr. Biol.* 19: 1806-10.

Herbinger, I., Papworth, S., Boesch, C. & Zuberbuhler, K. (2009). Vocal, gestural and locomotor responses of wild chimpanzees to familiar and unfamiliar intruders: a playback study. *Anim. Behav.* 78, 1389-96.

Janson, C. H. (1998). Experimental evidence for spatial memory in foraging wild capuchin monkeys, *Cebus apella*. *Anim. Behav.* 55, 1229-43.

Kawai, M. (1965). Newly-acquired pre-cultural behavior of the natural troop of Japanese monkeys on Koshima Islet. *Primates* 6, 1-30.

Köhler, W. (1921). Aus der Anthropoindenstation auf Teneriffa. V. Zur Psychologie des Schimpansen. *Sitzungsber. Preuss. Akad. Wiss.* 2, 686–92.

Kortlandt, A. & Kooij, M. (1963). Protohominid behaviour in primates. *Symp. Zool. Soc., Lond.* 10, 61.

Kummer, H. (1968). *Social Organization of Hamadryas Baboons.* Chicago, IL: University of Chicago Press.

Laland, K. N. & Janik, V. M. 2006. The animal cultures debate. *Trends Ecol. Evol.* 21, 542–7.

Martin, P. & Bateson, P. (1993). *Measuring Behaviour.* Cambridge: Cambridge University Press.

Müller, C. A. & Manser, M. B. 2007. 'Nasty neighbours' rather than 'dear enemies' in a social carnivore. *Proc. R. Soc. Lond.* B 274, 959–65.

Ouattara, K., Zuberbühler, K., N'goran, E. K., Gombert, J.-E. & Lemasson, A. (2009a). The alarm call system of female Campbell's monkeys. *Anim. Behav.* 78, 35–44.

Ouattara, K., Lemasson, A. & Zuberbühler, K. (2009b). Wild Campbell's monkeys concatenate vocalizations into context-specific call sequences. *Proc. Natl. Acad. Sci. USA* 106, 22026–31.

Pesendorfer, M. B., Gunhold, T., Shiel, N. *et al.* (2009). The maintenance of traditions in marmosets: individual habit, not social conformity? A field experiment. *PLoS One,* 4, e4472.

Price, E. E., Lambeth, S. P., Schapiro, S. J. & Whiten, A. 2009. A potent effect of observational learning on chimpanzee tool construction. *Proc. R. Soc. Lond.* B 276, 3377–83.

Ramsay, N. F. & Giller, P. S. (1996). Scent-marking in ring-tailed lemurs: responses to the introduction of 'foreign' scent in the home range. *Primates* 37, 13–23.

Rendall, D., Owren, M. J. & Ryan, M. J. 2009. What do animal signals mean? *Anim. Behav.* 78, 233–40.

Schel, A. M., Tranquilli, S. & Zuberbuhler, K. (2009). The alarm call system of two species of black-and-white colobus monkeys (*Colobus polykomos* and *Colobus guereza*). *J. Comp. Psychol.* 123, 136–50.

Scordato, E. S. & Drea, C. M. (2007). Scents and sensibility: information content of olfactory signals in the ringtailed lemur, *Lemur catta. Anim. Behav.* 73, 301–14.

Seyfarth, R. M., Cheney, D. L. & Marler, P. (1980). Monkey responses to three different alarm calls: evidence of predator classification and semantic communication. *Science* 210, 801–3.

Slocombe, K. E., Townsend, S. W. & Zuberbühler, K. (2009). Wild chimpanzees (*Pan troglodytes schweinfurthii*) distinguish between different scream types: evidence from a playback study. *Anim. Cog.* 12, 441–9.

Waser, P. M. (1975). Experimental playbacks show vocal mediation of intergroup avoidance in a forest monkey. *Nature* 255, 56–58.

Wedekind, C. & Furi, S. (1997). Body odour preferences in men and women: do they aim for specific MHC combinations or simply heterozygosity? *Proc. R. Soc. Lond.* B 264, 1471–9.

Whiten, A., Horner, V. & de Waal, F. B. M. (2005). Conformity to cultural norms of tool use in chimpanzees. *Nature* 437, 737–40.

Wittig, R. M. & Boesch, C. (2003). 'Decision-making' in conflicts of wild chimpanzees (*Pan troglodytes*): an extension of the Relational Model. *Behav. Ecol. Sociobiol.* 54, 491–504.

Wittig, R. M., Crockford, C., Seyfarth, R. M. & Cheney, D. L. (2007a). Vocal alliances in chacma baboons (*Papio hamadryas ursinus*). *Behav. Ecol. Sociobiol.* 61, 899–909.

Wittig, R. M., Crockford, C., Wikberg, E., Seyfarth, R. M. & Cheney, D. L. (2007b). Kin-mediated reconciliation substitutes for direct reconciliation in female baboons. *Proc. R. Soc. Lond. B* 274, 1109–15.

Zuberbühler, K., Cheney, D. L. & Seyfarth, R. M. (1999). Conceptual semantics in a nonhuman primate. *J. Comp. Psychol.* 113, 33–42.

USEFUL INTERNET SITES

www.praat.org and www.birds.cornell.edu/raven – Specialist software for creating auditory playback stimuli.

J. LAWRENCE DEW

12

Feeding ecology, frugivory and seed dispersal

INTRODUCTION

Frugivores are among the most diverse and abundant of tropical vertebrates. Studying them is thus a central part of tropical biology. Because fruit availability fluctuates throughout the year, most frugivores have variable diets that may include leaves, fruits, flowers and seeds, as well as insects or other animal matter. Identifying food plants and parts eaten is a critical part of this research, but can be complicated when you are studying canopy-dwelling animals in regions with diverse flora. Frugivores are often described as important ecological interactors (mutualists) because of their role as seed dispersers (Howe & Miriti, 2004). Studying plant and animal strategies simultaneously is no simple task, but doing so provides new perspectives on botany, feeding ecology, digestive physiology, coevolution and plant–animal interdependence, which may provide valuable tools for conservation (Lambert & Chapman, 2005; Link & Di Fiore, 2006; Barrera et al., 2008; Nunez-Iturri et al., 2008). Particularly in the tropics, endozoochory remains one area in which the dedicated natural historian still has much to reveal to the world of science. Here, I describe some of the methods and equipment required for animal observation, plant sample identification, and the analysis of faecal contents and feeding remains. I also give some practical advice on studying ranging, seed dispersal, gut passage times, seed germination and seedling survivorship in the field.

BASIC MATERIALS

Aside from the field biologist's basic kit of field clothes, notebook, compass and binoculars, the most important piece of gear for studying

Field and Laboratory Methods in Primatology: A Practical Guide, ed. Joanna M. Setchell and Deborah J. Curtis. Published by Cambridge University Press. © Cambridge University Press 2011.

seed dispersal is the simple plastic sandwich bag – or many such bags. These are ideal for collecting dung and seed specimens. Washing and re-using a soiled bag is not worth the mess or the risk of disease transmission (Chapters 1, 8), so bring a plentiful supply. I recommend cheaper bags that use small twists of wire for closure, because zipper bags are much more expensive. Specimen bags and vials sold by biological supply houses are even more expensive and I have not found them to be substantially better. You need permanent markers for labelling bags with the date and time the sample was collected. Keeping markers on a string tied to your belt or daypack may slow their inevitable disappearance in the field, but it is best to err on the side of abundance.

Another important piece of equipment is flagging tape to mark the trees in which your animals feed (another use for the permanent markers). Because tree size is an important determinant of habitat selection, and since trunk diameter correlates closely with fruit production, you need a measuring tape for measuring the diameters of feeding trees; a range finder allows accurate estimation of canopy heights and distances (Chapter 3). You will also need a good pair of callipers. For dietary description, you can use these to measure the fruits, fruitlets, and seeds of food plants on three axes, and calculate volume with the formula:

$$4/3\pi(\text{length}/2)(\text{width}/2)(\text{breadth}/2) \qquad (12.1)$$

If animals remove mouthfuls from larger fruits, or fruitlets from compound fruits, then measuring the sizes of the portions removed may be as important as measuring the sizes of entire fruits. Graduated cylinders are useful for measuring the volumes of irregularly shaped fruits and fruit parts by using water displacement (see Chapters 13 and 14 for more on measuring food mechanics and nutritional content).

Finally, a plant press and plant dryer, as well as paper bags of various sizes, will allow you to preserve specimens of food plants, particularly the fruits, which are typically rare in herbarium collections. Two companies that supply many of these tools and more are the Ben Meadows Company and Forestry Suppliers (see 'Useful Internet sites'). Press food plant specimens between layers of newspaper or blotting paper and corrugated cardboard and leave them in a dryer for a period of days until they are dessicated and non-perishable. Plant dryers typically use light bulbs or lamps as a heat source. Herbaria and university botany departments may have spare presses and can offer advice on constructing homemade dryers. Take a good supply of either newspaper or blotting paper to the field for these purposes.

SAMPLE IDENTIFICATION AND COLLABORATORS

Tropical botany remains a field with few guides and an immense diversity of taxa to learn. Some of the field ecologist's most important collaborators are the botanists who can help to identify plant specimens. If you are fortunate enough to work at a field site with local guides who know the indigenous flora, this will make the work much easier. Be aware, however, that local names may vary from region to region; I have found that the same local name can often refer to completely different plants in the same forest. You must therefore be prepared to collect multiple herbarium specimens of food plants for taxonomic identification. You can send these to a competent botanist in the country in which you work (usually affiliated with prominent universities), the national herbarium, and/or to a large international herbarium such as those at Kew Gardens (UK), The Chinese National Herbarium, The National Museum of Kenya, The South African National Botanical Institute, INPA (Instituto Nacional de Pesquisas da Amazônia, Brazil), the Muséum National d'Histoire Naturelle (France), the Smithsonian Institution, or the Missouri Botanical Garden (USA). Tropical plant taxonomy is a science of few textbooks, but three invaluable resources used by botanists worldwide are Gentry (1996), Letouzey (1986) and the library of edited volumes produced by the Foundation Flora Malesiana (Steenis, 1948–2001).

ANIMAL OBSERVATION

In studying seed dispersal, it is particularly important to be with the animals when they awaken at dawn (or dusk in the case of nocturnal species) because a large proportion of dispersed seeds are voided as the animals begin to stir upon waking. You should collect dung immediately, so a flashlight may be useful for collecting as many specimens as possible before the animals begin to travel. Collect dung from as many animals as possible to make the most complete list of food plants (a sort of scan sample of the group's diet). If the animals depart before you have finished collecting dung and seeds (a scenario which you can expect to experience quite often), then it is time to cut your losses and follow, because tracking a particular focal animal provides the most accurate data. Ideally, you should follow an individual from the time of food ingestion to the time of defecation to track gut passage time and dispersal distance – two key measures of dispersal ability. You may need to follow a group for several days in a row to determine these parameters. If animals are not easily recognizable as individuals then it may be useful to tag or mark them (Chapter 10).

Detailed observation of fruit and seed handling is obviously a key part of this research. Focal animal time sampling (Altmann, 1974; Martin & Bateson, 1993) is the standard data collection method (Chapter 11). However, observant ad libitum recording of feeding behaviours is also very important (Chapter 11). Keep track of the exact timing of feeding bouts for later determination of probable gut passage times. By recording the times and locations of each feeding bout and defecation over the course of many continuous hours you will, with luck, be able to deduce the rhythm of gut passage later. You may need to mark feeding trees quickly during follows and return to them later for measurement, botanical specimen collection and mapping (Chapter 3). Be sure to plan for such botanical data collection in addition to time spent following animals.

Fig. 12.1 shows the remarkable range of seed sizes that frugivorous primates can swallow. Fig 12.2 shows how complicated it can be to determine gut passage time in the wild. The two modes display the different passage times of seeds ingested at different times of day. Seeds in the second mode, at roughly 17 h, were ingested primarily in the afternoon and deposited under sleeping sites the following morning. The passage times and dispersal distances of frugivorous spider monkeys (*Ateles belzebuth belzebuth*) and woolly monkeys (*Lagothrix lagotricha poeppigii*) did not differ significantly in this study from Eastern Ecuador (Dew, 2001, 2008).

FAECAL CONTENTS ANALYSIS

Dung samples contain a wealth of biological information. You can manipulate specimens through their collection bags for items like large seeds, and dissect or wash them through a sieve to identify small seeds, seed fragments, fibre, plant parts and insect parts (Chapter 15). The rapidly passed dung of many frugivores can emerge from the animal with even delicate structures such as flower parts still intact. Counting and identifying gut-passed seeds is important, although estimating or recording simple presence/absence data may be more practicable for plants with large numbers of tiny seeds, such as

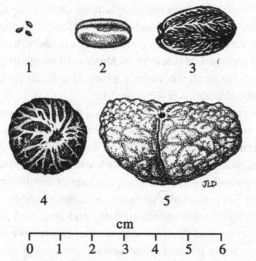

Fig. 12.1 A selection of seeds dispersed by Amazonian woolly monkeys and spider monkeys. Only a small proportion (16%) of dung samples from the two species contained small seeds, such as *Cecropia* seeds (#1), whereas at least 96% of samples contained large seeds. Both species dispersed large numbers of Inga seeds (#2). Seeds of *Spondias mombin* (#3) were the largest seeds passed by woolly monkeys, but spider monkeys dispersed seeds from *Ireartea deltoidea* (#4) and *Iryanthera* sp. (#5) as large as 27 mm in diameter. Illustration by J. L. Dew. Reprinted from Russo *et al.* (2005), Figure 1, p. 1023, with kind permission of Springer Science and Business Media.

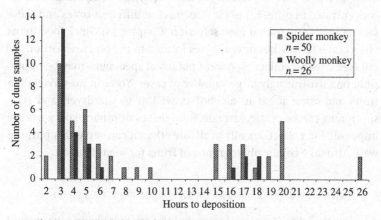

Fig. 12.2 Passage times of seeds dispersed by spider monkeys (*Ateles belzebuth belzebuth*) and woolly monkeys (*Lagothrix lagotricha poeppigii*) in the wild.

figs. Examine seeds closely for signs of damage, as the slightest nick in a seed coat can kill the plant embryo inside. These methods can also be useful in the examination of stomach contents, which, sadly, may be available sources of dietary data (Chapter 9). Along with dung samples, these can be preserved in alcohol or dried in a food dehydrator for later analysis in a nutritional laboratory (Chapter 14).

Be prepared to find a confusing array of seed taxa in dung, many of which will not match any plant you witnessed a focal animal eat. One of the first items of business will be to begin building a seed reference collection. With this you will be able to estimate the number of plant species eaten and dispersed even if you cannot identify all of them taxonomically. Keep examples of each seed morphotype in individual containers labelled with collection dates and specimen numbers. You can also store seed specimens in alcohol to preserve their original appearance. Seeds of some plant species can vary greatly in size and in shape, so it is wise to be conservative in 'lumping' apparent morphotypes rather than 'splitting' them until their identities can be verified with fruit specimens.

FOOD PLANTS AND FEEDING REMAINS

Tracking the changing availabilities of different plant resources is a key component of feeding ecology research (Chapter 3). It is important to note the stages of ripeness and the particular portions of the fruits, leaves, flowers, seeds, or pith swallowed by the animals, in contrast to those portions that are rejected. Plant secondary compounds may be concentrated in different plant structures at different times and these factors influence animal food selection (Chapter 14). Since food items often come from vines or epiphytes located in the obscure portions of tall tree crowns, collecting decent botanical specimens may be impossible in a frustratingly large number of cases. You can preserve fallen fruits and other foods in alcohol as well as in the dryer, and such specimens can be particularly useful if leaves or other plant parts are impossible to collect. Locally available ethanol can serve this purpose well. Also take colour photographs of fruits for identification.

RANGING, SEED DISPERSAL AND GUT PASSAGE

A well-mapped trail system is very helpful for measuring ranging and dispersal distances. This allows you to map sleeping sites, food plants and defecation sites with regard to their distances and positions relative to trail points (Chapter 4). Determining gut passage rates in the field is much more difficult than measuring them in a captive environment. When trying to identify the parent plant of a dispersed seed, one hopes to find a situation in which only one individual of a particular food plant species was visited, and that plant was visited only once during a particular day or period. The work is made more difficult when animals visit multiple individuals of the same fruiting plant species, or when the focal animal returns to a food plant multiple times during a follow. Of the various seeds collected from dung during a particular follow, there may be few or none that can indicate passage time. When animals return repeatedly to the same feeding tree, however, it may be possible to measure dispersal distance without knowing exact passage time. You can also use molecular genetic markers to identify the parent plants of dispersed seeds (Wang & Smith, 2002).

Frugivorous primates' gut passage rates are typically in the range of a few hours, and for most primate species, once you have followed a focal animal for one or more days, you can have some confidence that most voided seeds were ingested during the follow period (Fig. 12.2) (Milton, 1984; Julliot, 1996; Yumoto et al., 1999; Stevenson, 2000; Dew, 2001, 2008). If you have marked and mapped all fruit sources during a follow then you can determine passage rate and dispersal distance from the analysis of these bouts. It is wise to start conservatively by analysing the ingestion and defecation patterns of the second or third day of a continuous follow first. Sleeping sites are the best place to start backtracking through your data to determine the origins of dispersed seeds.

Large-bodied animals such as apes, some cercopithecine monkeys, and animals with highly folivorous diets may have considerably longer passage times than most frugivores (Milton, 1984; Lambert, 1999). When studying such species, as well as cathemeral animals, it may be impossible to follow seeds from ingestion to deposition. In such cases, you may be able to estimate minimum and maximum dispersal distances by analysing ranging data. It is simpler and also quite valuable to ascertain the proportions of dispersed seeds that fall in the 'hot zones' of high seed mortality under or near fruiting crowns with those that fall 15+ m away from any fruiting crown (Howe et al., 1985; Schupp, 1988; Terborgh et al., 1993). Likewise, you can describe dispersal sites with regard to canopy cover, substrate, microhabitat, and other factors likely to affect germination and survival (Rogers et al., 1998). It is also important to describe range usage accurately when studying seed dissemination patterns. If a study species uses the same resting sites repeatedly, for example, seed distribution may be heavily biased.

GERMINATION AND SEED SURVIVORSHIP TRIALS

Intact seed passage through the gut typically shortens the time to germination (Traveset, 1998; Dew & Wright, 1998). Plant species vary widely in time to germination, however, and this information is known for few tropical plants. You are likely to collect relatively few gut-dispersed seeds per plant species, and without large sample sizes the comparison of germination rates is statistically risky, but it is also important to know whether or not the gut-dispersed seeds of particular plants actually germinate following ingestion by particular frugivores. Seed germination trials therefore remain important in seed dispersal studies, and small sample sizes should not preclude you from collecting germination data.

If large quantities of gut-dispersed seeds are available from some plants, then it may be possible to compare the germination of gut-passed versus unpassed seeds for those taxa. This requires the collection of large quantities of unswallowed seeds that have not yet been infested by granivores such as weevils. Since the fruits fallen under a parent plant are often infested, this can be a tricky task.

The usual method for studying seed germination is to bury seeds in containers of damp sand or soil. Improvised local materials may be most practicable in field conditions, but some studies have used more precise laboratory setups with Petri dishes and moistened paper (Dew & Wright, 1998; Traveset, 1998). A laboratory building or screened area free of rodents and other pests is the best place to conduct undisrupted germination trials. Be sure to dissect all seeds at the end of the study to determine whether viable embryos remain. Seeds of some plants, such as *Spondias mombin* (Anacardiaceae), may not germinate for six months or more after deposition (Dew, 2001).

SEED GERMINATION AND SURVIVORSHIP IN THE WILD

A growing number of studies have examined the survivorship of gut-dispersed seeds under natural or semi-natural conditions (Janzen, 1982; Chapman, 1989; Estrada & Coates-Estrada, 1991; Feer, 1999; Andresen, 1999; 2001; Dew, 2001, 2008). Such studies can help determine the roles of secondary dispersers such as dung beetles and cache-hoarding rodents. You can place seeds in marked locations along transects and check them periodically for germination success, removal, and/or mortality (Blate et al., 1998; Andresen, 1999; Dew, 2001, 2008). Seed predation has been studied by enclosing subsets of seeds within wire cages that prevent their removal (Silman, 1996; Asquith et al., 1997; Brewer & Rejmanek, 1999; Wenny, 2000). Outside of such enclosures, you run into the difficulty of marking individual gut-dispersed seeds. Pierre-Michel Forget and others have pioneered a successful tracking method of marking large seeds with lengths of nylon thread (Forget et al., 2000; Forget & Wenny, 2004). Survival analysis is the standard statistical method for comparing time to germination and seed survivorship (Cox & Oakes, 1984; Lieberman, 1996).

CONCLUSIONS

Dietary ecology remains a field full of questions, particularly in the tropics. Thorough study can reveal the interdependence between species in new ways. Feeding ecologists excel at knowing their study animals well and scrutinizing behaviours in very fine detail. Knowledge of food characteristics, plant morphology, physiology and phenological patterns also provide valuable contributions to science. With thoughtful investigation, the interactions between frugivorous animals and the plants that sustain them may provide indispensable information to community ecology and conservation biology. Hopefully, the lessons learned will make a difference in the conservation of these spectacular species and their habitats.

REFERENCES

Altmann, J. (1974). Observational study of behavior: sampling methods. Behaviour 49, 227–65.

Andresen, E. (1999). Seed dispersal by monkeys and the fate of dispersed seeds in a Peruvian rain forest. Biotropica 31, 145–58.

Andresen, E. (2001). Effects of dung presence, dung amount and secondary dispersal by dung beetles on the fate of Micropholis guyanensis (Sapotaceae) seeds in Central Amazonia. J. Trop. Ecol. 17, 61–78.

Asquith, N. M., Wright, S. J. & Clauss, M. J. (1997). Does mammal community composition control recruitment in neotropical forests? Evidence from Panama. Ecology 78, 941–6.

Barrera, Z. V. A., Zambrano, M. J. & Stevenson, P. R. (2008). Diversity of regenerating plants and seed dispersal in two canopy trees from Colombian Amazon forests with different hunting pressure. Rev. Biol. Trop. 56, 1531–42.

Blate, G. M., Peart, D. R. & Leighton, M. (1998). Post-dispersal predation on isolated seeds: a comparative study of 40 tree species in a Southeast Asian rainforest. *Oikos* 82, 522–38.

Brewer, S. W. & Rejmanek, M. (1999). Small rodents as significant dispersers of tree seeds in a Neotropical forest. *J. Veg. Sci.* 10, 165–74.

Chapman, C. A. (1989). Primate seed dispersal: the fate of dispersed seeds. *Biotropica* 21, 148–54.

Cox, D. R. & Oakes, D. (1984). *Analysis of Survival Data*. London, New York: Chapman and Hall.

Dew, J. L. (2001). Synecology and seed dispersal in woolly monkeys (*Lagothrix lagotricha poeppigii*) and spider monkeys (*Ateles belzebuth belzebuth*) in Parque Nacional Yasuni, Ecuador. Unpublished PhD thesis, University of California, Davis.

Dew, J. L. (2008). Spider monkeys as seed dispersers. In *Spider Monkeys: The Biology, Behavior and Ecology of the Genus Ateles*, ed. C. Campbell, pp. 155–84. New York: Kluwer Academic Publishers.

Dew, J. L. & Wright, P. C. (1998). Frugivory and seed dispersal by four species of primates in Madagascar's rastern rainforest. *Biotropica* 30, 425–37.

Estrada, A. & Coates-Estrada, R. (1991). Howler monkeys (*Alouatta palliata*), dung beetles (*Scarabaeidae*) and seed dispersal: ecological interactions in the tropical rain forests of Los Tuxtlas, Mexico. *J. Trop. Ecol.* 7, 459–74.

Feer, F. (1999). Effects of dung beetles (Scarabaeidae) on seeds dispersed by howler monkeys (*Alouatta seniculus*) in the French Guianan rain forest. *J. Trop. Ecol.* 15, 129–42.

Forget, P.-M. & Wenny, D. (2004). How to elucidate seed fate? A review of methods used to study seed removal and secondary seed dispersal, In *Seed Fate: Predation, Dispersal and Seedling Establishment*, ed. P.-M. Forget, J. E. Lambert, P. E. Hulme & S. B. Vander Wall, pp. 351–62. Wallingford, UK: CABI International.

Forget, P.-M., Milleron, T., Feer, F., Henry, O. & Dubost, G. (2000). Effects of dispersal pattern and mammalian herbivores on seedling recruitment for *Virola michelii* (Myristicaceae) in French Guiana. *Biotropica* 32, 452–62.

Gentry, A. H. (1996). *A Field Guide to the Families and Genera of Woody Plants of Northwest South America (Colombia, Ecuador, Peru), with Supplementary Notes on Herbaceous Taxa*. Chicago, IL: University of Chicago Press.

Howe, H. F. & Miriti, M. N. (2004). When seed dispersal matters. *Bioscience* 54, 651–60.

Howe, H. F., Schupp, E. W. & Westley, L. C. (1985). Early consequences of seed dispersal for a neotropical tree (*Virola surinamensis*). *Ecology* 66, 781–91.

Janzen, D. H. (1982). Removal of seeds from horse dung by tropical rodents: influence of habitat and amount of dung. *Ecology* 63, 1887–990.

Julliot, C. (1996). Seed dispersal by the red howler monkey (*Alouatta seniculus*) in the tropical rain forest French Guiana. *Int. J. Primatol.* 17, 239–58.

Lambert, J. E. (1999). Seed handling in chimpanzees (*Pan troglodytes*) and redtail monkeys (*Cercopithecus ascanius*): implications for understanding hominoid and cercopithecine fruit-processing strategies and seed dispersal. *Am. J. Phys. Anthropol.* 109, 365–86.

Lambert, J. E. & Chapman, C. A. (2005). The fate of primate dispersed seeds: deposition pattern, dispersal distance, and implications for conservation. In *Seed Fate: Predation, Dispersal and Seedling Establishment*, ed. P.-M. Forget, J. E. Lambert, P. E. Hulme & S. B. Vander Wall, pp. 137–50. Wallingford, UK: CABI International.

Letouzey, R. (1986). *Manual of Forest Botany. Tropical Africa*, Vols. I, II, III. (transl. R. Huggett). Nogent-sur-Marne, France: Centre de Technique Forestier Tropical.

Lieberman, D. (1996). Demography of tropical tree seedlings: a review. In *The Ecology of Tropical Forest Seedlings*, ed. M. D. Swaine, pp. 141–8. Paris: Parthenon Publishing Group.

Link, A. & Di Fiore, A. (2006). Seed dispersal by spider monkeys and its importance in the maintenance of neotropical rain-forest diversity. *J. Trop. Ecol.* 22, 235–46.

Martin, P. R. & Bateson, P. P. G. (1993). *Measuring Behaviour: An Introductory Guide.* Cambridge, New York: Cambridge University Press.

Milton, K. (1984). The role of food-processing factors in primate food choice. In *Adaptations for Foraging in Nonhuman Primates*, ed. P. S. Rodman & J. G. H. Cant, pp. 249–79. New York: Columbia University Press.

Nunez-Iturri, G., Olsson, O. & Howe, H. F. (2008). Hunting reduces recruitment of primate-dispersed trees in Amazonian Peru. *Biol. Conserv.* 141, 1536–46.

Rogers, M. E., Voysey, B. C., McDonald, K. E., Parnell, R. J. & Tutin, C. E. G. (1998). Lowland gorillas and seed dispersal: the importance of nest sites. *Am. J. Primatol.* 45, 45–68.

Russo, S. E., Campbell, C. J., Dew, J. L., Stevenson, P. R. & Suarez, S. A. (2005). A multi-forest comparison of dietary preferences and seed dispersal by *Ateles* spp. *Int. J. Primatol.* 26, 1017–37.

Schupp, E. W. (1988). Seed and early seedling predation in the forest understory and in treefall gaps. *Oikos* 51, 71–8.

Silman, M. R. (1996). Regeneration from seed in a neotropical rain forest. Unpublished PhD thesis, Duke University, Durham, NC.

Steenis, C. G. G. J. van (ed.) (1948–2001). *Flora Malesiana, Being an Illustrated Systematic Account of the Malaysian Flora.* Vols. 1–13. Dordrecht: Kluwer Academic Publishers.

Stevenson, P. R. (2000). Seed dispersal by woolly monkeys (*Lagothrix lagothricha*) at Tinigua National Park, Colombia: dispersal distance, germination rates, and dispersal quantity. *Am. J. Primatol.* 50, 275–89.

Terborgh, J., Losos, E., Riley, M. P. & Riley, M. B. (1993). Predation by vertebrates and invertebrates on the seeds of five canopy tree species of an Amazonian forest. *Vegetatio* 107–108, 375–86.

Traveset, A. (1998). Effect of seed passage through vertebrate frugivores' guts on germination: a review. *Perspect. Plant Ecol. Evol. Syst.* 1, 151–90.

Wang, B. C. & Smith, T. B. (2002). Closing the seed dispersal loop. *Trends Ecol. Evol.* 17, 379–86.

Wenny, D. G. (2000). Seed dispersal, seed predation, and seedling recruitment of a neotropical montane tree. *Ecol. Monogr.* 70, 331–51.

Yumoto, T., Kimura, K. & Nishimura, A. (1999). Estimation of retention times and distances of seed dispersed by two monkey species, *Alouatta seniculus* and *Lagothrix lagotricha*, in a Colombian forest. *Ecol. Res.* 14, 179–91.

USEFUL INTERNET SITES

www.benmeadows.com and www.forestry-suppliers.com – for equipment.

PETER W. LUCAS, DANIEL OSORIO, NAYUTA YAMASHITA,
JONATHAN F. PRINZ, NATHANIEL J. DOMINY AND
BRIAN W. DARVELL

13

Dietary analysis I: food physics

INTRODUCTION

This chapter and the next focus on measurements of the physical and chemical attributes of potential foods that primates select or reject. The major reason for analysing primate diets in this manner is to understand the basis for their food choice. Observing primates as they feed quickly raises questions in the observer's mind about the possible foraging strategies that the animals might be following in order to survive. How do primates distinguish food from what is otherwise scenery? Can we measure the attributes of potential foods in the form in which primates are actually sensing them? What do primates get out of the foods they choose and are their choices, based on sensory capabilities, optimal in terms of nutrients? Tests of hypotheses that address these questions will require objective dietary analysis (e.g. for colour: Osorio et al., 2004). It is important to tailor your measurements to the questions being asked.

The physicochemical characteristics of foods may form important sensory cues for their detection, selection and subsequent processing by primates, but all these characteristics are affected to some extent by specimen storage. Physical characteristics, such as colour, geometry and mechanical properties, may change drastically and rapidly, so it is often important and sometimes vital to make measurements almost immediately, while the specimen is fresh. The alternative of drying specimens for later chemical analysis not only involves a substantial time lag between fieldwork and subsequent access to a laboratory, but can also lead to inaccurate results. Thus, a major function of these two chapters is to describe the practical application of a new range of field techniques that can produce results in situ. A full field laboratory would be the ideal situation, but in reality many

Field and Laboratory Methods in Primatology: A Practical Guide, ed. Joanna M. Setchell and Deborah J. Curtis. Published by Cambridge University Press. © Cambridge University Press 2011.

scientists can only dream of such facilities. Thus, wherever possible, we also include field options that are possible on a smaller budget. Otherwise, for chemistry, we describe briefly what can be achieved 'back home' in the laboratory. Overall, field workers should think broadly and prepare well before setting out because many food properties can now be estimated synchronously with the period of observation. It is wise, though, to remember that although many things can be done, only one thing is essential – the accurate observation of primate feeding.

The current high rates of tropical deforestation, coupled with the vulnerability of primates to hunting even in intact forests, means that many first studies of primate ecology will also be the last for that particular species and location. Primate researchers therefore have an unusually strong obligation to target their measurements at future generations of primatologists, rather than simply their thesis examiners or journal referees. The most important step towards making present-day measurements 'upwardly compatible' with future questions is to document precisely how you selected, collected, stored, processed and analysed your samples, even to the extent of including ancillary data that might ordinarily be judged as irrelevant. Too many published data are unusable because the methods are inadequately described. Our approach here is meant to be positive and pragmatic: we want to encourage primatologists to gain much more information from their field studies while still on site.

BASIC EQUIPMENT

'In the field' does not just mean 'in the shade': you need a roofed enclosure. The majority of the tests we describe require a power supply from solar panels, generators or (rather rarely available) a connection to the mains. The tests can be conducted with rechargeable batteries that require intermittent access to a power supply for recharging and everything can be run off, or recharged using, solar panels. Basic equipment can be expensive and we present lower-cost options whenever possible. We assume that most researchers will have access to a laptop computer in the field. Adequate space for operating the equipment should also be taken into account.

Researchers interested in testing mechanical properties of foods will need to invest in a portable tester (see, for example, Darvell *et al.*, 1996). Although you can conduct toughness tests without a computer, other tests that measure a force–displacement gradient require one with the appropriate software loaded. We are not aware of lower-cost alternatives than that described by Darvell *et al.* (1996), production of which is lapsing (although a new version is being built), but commercial testers have been used successfully in the field (Overdorff & Strait, 1998). The accurate measurement of colour requires a spectrometer, which can also be used for quantifying leaf chemistry, as described in the next chapter. We describe additional equipment in the relevant sections. Finally, careful laboratory notes for both mechanical and chemical tests are essential, including full details of all tests conducted.

TESTS

All tests follow directly from observations of the exact plant part that is consumed. Mechanical properties, colour, and nutrients change throughout the developmental cycle of a plant, so it is critical to identify the developmental stage of a food accurately.

Food geometry

Aspects of food size and shape are important determinants of primate acceptance, affecting visibility at long range. Depending on the manual dexterity and incisal form of a primate, the size and shape of food items also affect the method of ingestion. Dimensional measurements are straightforward with mechanical callipers and a portable battery-powered analytical balance (accurate to ±0.01g; e.g. a Mettler Toledo JL1502GA03 Class 2 'jeweller's balance' for approx. US$600) can be used to obtain food mass. In addition, digital cameras (Chapter 17) provide simple electronic records of basic form and colour. Downloading these images to a computer allows you to use image analysis software to quantify shape.

Food colour

The detection of food colour at long distances (say 10 m or more) is likely to provide an important sensory cue for primates (Dominy *et al.*, 2001). At shorter ranges, colour may well be useful for identifying the developmental stage of fruit and leaves. Primate colour vision is very variable, both within and between species (Jacobs, 1996). Although it has long been assumed that such differences are important for foraging primates, there were almost no field data to support this until recently. Now simple field techniques have begun to change this state of affairs, and we expect many advances in the next few years. It is fairly easy to record the reflectance of an object with a small computer-based spectroradiometer. There are two aspects to the coloration of most objects: a diffuse, spectrally selective reflection due to chemical pigmentation and a more directional, non-spectrally selective specular reflection (shininess or gloss). To record reflectance spectra, it is conventional to use a viewing geometry that minimizes specular reflection, both because such measurements are more reproducible and because pigment reflectance probably relates more closely to the principal attraction for a primate. Illumination can either be directional or diffuse. Diffuse sources (i.e. integrating spheres) are widely used for colorimetry in industry (and can be used in biology, see Lucas *et al.*, 1998), but though accurate, they are still extremely expensive. Directional lighting is far more usual in work on animal colour vision and is cheaper. Lucas *et al.* (2001) describe a set up that includes a fibre-optic spectrometer. The Ocean Optics USB2000+UV-VIS is a cheap option, but it requires a laptop computer to process its signals. The latest spectrometers from Ocean Optics include a data-logger ('Jaz'), and are thus completely standalone. A fibre-optic light source is required (e.g. USB-DT Deuterium Tungsten Light Source from Ocean Optics), and either a home-made lightproof specimen box that arranges the lighting at a fixed angle to the specimen surface or a reflectance probe (Endler & Mielke, 2005). Orienting the specimen is easy for flat leaves, difficult for curved shiny fruit. The reflectance must be referred to a white standard. For this, compacted barium sulphate powder is cheap, whereas commercial standards are robust but expensive (Ocean Optics WS-1-SL diffuse reflection standard).

It is relatively straightforward to predict the responses of the primate cone photoreceptors from the spectra of your specimens since the spectral sensitivities of a considerable number of species have now been determined. This provides the information that the eye makes available to the brain. With care, you can predict whether a pair of colours will be discriminable for a given animal – e.g. a food item as seen against a leaf background, or adjacent fruits or leaves at different developmental stages (examples in Fig. 13.1). The spectra of food items alone are meaningless: you need to record food, non-food and background. Typical backgrounds are foliage and bark, but earth, water, rock and sky are also possible.

Several papers have presented and reviewed analytical methods for measuring colour (Regan *et al.*, 2001; Kelber *et al.*, 2003; Stevens

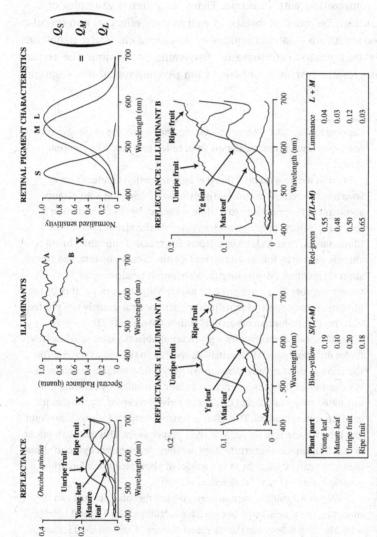

Fig. 13.1 (For legend, see p. 242.)

et al., 2009). Figure 13.1 illustrates the general approach. Ideally, object reflectance, referenced to a standard, would be 'lit' as it would be for a foraging primate. However, natural lighting, especially in a forest, is neither highly directional nor completely diffuse and varies in its spectral composition with direction. Figure 13.1 shows examples of two illuminants recorded in forests, as well as their effect on non-specular reflection: this is small but noticeable. Results on chromatic signals are for a trichromatic catarrhine or platyrrhine and assume the retinal photoreceptor responses obtained from psychophysical investigations

---

Caption for Fig. 13.1 Colour processing: the estimation of photoreceptor responses and chromatic signals from reflectance and illumination spectra.

Top row: Relative quantal flux (i.e. light intensity) as a function of wavelength incident on the eye is given by multiplying a reflectance spectrum (left, referenced to a white standard) by the illumination spectrum (centre) at each wavelength across the visible spectrum. Illuminant A, from Yakushima, Japan, is unshaded sunlight while B is slightly shaded by foliage and is from Kibale. Relative quantal catches, of short (**S**) medium (**M**) and long (**L**)-wavelength sensitive cone photoreceptors ($Q_{S,M,L}$) are given by multiplying the incident flux by the photoreceptor's spectral sensitivity (right). Spectral sensitivities plotted here are from a human observer (Smith & Pokorny, 1972).

Middle row: Quantal flux for a given set of objects varies substantially under different natural illumination spectra. In the visual system, this variation is substantially 'discounted' by colour constancy mechanisms. We can model the consequences of colour constancy by dividing the estimated receptor quantal catches of a given receptor, by the quantal catch to some standard. This gives estimated outputs for short, medium and long wavelength cones: S, M and L respectively. This standard might be the illumination spectrum itself, or some 'average background'. This algorithm can be thought of as a model of photoreceptor adaptation (Osorio & Vorobyev, 1996; Kelber *et al.*, 2003).

Lower row: Chromatic signals derived from receptor outputs can be modelled in various ways, but we have calculated $S/(L+M)$ and $L/(L+M)$ after Macleod & Boynton (1979). An advantage of this method is that $S/(L+M)$ represents the blue-yellow chromatic signal available to a dichromatic mammal, while the red-green parameter, $L/(L+M)$, is available only to trichromats. Bivariate plots of $S/(L+M)$ and $L/(L+M)$ are useful to show separations but this does not predict their discriminability (Osorio & Vorobyev, 1996; Kelber *et al.*, 2003). The parameter $(L+M)$ represents luminance.

of humans (Wysecki & Stiles, 2000). Where receptor sensitivities for other primates differ significantly from those of humans, they can be estimated from photoreceptor spectral absorbance functions, taking account of light absorbance within ocular media (e.g. the lens and macular pigment).

Diurnal vision in primates is restricted to wavelengths between 400 and 700 nm. Although diurnal primates probably cannot see into the ultraviolet (UV) region (350–400 nm), many other animals can (Jacobs, 1992), and if you are interested in questions about food selection by primates *vis-à-vis* their competitors, then it might be appropriate to measure UV reflectance (Bennett *et al.*, 1994; Altshuler, 2001). In nocturnal primates, the tapetum increases sensitivity of the retina in dim light, but it also fluoresces, permitting them to use a portion of the ultraviolet spectrum when this is present (Martin, 1990).

Given that primate species vary in their colour vision, you cannot rely on human perception to describe spectral stimuli. One low-budget alternative used in several published studies (semi-quantified although retaining some subjectivity) is to use Munsell colour cards (approx. US$100), to match food and background samples. These are limited, particularly in range of brightness, but are far superior to the simple descriptive colour categories commonly found in the ecological literature. An alternative is software that deciphers colour coordinates from high-resolution digital camera photographs (see Useful Internet sites). For those with scant funds, use of basic colour terms has shown trends in the fruit choice of different primate species that correspond to interspecific variations in colour vision (Dominy, 2004).

Surface texture of foods

Many potential plant food items have irritant thorns, spines or hairs that can reduce feeding rates or deter primates completely. Leaves in particular can contain amorphous opaline silica or calcium oxalate crystals, which can wear teeth appreciably if ingested regularly. Silica is surprisingly common on the surfaces of leaves eaten by primates. The Moraceae, species of which are ubiquitous in primate diets, provide a prime example (Fig. 13.2). Thus, estimation of the abrasiveness of leaves might be of interest. A recent laboratory method to ascertain the wear effect of grasses has been described by Hammond and Ennos (2000) and adapted to a field setting by JFP and NJD (Fig. 13.2), but requires development. Scratching could be analysed by examining tooth microwear, but a simpler solution is probably just to take the

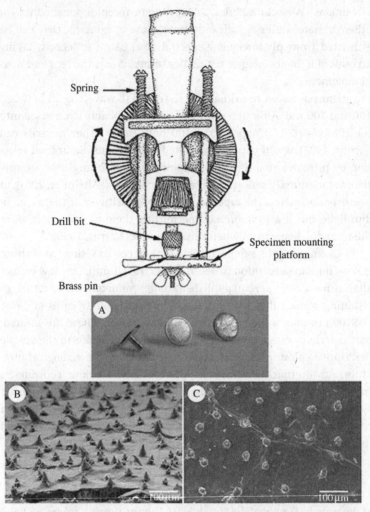

Spring

Drill bit

Specimen mounting platform

Brass pin

Fig. 13.2 Abrasion testing. A toy minidrill, modified so that a brass pin, shown in A, inserted in the drill bit rotates against a food specimen to test its abrasiveness. A standard load is imposed by two springs. The specimen-mounting platform has a holder for clamping food specimens such as leaves. The pin has been pre-coated with nail varnish. You can turn the drill a set number of revolutions (e.g. three) then remove and inspect the pin for scratches to the varnish, as seen in A. You can save pins for quantitative wear analysis. B and C are scanning electron micrographs of *Poulsenia armata* (Moraceae) leaves (from Barro Colorado Island, Panama), showing the surface armour that does the damage.

dried food back to the laboratory and quantify the density of features under a scanning electron microscope (Fig. 13.2; Lucas & Teaford, 1995).

Mechanical properties of foods

The physical resistance of foods to being ingested, chewed and swallowed probably forms an important factor in their acceptance or rejection by primates. To simplify understanding, this section treats foods as though they were real solids. Of course, they are not and there is an extensive theory of cellular mechanics that deals with foods, such as plant tissues, wherein a trellis of cell walls traps fluid (Ashby & Gibson, 1997).

The most important properties of solid foods are toughness, Young's modulus and yield stress. Young's modulus and yield stress can be measured under force control (using weights as 'dead' loads, and measuring the effect on specimen dimensions), but toughness can only be measured under displacement control (i.e. the dimensions of the specimens are systematically changed, while monitoring the effect on the force as the dependent variable). All these properties are directional and the amount of information required to fully characterize a food object depends on its heterogeneity. This level of complexity, though, is not the place to start, certainly not in the field, and any measure of one or more of these food properties could go a long way towards explaining food choices. In fact, groupings of two of these properties can explain the general basis of plant defence against fracture (Lucas et al., 2000), defences that a primate must overcome in order to eat these items. The square root of the ratio of toughness to Young's modulus describes what is usually referred to in lay terms as 'toughness', while the square root of their product is a quantity that most would denote as 'hardness'. Note, though, that any specific analysis typically requires more information than this, with a variable mix of friction ('Other possibilities', below), indentation (controlled by the yield stress) and fracture. No one has yet attempted measurements of all these in the field.

Much less is known about how the mechanical properties of foods are sensed by primates than about food attributes such as colour. Even in humans, the branch of food science that deals with mechanics (texture studies) has no generally accepted methodology. Increasingly, however, the influence of materials science is being felt and we strongly advocate this fundamental approach. However, the learning curve for mechanical tests is steep and you will need to read and practise, and perhaps collaborate with someone expert in this area. Vincent (1992) is invaluable as a 'cookbook' for methods and Ashby & Jones (1996, 1998) are superlative general introductions to the science behind materials testing. Darvell *et al.* (1996) have described a portable tester, which now has provision for jigs for many types of mechanical test (Lucas *et al.*, 2001). Supplementary equipment includes callipers, a thickness gauge (particularly useful for measuring leaf thickness) and tools for making test specimens such as cork borers (perhaps US$500 in all). The best way to preserve plants in the field in the state in which they were eaten is probably to place them in plastic bags with damp paper towels to preserve moisture (i.e. maintain 100% relative humidity but without permitting osmotic absorption).

Young's modulus

The Young's or elastic modulus of a material is defined as the ratio of stress to strain during the early stage of its deformation under load. It is a measure of the stiffness of a food and calculated from the early (greatest) slope of a force–displacement diagram. Young's modulus can be measured in bending, compression or tension tests, but we recommend the compression of cylinders of foods as the easiest (described in Fig. 13.3). In any type of test, there may be an initial 'toe' at the start of the test where some slack is taken up in tension or where small geometrical errors are being corrected (Fig. 13.3). In plant tissues, ignore the toe when measuring the gradient, but remember that many soft animal tissues have curved stress–strain relations even at low loads, especially in tension. Knowledge of limits to the dimensional ratios of specimens in tests will avoid errors. In a bending test, the length of the specimen beam should be >16 times its depth, while in a cylindrical compression test, the height : diameter ratio should be standardized to about 2 for comparability (but definitely <10 to avoid buckling). Both compression and tension tests are subject to specimen 'end effects'. In a tension test, compression of the specimen by the grips may affect the modulus estimate (the solution is to keep the length : thickness ratio high); in a compression test, friction of the food against the compressing platform does the same. A strip of PTFE-coated adhesive tape or a little mineral oil attached to the compression plates can help to prevent this.

Toughness

Toughness, the energy expended per unit area of crack growth, is probably the vital mechanical element in the resistance of most foods

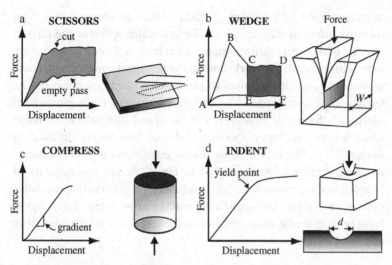

Fig. 13.3 Mechanical testing.

(a) A scissors test gives toughness estimates for foods that are difficult to shape or in the form of rods or sheets. Application of a force (e.g. to the handles) causes the scissor blades to act like double wedges producing a crack that tracks along the fine dotted line as the scissors are closed. This produces the force–displacement trace marked as 'cut'. The area below this curve gives the work done during the test. However, you need to deduct friction, largely work done by the scissors blades against each other: you can do this simply by running an empty pass over the identical path used to cut the specimen. The work done during the cut, minus the work done by the blades against themselves gives the work done on the specimen (shaded in the figure). Dividing this work by the fractured cross-sectional area (using callipers, etc.) gives the toughness estimate.

(b) A sharp wedge is forced through a block of tissue to measure toughness. The loading curve moves through points ABCD. From A, the force builds with indentation until cracking at B. This is unstable at first but settles and, from C, further movement of the wedge between C and D makes the crack run just ahead of the wedge tip. Reversing the wedge back to position A, then running it through the cracked specimen produces a friction curve (often negligible) labelled AEF. The area bounded by BCDE gives the work done to produce the crack. The physical crack area (shaded) is given by the depth of wedge penetration (distance CD) multiplied by the width, w. Of the block dimensions, only the width needs to be cut at all accurately.

(c) A cylinder of tissue is compressed between plates to obtain the force–displacement gradient at small deformations. The force–displacement gradient at small deformations is converted into Young's to consumption by primates. In plant tissues, toughness arises from the structure and dimensions of the cell walls, whereas the characteristics of the cuticle are all-important in insects. To determine toughness accurately, all the work done in the test must contribute to crack growth. Usually, this requires you to notch a specimen, measure the length of the notch accurately, and ensure that the crack grows from the tip of the notch. However, you can avoid this palaver by using scissors tests on floppy specimens shaped like sheets or rods, or wedge tests on blocks, which are simple yet effective tests of toughness (Lucas & Pereira, 1990; described in Fig. 13.3). For the majority of primate foods, a portable device is sufficiently rigid to withstand deformation itself (and thus avoid inaccuracy) during a test. Some foods, however (e.g. seeds), must be taken home and tested with a laboratory universal tester.

Indentation

Indentation tests are some of the oldest in the mechanical canon and have many uses in field biology. They generally involve pressing a conical, spherical or pyramidal spike into an object. The force that produces a permanent mark of given surface area is called the hardness of the object (Fig. 13.3). Such tests involve contacts similar to those of teeth on food particles, and this imitative aspect has led to the use of simple and cheap instruments like durometers in ecology. Although food hardness is an undeniably important aspect of primate ecology, measurements are rarely related to fundamental properties and vary from the descriptive to the quantitative. For example, Boubli (1999) describes fruits as hard (like a walnut), medium (like an avocado), or soft (like a grape). However, for those with sparse funds, even this level of resolution has proved informative for understanding the ecological specializations of black-headed uakaris (*Cacajao*

Caption for Fig. 13.3 (cont.)

modulus by first dividing the force by the original cross-sectional area (the area of the black top of the cylinder) and then the displacement by the original cylinder height. Indentation is shown in D. Indentors can be made in various shapes. The figure shows a hemispherical indentor, but a Knoop pyramid is probably best. The hardness of a specimen is the force divided by the area of indentation, where the area is measured in the plane of the specimen, in this case, from measuring the diameter, d.

melanocephalus), primates adapted to forests with unusually well-protected fruits. Other studies, making quantitative measures with a variety of instruments, have shown the importance of food mechanics to dietary divergence and species coexistence (Kinzey & Norconk, 1993; Yamashita, 1996, 2000; Overdorff & Strait, 1998). Recent developments may change this and elevate indentation to a vital field technique (Lucas *et al.*, 2010). If you attach an indenter to a device with displacement control, such as the Darvell tester, it is possible to sense the depth of indentation, both to measure the hardness of a material during loading and the elastic modulus from the rebound of the deepest part of the indentation as the indenter is withdrawn.

Hardness reflects the yield stress, the stress at which a mark becomes permanent (Fig. 13.3). In most real solids, where volume is preserved during indentation, hardness values are around three times that of the yield stress (Ashby & Jones, 1996) and this relationship probably applies to the scratching of tooth enamel by abrasive particles such as the plant silica described above. However, in cellular solids such as most foods, this will not be so. The lightest touch with an indenter produces a recoverable dent. In a true solid, this is an indirect measure of Young's modulus, but edible tissues are not true solids and the meaning is somewhat different. Resistance to the lightest touch in thin-walled plant tissues is likely to be due to the turgidity of cellular contents, a feature ingeniously exploited by the Outwater ball tonometer to measure the internal cell pressure (Lintilhac *et al.*, 2000). Tonometers may have applications in primatology since moisture is probably important in food choice. On heavier loading, turgid cells will burst and the tissue is likely to collapse in on itself, effectively with a loss of volume. When this happens, hardness is about equal to yield stress (Wilsea *et al.*, 1975). However, the cells of many ripe fruits slide across each other, and tissues such as seed shells are often very dense woods (Lucas *et al.*, 2000), and indentation results in less volume reduction.

One specific application of indentation tests is to estimate the bite forces that primates use. When they feed, primates often drop food parts, either deliberately (e.g. fruit peel) or accidentally. Examination of these food fragments can reveal a substantial amount of information about how a food was processed. In particular, when mammals feed on harder foods, indentations produced by tooth cusps can usually be seen on or around the fracture line. Peak forces are usually produced during these indentations. Lucas *et al.* (1994) and Hill *et al.* (1995) provide a method for establishing the magnitude of forces from an examination of these indentations.

OTHER POSSIBILITIES

In addition to reflectance, it might be worth measuring the transmission of light through young leaves. Their translucence is very obvious when they are seen against the light, which is how some primates have been observed to feed on them. However, because leaves scatter light as well, the accurate measurement of transmission needs an extended light source (e.g. a light box or uniformly lit white paper) rather than a spectrophotometric set-up such as that described in the next chapter.

Fruits eaten by higher primates, i.e. anthropoids with spatulate incisors, are often characterized by outer tissues (which botanists call rinds) that peel cleanly from the underlying flesh at ripeness. The work done in peeling is work done against adhesion (Kendall, 2001). The work of adhesion is a quantity analogous to toughness and can be estimated roughly and easily with a mechanical tester. The stickiness of a food is also a matter of adhesion. It is important: many primate foods contain latex, but we cannot offer any obvious test of its effectiveness.

The surfaces of food objects have been very poorly investigated. The shininess of object surfaces (i.e. the specular component of object reflectance) could be very important for attracting the attention of a primate to a fruit and might be worth measuring systematically. Surfaces are important for mechanical action too. Friction helps hands to grip branches that contain potential food items, but when teeth are used to remove tree bark or soils (geophagy), friction simply raises the forces that need to be exerted to obtain fracture and this adds to tooth wear. Field devices that measure the coefficient of friction may be needed for the analysis of such actions. Resistance to being gripped also lies in the domain of friction, depending not only on the surface of the animal or plant part being gripped, but on the surfaces of the hands or teeth that are gripping them. Tooth enamel is very smooth, but intra-oral friction is not always low: one of the major actions of tannins in plant foods is to reduce the lubricating effect of saliva (the more so in primates that have tannin-specific binding proteins), allowing tannin levels to be detected by an increase in friction between oral tissues (Prinz & Lucas, 2000). This characteristic is often referred to as 'astringency' in food texture studies.

This chapter has only scratched the surface of food physics. Almost ignored, this subject offers enormous potential for explaining many aspects of primate food choice. We have limited the discussion to solid foods, but for the sake of completeness, gums and flower nectar are, of course, food items. They are viscous liquids, and the ripening of fruit flesh often involves considerable flow. Investigations of these foods require viscometers of some sort, a large variety of which are available. We have not yet tried any of these in the field.

ACKNOWLEDGEMENTS

Camilla Pizano drew the abrasion tester shown in Fig. 13.2; Johnny Leung arranged the figure.

REFERENCES

Altshuler, D. L. (2001). Ultraviolet reflectance in fruits, ambient light composition, and fruit removal in a tropical forest. *Evol. Ecol. Res.* 3, 767–78.

Ashby, M. F. & Gibson, L. J. (1997). *Cellular Solids: Structure and Properties.* Cambridge: Cambridge University Press.

Ashby, M. F. & Jones, D. R. H. (1996). *Engineering Materials 1.* (2nd edn) Oxford: Butterworth Heineman.

Ashby, M. F. & Jones, D. R. H. (1998). *Engineering Materials 2.* (2nd edn) Oxford: Butterworth Heineman.

Bennett, A. T. D., Cuthill, I. C. & Norris, K. J. (1994). Sexual selection and the mismeasure of color. *Am. Nat.*, 144, 848–60.

Boubli, J. P. (1999). Feeding ecology of black-headed uacaris (*Cacajao melanocephalus melanocephalus*) in Pico da Neblina National Park, Brazil. *Int. J. Primatol.* 20, 719–49.

Darvell, B. W., Lee, P. K. D., Yuen, T. D. B. & Lucas, P. W. (1996). A portable fracture toughness tester for biological materials. *Meas. Sci. Technol.* 7, 954–62.

Dominy, N. J. (2004). Color as an indicator of food quality to anthropoid primates: ecological evidence and an evolutionary scenario. In *Anthropoid Origins*, ed. C. R. Ross & R. F. Kay, pp. 615–44. New York: Kluwer.

Dominy, N. J., Lucas, P. W., Osorio, D. & Yamashita, N. (2001). The sensory ecology of primate food perception. *Evol. Anthropol.* 10, 171–86.

Endler, J. A. & Mielke, P. W. (2005). Comparing entire colour patterns as birds see them. *Biol. J. Linn. Soc.* 86, 405–31.

Hammond, T. A. & Ennos, A. R. (2000). Mechanical testing of the abrasiveness of grass leaves. In *Plant Biomechanics 2000*, ed. H.-C. Spatz T. Speck, pp. 535–40. Stuttgart: Georg Thieme Verlag.

Hill, D. A., Lucas, P. W. & Cheng, P. Y. (1995). Bite forces used by Japanese macaques (*Macaca fuscata yakui*) on Yakushima Island, Japan to open aphid-induced galls on *Distylium racemosum. J. Zool.* 237, 57–63.

Jacobs, G. H. (1992). Ultraviolet vision in vertebrate. *Am. Zool.* 32, 544–54.

Jacobs, G. H. (1996). Primate photopigments and primate color vision. *Proc. Nat. Acad. Sci. USA* 93, 577–81.

Kelber, A., Vorobyev, M. & Osorio, D. (2003). Animal colour vision – behavioural tests and physiological concepts. *Biol. Rev.* 78, 81–118.

Kendall, K. (2001). *Molecular Adhesion and its Applications: The Sticky Universe.* New York: Kluwer/Plenum Press.

Kinzey, W. G. & Norconk, M. A. (1993). Physical and chemical properties of fruit and seeds eaten by *Pithecia* and *Chiropotes* in Surinam and Venezuela. *Int. J. Primatol.* 14, 207–27.

Lintilhac, P. M., Wei, C., Tanguay, J. J. & Outwater, J. O. (2000). Ball tonometry: a rapid, non-destructive method for measuring cell turgor pressure in thin-walled plant cells. *J. Plant Growth Regul.* 19, 90–7.

Lucas, P. W. & Pereira, B. (1990). Estimation of the fracture toughness of leaves. *Func. Ecol.* 4, 819–22.

Lucas, P. W. & Teaford, M. F. (1995). Significance of silica in leaves eaten by long-tailed macaques (*Macaca fascicularis*). *Folia Primatol.* 64, 30–6.

Lucas, P. W., Peters, C. R. & Arrandale, S. (1994). Seed-breaking forces exerted by orang-utans with their teeth in captivity and a new technique for estimating forces produced in the wild. *Am. J. Phys. Anthropol.* 94, 365–78.

Lucas, P. W., Darvell, B. W., Lee, P. K. D., Yuen, T. D. B. & Choong, M. F. (1998). Colour cues for leaf food selection by long-tailed macaques (*Macaca fascicularis*) with a new suggestion for the evolution of trichromatic colour vision. *Folia Primatol.* 69, 139–52.

Lucas, P. W., Turner, I. M., Dominy, N. J. & Yamashita, N. (2000). Mechanical defences to herbivory. *Ann. Bot.* 86, 913–20.

Lucas, P. W., Beta, T., Darvell, B. W. *et al.* (2001). Field kit to characterize physical, chemical and spatial aspects of potential foods of primates. *Folia Primatol.* 72, 11–15.

Lucas, P. W., Constantino, P. J., Chalk, J. *et al.* (2010). Indentation as a technique to assess the mechanical properties of fallback foods. *Am. J. Phys. Anthrop.* 140, 653–52.

Macleod, D. I. A. & Boynton, R. M. (1979). Chromaticity diagram showing cone excitation by stimuli of equal luminance. *J. Opt. Soc. Am.* 69, 1183–6.

Martin, R. D. (1990). *Primate Origins and Evolution.* Princeton, NJ: Princeton University Press.

Osorio, D. & Vorobyev, M. (1996). Colour vision as an adaptation to frugivory in primates. *Proc. R. Soc. Lond.* B 263, 593–9.

Osorio, D., Smith, A. C., Vorobyev, M. & Buchanan-Smith, H. M. (2004). Detection of fruit and the selection of primate visual pigments for color vision. *Am. Nat.* 164, 696–708.

Overdorff, D. J. & Strait, S. G. (1998). Seed handling by three prosimian primates in southeastern Madagascar: implications for seed dispersal. *Am. J. Primatol.* 45, 69–82.

Prinz, J. F. & Lucas, P. W (2000). Saliva tannin interactions. *J. Oral Rehabil.* 27, 991–4.

Regan, B. C., Julliot, C., Simmen, B. *et al.* (2001). Fruits, foliage and the evolution of primate colour vision. *Phil. Trans. R. Soc. Lond.* B 356, 229–83.

Smith, V. C. & Pokorny, J. (1972). Spectral sensitivity of colorblind observers and the cone pigments. *Vis. Res.* 12, 2059–71.

Stevens, M., Stoddard, M. C. & Higham, J. P. (2009). Studying primate color: towards visual system dependent methods. *Int. J. Primatol.* 30, 893–917.

Vincent, J. F. V. (1992). *Biomaterials – A Practical Approach.* Oxford: IRL Press.

Wilsea, M., Johnson, K. L. & Ashby, M. F. (1975). Indentation of foamed plastics. *Int. J. Mech. Sci.* 17, 457–60.

Wysecki, G. & Stiles, W. S. (2000). *Color Science.* New York: John Wiley.

Yamashita, N. (1996). Seasonality and site specificity of mechanical dietary patterns in two Malagasy lemur families (Lemuridae and Indriidae). *Int. J. Primatol.* 17, 355–87.

Yamashita, N. (2000). Mechanical thresholds as a criterion for food selection in two prosimian species. In *Plant Biomechanics 2000*, ed. H.-C. Spatz & T. Speck, pp. 590–5. Stuttgart: Georg Thieme Verlag.

USEFUL INTERNET SITES

http://chrometrics.com/ – for obtaining colour coordinates from raw photographic images.

http://vischeck.com/ – for a subjective impression of how images appear when processed for dichromacy (pick 'protanope' to see this).

http://www.gwu.edu/~hebdp/fieldtech/ – for information on mechanical tests.

PETER W. LUCAS, RICHARD T. CORLETT,
NATHANIEL J. DOMINY, HAFEJEE C. ESSACKJEE,
PABLO RIBA-HERNANDEZ, LAWRENCE RAMSDEN,
KATHRYN E. STONER AND NAYUTA YAMASHITA

14

Dietary analysis II: food chemistry

INTRODUCTION

The preceding chapter introduced dietary analysis and discussed physical aspects of potential foods as they might influence feeding behaviour. Here, we deal with chemical aspects of potential foods. From the outset though, we should point out that attempts to explain the influence of chemical factors on primate nutrition, and the dietary factors that promote or deter the uptake of nutrients, are limited by our understanding of how the primate gut operates. It is unclear what the optimal dietary requirements are even for humans. Gut research is developing on both theoretical (see, for example, Jumars, 2000) and practical levels (Dominy et al., 2004), but it is important to point out that the effective rate of uptake is not simply a question of enzymatic action. The quantity of plant fibre that a primate ingests has a major influence on the rate of passage of food through the gut and thus digestibility (Lambert, 2002). Variable gut populations of microorganisms and parasites also play a large positive or negative role, particularly in relation to specializations in the stomach or large intestine. The situation is even less clear when it comes to chemical compounds that act as feeding deterrents, toxins or anti-nutritional factors. These have largely been bred or processed out of the agricultural products on which humans feed, so they have received relatively little attention in food science. Currently, these chemicals are often assessed by crude measures, such as total phenolics, which, of necessity, ignore the wide range of variation within each class of chemicals in the nature and intensity of the biological effect.

Field chemistry tests in primatological studies have often been very basic, employing 'presence/absence' tests for compounds such as alkaloids (Waterman, 1984), cyanogenic glycosides (Glander et al., 1989) and phenolics (Simmen & Sabatier, 1996). There is nevertheless

Field and Laboratory Methods in Primatology: A Practical Guide, ed. Joanna M. Setchell and Deborah J. Curtis. Published by Cambridge University Press. © Cambridge University Press 2011.

a lot of scope for such tests: Mowry *et al.* (1996), for example, is note-worthy for the innovative use of Nestle Lactogen milk-mix to screen for tannins. Others have used semi-quantitative tests, like pH papers to assess acidity (Ungar, 1995) and refractometers to measure sugars in flowers or fruits (White & Stiles, 1985). However, portable technology has opened the door to a much wider range of accurate quantitative field tests, e.g. colorimetric assessment of chemical reactions for several extracted food ingredients (briefly described here). All these tests can use fresh, rather than dried, material, allowing you to express results as concentrations (akin to how they might be sensed by a primate) rather than on a dry mass basis (which relates more to nutritional gain). Very importantly, they give rapid results. Though quantified, these new tests still do not provide what is, in the end, necessary: they do not tell you *exactly* what is in the food item. Instead, they generally refer test results to a standard, describing the results as 'equivalents', a term assuming that the extract contains components with a chemical behaviour equivalent to the standard. Should you want to know more, bring the food home for more detailed laboratory analysis. Bear in mind that the acceptance/rejection of potential foods depends on sensory perceptions that are not one-to-one with nutritional quality and it should not be assumed that a detailed understanding of the latter would tell you anything about foraging decisions.

We limit this chapter to the analysis of plant parts because that is the area in which most research has been directed, and divide it into field techniques, which require a minimal set-up, and laboratory techniques, for which we provide only a point of reference. Laboratory studies of the nutritional composition of invertebrates (Barker *et al.*, 1998) and exudates (Smith, 2000) indicate that field-adapted techniques would also be valuable, particularly as insects may provide some important amino acids (Hladik, 1977). Analysis of vertebrate meat consumed by primates is straightforward because the major nutritional components are highly digestible and the variation between species is small in comparison with plant materials. Invertebrates, such as insects, are more problematic because they are less digestible and much more variable in composition, both between species and between life cycle stages (Bell, 1990).

BASIC EQUIPMENT

An adequate dust-free space is a must; electricity, or the means to recharge batteries, is also required. Depending on the tests to be performed, you will need a stovetop or hotplate and refrigeration. A method of reducing the particle size of plant parts prior to testing is essential. A tissue homogenizer (e.g. Tissue Tearor, Dremel, approx. US$600) is ideal, but you could also use a pestle and mortar, although this is slow and difficult to control. You will also need a spectrometer (Chapter 13) and cuvette holder (e.g. Ocean Optics CUV-FL-DA, which attaches easily to light sources and spectrometers, US$461). General requirements for accurate field spectrometry start with good water quality. You can use rainwater, but it requires filtration. Both ceramic and membrane filters are good. Low-cost deionizers are available to produce dH_2O (approx. US$500 for a range of choices), invaluable if you wish to measure ion concentrations in food samples later (e.g. Ca^{2+} concentrations in figs: O'Brien *et al.*, 1998). Bottled water is also a possibility, but it should be distilled and should not contain mineral additives. It is best to use quartz cuvettes, although they are expensive (approx. US$50 each).

We give the chemicals required for individual tests in the relevant sections. Bear in mind that some solvents may be difficult to locate in some countries, in which case you will need to import them. You only need small quantities, and these are difficult to price. Additional equipment includes a good analytical balance (±0.01 g; Chapter 13), adjustable micropipettes (e.g. Gilson, approx. US$200 each), glassware, plasticware, gloves and related accessories such as plastic 'eppendorff' microtubes (1.5 ml) and 2–3 microtube racks. Altogether, budget about US$1000 for consumables.

FOOD PRESERVATION AND EXTRACTION

Most researchers have dried plant food for transport to a laboratory. Drying samples at high temperature (>50 °C) or in direct sunlight is never a good idea because many constituents of plant parts are irreversibly altered by such treatment. If necessary, we recommend the low-temperature drying techniques of Wrangham *et al.* (1998). For small amounts of material, rapid drying with silica gel or molecular sieves in the dark at room temperature is probably best. You can also use silica gel to keep material dry. Field extraction from fresh material is usually preferable to laboratory extraction from dried material and there is evidence for considerable change in some components, even when you take great care (e.g. tannins: Hagerman, 1988; Orians, 1995). We have used 50% methanol : 50% dH_2O (hereafter called 50% methanol) because most nutrients are sufficiently soluble in it to provide a balanced extract, but you should bear in mind that methanol is highly toxic (see below for general precautions).

To extract from fresh material, cut the food sample into pieces of *c*.1 mm in length (incidentally, approximately the degree of comminution that chewing achieves). Weigh about 0.1 g of sample accurately, and transfer it into a homogenizing tube with 5 ml increments. Add 50% methanol to a final volume of 5 ml and lower the tip of the homogenizer into the suspension. Operate at low speed first before shifting to high speed. Administer for 30 s to 2 min using an up–down dabbing motion, and then allow the homogenate to stand. The longer it stands in the solvent, the greater the extraction is likely to be. It is important to standardize the time period for all extractions – we chose 1–2 min for our work. Withdraw the supernatant into a 5 ml syringe. Fit the syringe with a Luerlock and a plastic filter holder containing a glass fibre filter (1.6 μm pore size, Type 1, Millipore, USA). Engage the syringe plunger, press gently to direct the extract through the filter and into a 1.5 ml eppendorff microcentrifuge tube, filling this up to about 10% from the top. Label the tube. In the case of liquid samples such as nectar, you can extract the sample from flowers by using a pre-weighed syringe, and pour it into an eppendorff tube with 5 ml of 50% methanol. Refrigerate the extracts produced if at all possible. All chemical tests depend on these extracts, so at minimum you can store them and conduct the remainder of the tests back in a laboratory. If you dry plant tissues then you will need a different extraction procedure and spectrophotometric measurements will be less accurate.

General safety tips: Wear gloves when handling chemicals. Store chemicals away from direct sunlight and label reagent bottles with both the reagent name and date. In field sites, it is very likely there will be people around who will not understand what you are doing or know the danger posed by some of these chemicals, so be very vigilant and lock up chemicals when not in use. Do not dispose of chemicals carelessly. If you store chemical waste, be careful not to produce an inflammable mass. Keep your equipment (e.g. spatulas) clean and dry. Remember some important contamination precautions – never pipette directly from your stock reagent. Dispense what is needed into a separate aliquot. The bottom line is: be clean, be thorough and be safe.

FIELD TESTS

We present three examples of field tests, for protein, phenolics and tannins, in some detail to show exactly what can be achieved in the field and suggest other possibilities below. Only the protein test can be conducted without a fridge, as some of the stock and buffer solutions employed in the phenolics and tannin tests require low-temperature storage.

Protein

Proteins react with a dye called Coomassie Brilliant Blue, producing a strong blue colour, the absorbance of which is measured spectro-photometrically at 595 nm. The test is very simple and a typical stand-ard curve uses Bovine Serum Albumin (BSA), a protein chosen simply because it gives a good linear relation between absorbance and BSA concentration. Results are reported as % BSA equivalents. This method follows Read and Northcote (1981). To prepare a 100 ml stock solution of Coomassie Blue, weigh out 0.33 g of Coomassie blue G (Sigma). Dissolve the dye in a 2:1 volume by volume of phosphoric acid (85%) and ethanol (100%). Filter through a Whatman's No. 1 paper and store at environmental temperature. Protect the stock solution from light, e.g. by wrapping the reagent bottle in aluminium foil. You may be able to take ready-made stock solution with you to the field. To prepare the diluted working protein reagent (100 ml), dispense 3 ml of the concentrated stock solution into a 100 ml measuring cylinder. Next, add 8 ml of phosphoric acid (85%) and 3.75 ml of ethanol (100%). Make up the final volume to 100 ml with dH$_2$O. Filter twice through a Whatman's No. 1 paper and store at room temperature. The resulting solution is reddish-brown in colour. You will need just 1 ml of working reagent per test, so prepare only the amount that you will need for a day or two of work.

To perform the tests, you need eppendorf tube racks, a 200 μl and a 1 ml adjustable micropipette with disposable tips, 50% methanol, working protein reagent, eppendorf tubes containing extracts, and an equal number of empty eppendorff tubes plus one extra for the blank and spectrometry equipment (Lucas et al., 2001). First, align the extracts that you plan to analyse along one row of an eppendorff rack. Place an empty eppendorf opposite each extract, reserving one for a blank (a solution containing everything except the extract, which is replaced by 50% methanol). Using a new, sterile pipette tip for each extract, dispense 50 μl of each extract into the respective eppendorf tube (the blank receives only 50 μl of 50% methanol). Add 1 ml of the working staining reagent into each. Cap and mix the tubes by gently inverting them. Allow the tubes to stand for 5 min at room temperature. A blue colour will develop almost immediately. To perform the spectroscopic measurement you first need to measure the electronic activity of the detectors in dark conditions. This is called the 'dark current' and should be deducted from every further measurement taken. You then need to measure the incident light with the light source switched on and everything in place except a solution of sample or blank. Then measure the absorbance of the blank by taking 1 ml from the microtube, filling the cuvette and placing it in the transmittance stage. For this test, the absorbance for the blank can be as high as 0.5. Empty the cuvette completely by using the pipette. It is possible to evacuate every noticeable drop by doing this, which avoids the need to wash the cuvette after each absorbance reading. Invert the cuvette over absorbent paper and shake and wipe the pipette clean. Measure each sample identically to the procedure for the blank. Dilute any sample absorbance above 1.0 to get an accurate reading. To ensure the quality of the results, we suggest periodically re-testing the blank.

Phenolics

These compounds are the parent group of tannins, distinguished by the presence of six-membered carbon rings with hydroxyl (OH) groups, and presumed to have a variety of defence functions in plants (Waterman & Mole, 1994). In the following test, the phenolics present in a plant sample oxidize potassium ferricyanide to produce ferrous ions. These ions then react with ferric chloride in HCl to produce a Prussian Blue complex, the absorbance of which is measured spectrophotometrically at 700 nm. The stabilizer is an attempt to prevent precipitation of Prussian Blue by increasing the viscosity of the solution. It is only partially successful and you should adopt strict time protocols. The standard is gallic acid, a simple phenolic acid with a low molecular mass. Too small a molecule to precipitate protein, gallic acid is an example of a phenolic compound that is not a tannin. We do not advise running more than 50 samples at a time, as it requires considerable practice to build up to that level. The amount of phenolics is determined at 700 nm. The test follows Hagerman (see Useful Internet sites), converted to a microassay (Lucas et al., 2001). Note that this measures the reducing power of phenolics, not the overall concentration (Appel et al., 2001). First, you need to make 100 ml 0.1 M HCl. This chemical is very corrosive and dangerous. [Never add water to concentrated acid. Always add acid to water.] Measure 100 ml of dH_2O

in the large measuring cylinder and transfer into a 500 ml reagent bottle. Next, carefully add 830 µl of concentrated HCl. Cap the bottle and store at room temperature. Secondly, make 0.02 M $FeCl_3$, in 0.1 M HCl (100 ml). Weigh out 0.32 g of $FeCl_3$ and transfer to a large reagent bottle containing 100 ml of 0.1 M HCl. Label the bottle, e.g., as 'CL', date it and store at room temperature. Thirdly, make 0.0016 M $K_3Fe(CN)_6$ (100 ml). Measure 100 ml of dH_2O in the large measuring cylinder and transfer to a 500 ml reagent bottle. Weigh out 0.53 g of $K_3Fe(CN)_6$ and transfer to the 100 ml of dH_2O in the large reagent bottle. Label the bottle, e.g., as 'CN', date it, and store at room temperature. To make the stabilizer (1% gum arabic solution), measure 80 ml of dH_2O in the large measuring cylinder and add it to the conical flask. Weigh out 1.00 g of gum arabic and transfer it to the flask. Heat the solution to boiling so as to dissolve the gum arabic completely. Filter the solution through a Whatman's No. 1 filter paper into another flask. This will take time, since the gum solution is very viscous. Once complete, make up the final volume of filtrate to 100 ml with dH_2O. Store in a small reagent bottle at 4 °C. It should be usable for about a week.

Set up the extracts again along the back row of an eppendorff rack, leaving the space furthest to the left free. Load the entire front row with empty eppendorf tubes. The first tube will be for the blank. Add 300 µl of deionized water to each tube. Add 10 µl of 50% methanol to the 'blank' eppendorf. This is a very small volume to pipette: make sure the pipette tip touches the surface of the water in the microtube while the pipette plunger is fully depressed. Retract the pipette before releasing the plunger. Remove the pipette tip and reserve it for methanol only. Add 10 µl of the sample extract from the tube in the back row to the empty tube placed in front of it using a new tip for each extract. Add 100 µl of CN to each of the front row mixtures, working evenly from left to right. Set a stopwatch to 15 min, put a new tip on the 200 µl pipette and start the watch. Add 100 µl of CL to each of the front row mixtures, working at an even pace from left to right. When the stopwatch goes off, add 1 ml of stabilizer to each tube on the front row of the rack, again working at the same even pace from left to right. The solutions should now be relatively stable for long enough to read the absorbance. If the Prussian Blue complex precipitates in spite of the stabilizer, or the absorbance is >1.3, we advise repeating the assay with a diluted sample. A two-fold dilution with 50% methanol is normally sufficient. Remember to take the dilution factor into account for the determination of the final concentration.

Tannins

Tannins precipitate proteins. A simple method based on this mechanism, devised by Hagerman (1987), works well for field estimation. Plant extracts, loaded into wells of protein-containing agarose gel, combine with the protein to form a whitish precipitate. The following chemicals are required for the preparation of the buffer solution reagents: dH_2O, 2 M sodium hydroxide, glacial acetic acid and ascorbic acid, while agarose and bovine serum albumin (BSA) form the gel. Note that BSA is very prone to fungal contamination in the absence of refrigeration. In addition to items listed under 'Protein' and 'Phenolics', you will need a dry pH meter (e.g. Sentron NV, Netherlands), a 500 ml plastic measuring cylinder, a thermometer, a 25 ml wide-mouthed pipette with a pipette filler and eight or more plastic Petri dishes, the lids of which are fitted with 5 mm diameter acrylic pins (Fig. 3 in Lucas *et al.*, 2001). These pins will make wells during gel setting and are used for the loading of plant extracts. You will also need standard lids without pins. On the undersurface of the Petri dishes, just beside the future locations of each well, label the wells anti-clockwise 1–6 with a water-proof marker pen. To make the buffer, weigh 5.3 g of ascorbic acid and add this to a 500 ml glass beaker. Dispense 400 ml dH_2O to the beaker and swirl to dissolve. Add 2.43 ml of glacial acetic acid and mix well with a spatula. Make a 2 M solution of NaOH by dissolving 8 g of NaOH in 100 ml of dH_2O and use this to bring the pH to 5.0. Transfer the pH 5.0 solution into the cylinder and make up the volume to 500 ml with dH_2O. This buffer can be stored for one week at 4 °C. To make an agarose solution sufficient for 8 Petri dishes, dispense 200 ml of the buffer into a 250 ml conical flask. Weigh out 2.0 g of agarose and 0.02 g of BSA. Add the agarose to the buffer and boil with constant stirring until completely dissolved. Monitor the cooling of the agarose solution to 40–45 °C with a thermometer (this temperature range is critical: agarose will gel/solidify at temperatures lower than 40 °C, while denaturation of BSA occurs at temperatures higher than 45 °C). Add the BSA and swirl gently to dissolve it completely. Now working rapidly yet smoothly, withdraw 25 ml of solution using a pipette equipped with filler (Fig. 3 in Lucas *et al.*, 2001) and dispense this into each dish. It is very important to avoid air bubbles and you need to practise. The trick is to empty the pipette gently on the side of the Petri dish. Cover each dish with the modified lids. Allow the agarose to gel at room temperature. When cool, remove the modified lids cautiously to expose the wells while avoiding cracks. Cover with a standard Petri dish lid. Store upside down to avoid condensation in the wells. Be certain that the wells are free of condensation before loading the samples. You can carefully aspirate the wells with a Pasteur pipette, taking sufficient care not to puncture them, or, alternatively, store them upside down for a short period to allow moisture to drain and dry out. Within two hours, you can start loading 40 µl of your chosen sample extract into each well. Move clockwise when doing this, keeping a record of which extract goes in which numbered well. Replace the standard lid, seal the dishes with clingfilm or Parafilm M® (to preserve moisture and to limit microbial contamination) and incubate for 96 h at room temperature (although the precipitation ring tends to develop within 36 h). We express tannin concentrations as equivalents to six-point standard curves based on crude quebracho tannin (gift of Dr A. E. Hagerman). Although the ring edge is sometimes difficult to discern, we use callipers to measure ring diameters. You could use a digital camera mounted on a tripod and a ruler for scale for computer analysis.

LABORATORY CHEMISTRY

The advantages of laboratory over field analyses are the much wider range of chemical components that can be measured with potentially greater accuracy. Fast and accurate instrumental techniques for analysis have replaced much traditional wet chemistry over the last two decades. It is also possible to use commercial analysis services, for example those that assess animal feed. These will do some routine analyses faster, cheaper and more accurately than an inexperienced researcher and/or an under-equipped laboratory. A huge range of non-routine analyses is also available commercially to the researcher with sufficient funds.

Fibre

In food science, 'fibre' includes all the digestion-resistant components of plant materials. The most useful measure of fibre content is Neutral Detergent Fibre (NDF) – that which remains after the sample is boiled for an hour in neutral detergent solution. NDF includes cellulose and lignin, plus most hemicelluloses and is quick, cheap and easy to determine with the appropriate equipment. The standard procedure leaves a variable amount of digestible starch that can be removed by pretreatment with amylase. Acid Detergent Fibre (ADF) is basically NDF without hemicelluloses. Specialized laboratories can run the sequence

NDF–ADF–lignin, allowing a segregation of all components of the cell wall (Wrangham *et al.*, 1998).

Protein and amino acids

Crude protein has traditionally been estimated from the nitrogen content of samples measured by the Kjeldahl method. This is still widely used, but requires special glassware and produces corrosive fumes and toxic waste products. Micro-Kjeldahl equipment requires less than 0.25 g of sample material and is semi-automated. The now commonly used Dumas combustion method gives a nitrogen reading 25% higher than the Kjeldahl method (Simonne *et al.*, 1998).

Total nitrogen has traditionally been converted into protein by multiplying by a factor of 6.25, derived from studies with animal proteins. Numerous studies have shown that this overestimates protein in plant materials because a variable amount of nitrogen is in non-protein forms (Conklin-Brittain *et al.*, 1999), some of which are nutritionally valuable (e.g. free amino acids) whereas others may have a neutral or negative impact on nutrition (e.g. nitrogen-based plant secondary chemicals). A similar problem occurs with the analysis of insects, since the chitin component of the exoskeleton contains 6.9% nitrogen (Bell, 1990). Simply using a lower conversion factor may be adequate for a rough estimate of the protein content in the diet (Conklin-Brittain *et al.*, 1999). Amino acid analysers can measure their individual concentrations in proteins, which is important in assessing protein quality (plant proteins often have a seriously unbalanced amino acid composition).

Non-structural carbohydrates

Wet chemistry methods are available for all aspects of carbohydrate analysis (Chaplin & Kennedy, 1994) but it takes considerable skill to apply these accurately to complex, unknown mixtures of carbohydrates and other organic materials. High-Pressure Liquid Chromatography (HPLC) has largely replaced them for quantitative analysis (Simmen & Sabatier, 1996; Reynolds *et al.*, 1998). Different combinations of columns, solvents, detectors and standards are needed for different carbohydrates, so an analysis targeted at glucose, fructose and sucrose, for instance, may overlook some less common sugars in the sample. However, it is glucose, fructose and sucrose that plants generally produce in bulk and which seem to play an exceedingly important role in primate food choice, particularly for fruits (Riba-Hernández *et al.*, 2005). Sucrose is the transport sugar, the splitting of which produces equal quantities of fructose and glucose. However, sucrose is not always present in fruits, particularly those targeted at birds, which appear to have a general preference for hexose sugars (Lotz & Schondube, 2006). Primates can often detect sucrose in the mouth at lower concentrations than the other sugars and this may relate to fruit consumption (see, for example, Laska *et al.*, 1999). The importance of this, though, depends on how sugars are perceived. Although humans appear to possess only one sweetness signal (Breslin *et al.*, 1994), there is now strong evidence that non-structural carbohydrates are also sensed orally by a second, as yet unknown set of receptors (Chambers *et al.*, 2009).

Sugars increase during fruit development. In leaves, they may increase with sun exposure (Ganzhorn, 1995) and vary with time of day (Ganzhorn & Wright, 1994). Indeed, it has been claimed that chimpanzees (*Pan troglodytes*) may prefer young leaves to mature leaves due to significantly higher levels of glucose (Reynolds *et al.*, 1998). Even ants can be relatively sugar-rich (Hladik, 1977).

With the exception of many fruits, plant parts usually contain more starch than sugars. Raw starch is often assumed to be as tasteless and unattractive to other primates as it is to humans. However, some primates respond to long-chain saccharides (Laska *et al.*, 2001) and catarrhines have a salivary amylase that may be able to break raw starch down rapidly to release tasty sugars (Perry *et al.*, 2007), presumably providing that the starch is not present in an insoluble granular form.

Lipids

These are defined by their solubility in organic solvents and are usually extracted from plant materials in a Soxhlet apparatus (Gunstone, 1996). Crude fat (or ether extract) is defined as the lipid extracted with anhydrous ether (which is flammable and explosive). Other solvents, such as petroleum ether or hexane, are less dangerous, but give slightly different results. Crude fat is a complex mixture of substances including plant cuticles, which are not digested (even by ruminants) and you could potentially analyse these in the faeces to indicate plant composition in the diet. Methods for measuring fatty acids are given by Chamberlain *et al.* (1993) and Simmen and Sabatier (1996).

Detailed lipid analysis is probably only worth while if fresh material is available because plant lipids are very sensitive to oxidation under normal storage and preparation conditions. You can store seeds intact as long as they remain alive. Thin-layer chromatography is used to separate the total lipids into the major classes of compounds. Fatty acid composition is then determined by gas chromatography.

Secondary compounds

Laboratory determinations of terpenoids have been reported by Wrangham *et al.* (1998). For accurate identification, these must be extracted and quantified individually, usually by gas–liquid or gas chromatography (GLC; GC). Saponins are another widespread and potentially interesting group of compounds known for their toxicity. You can identify and quantify individual saponins by HPLC coupled with mass spectrometry (Marston *et al.*, 2000). Waterman and Mole (1994) describe a simple spectrophotometer-based haemolysis technique that is only suitable if the extract is tannin-free. Alkaloids are so heterogeneous as to lack even a satisfactory definition. They are less widespread in plant material than the other two major classes of secondary chemicals, but can be crucial in feeding decisions when present. Although there are more or less reliable general tests to screen for the presence of alkaloids (Harborne, 1998), they must be quantified individually, using GLC or HPLC. In the recent human food literature, interest in secondary compounds has focussed on their antioxidant capacity, because of the possible role of antioxidants in human health. A variety of techniques are currently used to characterize 'total antioxidant capacity', based on different chemical mechanisms (Pérez-Jiménez *et al.*, 2008).

OTHER POSSIBILITIES

The major limit on possible field tests is the time needed to develop them, but those involving extraction are always going to be limited. Lipid analysis, for example, would be almost impossible in the field. Those compounds can certainly be stained, however, in freehand thin sections. Commercial microscopes are available that connect to a computer (e.g. Veho VMS-001) and permit quantification in two dimensions using image analysis software. When extraction can be avoided, possibilities are greater. A quantitative test for alkaloids and cyanide, for example, seems possible, the latter by adapting a small commercial kit (cyanide test kit – Cole-Parmer). Others have described methods for measuring mineral content (Silver *et al.*, 2000; Remis *et al.*, 2001). Ion concentrations can be measured with ion-specific electrodes that can be attached to pH meters or by thin films based on chitin derivatives (Schauer et al., 2004). The antioxidant potential of vitamin C (Milton & Jenness, 1987) and other compounds has also received attention (Leitão et al., 1999). Terpenoids as a whole raise many problems, particularly the low-molecular-mass volatile mono-terpenes, which seem to be released through the plant cuticle and could be sensed by smell. The quantification of scent chemistry is now practicable, using headspace techniques for field collection (Pettersson & Knudsen, 2001). A further development, portable mass spectrometry, will assist in analysis in the field.

In the laboratory, you can use near-infrared reflectance spectrom-etry on cut food surfaces to estimate broad suites of food components (Felton et al., 2009). This technique could actually be implemented in the field, since some spectrometers can operate in adverse environ-ments (e.g. Bruker MATRIX-F FT-NIR series). A prerequisite, though, would be the establishment of statistical relationships between spectra and a set of laboratory analyses of components of interest to produce calibration equations (as described in ASTM Standard E1655-00, 1995). These then allow the 'decomposition' of any given spectrum into its probable constituents. On the horizon, however, lies potentially the most powerful technique of all, electrospray ionization (Møller et al., 2007), which could potentially analyse exactly what is in a food.

REFERENCES

Appel, H. M., Governor, H. L., D'Ascenzo, M., Siska, E. & Schultz, J. C. (2001). Limitations of Folin assays of foliar phenolics in ecological studies. *J. Chem. Ecol.* 27, 761–78.

ASTM Standard E1655-00 (1995). *Standard Practices for Infrared Multivariate Quantitative Analysis*: West Conshohocken, PA. American Society for Testing and Materials.

Barker, D. Fitzpatrick, M. P. & Dierenfeld, E. S. (1998). Nutrient composition of selected whole invertebrates. *Zoo Biol.* 17, 123–34.

Bell, G. P. (1990). Birds and mammals on an insect diet: a primer on diet compo-sition analysis in relation to ecological energetics. *Stud. Avian Biol.* 13, 416–22.

Breslin, P. A. S., Kemp, S. & Beauchamp, G. K. (1994). Single sweetness signal. *Nature* 369, 447–8.

Chamberlain, J., Nelson, G. & Milton, K. (1993). Fatty acid profiles of major food sources of howler monkeys (*Alouatta palliata*) in the neotropics. *Experientia* 49, 820–4.

Chambers, E. S., Bridge, M. W. & Jones, D. A. (2009). Carbohydrate sensing in the human mouth: effects on exercise performance and brain activity. *J. Physiol.* 587, 1779–94.

Chaplin, M. F. & Kennedy, J. F. (1994). *Carbohydrate Analysis: A Practical Approach*. Oxford: IRL Press.

Conklin-Brittain, N. L., Dierenfeld, E. S., Wrangham, R. W., Norconk, M. & Silver, S. C. (1999). Chemical protein analyses: a comparison of Kjeldahl crude protein and total ninhydrin protein from wild, tropical vegetation. *J. Chem. Ecol.* 25, 2601-22.

Dominy, N. J., Davoust, E., & Minekus, M. (2004). Adaptive function of soil consumption: an *in vitro* study modeling the human stomach and small intestine. *J. Exp. Biol.* 207, 319-24.

Felton, A. M., Felton, A., Raubenheimer, D. *et al.* (2009). Protein content of diets dictates the daily energy intake of a free-ranging primate. *Behav. Ecol.* 20, 685-90.

Ganzhorn, J. U. (1995). Low-level forest disturbance effects on primary production, leaf chemistry, and lemur populations. *Ecology* 76, 2084-96.

Ganzhorn, J. U. & Wright, P. C. (1994). Temporal patterns in primate leaf eating: the possible role of leaf chemistry. *Folia Primatol.* 63, 203-8.

Glander, K. E., Wright, P. C., Seigler, D. S., Randrianasolo, V. & Randrianasolo, B. (1989). Consumption of a cyanogenic bamboo by a newly discovered species of bamboo lemur. *Am. J. Primatol.* 19, 119-24.

Gunstone, F. D. (1996). *Fatty Acid and Lipid Chemistry.* London: Blackie Academic & Professional.

Hagerman, A. E. (1987). Radial diffusion method for determining tannin in plant extracts. *J. Chem. Ecol.* 13, 437-49.

Hagerman, A. E. (1988). Extraction of tannin from fresh and preserved leaves. *J. Chem. Ecol.* 14, 453-61.

Harborne, J. B. (1988). *Phytochemical Methods*, 3rd edn. London: Chapman & Hall.

Hladik, C. M. (1977). Chimpanzees of Gabon and chimpanzees of Gombe: some comparative data on the diet. In *Primate Ecology: Studies of Feeding and Ranging Behaviour in Lemurs, Monkeys and Apes*, ed. T. H. Clutton-Brock, pp. 481-501. London: Academic Press.

Jumars, P. A. (2000). Animal guts as ideal chemical reactors: maximizing absorption rates. *Am. Nat.* 155, 527-43.

Lambert, J. E. (2002). Digestive retention times in forest guenons with reference to chimpanzees. *Int. J. Primatol.* 26, 1169-85.

Laska, M., Schüll, E. & Scheuber, H.-P. (1999). Taste preference thresholds for food-associated sugars in baboons (*Papio hamadryas anubis*). *Int. J. Primatol.* 20, 25-34.

Laska, M., Kohlmann, S., Scheuber, H., Salazar, L. T. H. & Luna, E. R. (2001). Gustatory responsiveness to polycose in four species of nonhuman primates. *J. Chem. Ecol.* 27, 1997-2011.

Leitão, G. G., Mensor, L. L., Amaral, L. F. G. *et al.* (1999). Phenolic content and antioxidant activity: A study on plants eaten by a group of howler monkeys (*Alouatta fusca*). In *Plant Polyphenols 2: Chemistry, Biology, Pharmacology, Ecology*, ed. G. G. Gross, R. W. Hemingway & T. Yoshida, pp. 883-95. New York: Kluwer Academic.

Lotz, C. N. & Schondube, J. E. (2006). Sugar preferences in nectar- and fruit-eating birds: behavioral patterns and physiological causes. *Biotropica* 38, 3-15.

Lucas, P. W., Beta, T., Darvell, B. W. *et al.* (2001). Field kit to characterize physical, chemical and spatial aspects of potential foods of primates. *Folia Primatol.* 72, 11-5.

Marston, A., Wolfender, J. L. & Hostettmann, K. (2000). Analysis and isolation of saponins from plant material. In *Saponins in Food, Feedstuffs and Medicinal Plants*, ed. W. Oleszek & A. Marston, pp. 1-12. Dordrecht: Kluwer Academic.

Milton, K. & Jenness, R. (1987). Ascorbate content of neotropical plant parts available to monkeys and bats. *Experientia* 43, 339-42.

Møller, J. K. S., Catharino, R. R. & Eberlin, M. N. (2007). Electrospray ionization mass spectrometry fingerprinting of essential oils: spices from the Labiatae family. *Food Chem.* 100, 1283–8.

Mowry, C. B., Decker, B. S. & Shure, D. J. (1996). The role of phytochemistry in dietary choices of Tana River red colobus monkeys. *Int. J. Primatol.* 17, 63–84.

O'Brien, T. G., Kinnaird, M. F., Dierenfeld, E. S. *et al.* (1998). What's so special about figs: a pantropical mineral analysis. *Nature* 392, 668.

Orians, C. M. (1995). Preserving leaves for tannin and phenolic glycoside analyses: a comparison of methods using three willow taxa. *J. Chem. Ecol.* 21, 1235–43.

Pérez-Jiménez, J., Arranz, S., Tabernero, M. *et al.* (2008). Updated methodology to determine antioxidant capacity in plant foods, oils and beverages: extraction, measurement and expression of results. *Food Res. Internat.* 41, 274–85.

Perry, G. H., Dominy, N. J., Claw, K. G. *et al.* (2007). Diet and the evolution of human amylase gene copy number variation. *Nat. Genet.* 39, 1256–60.

Pettersson, S. & Knudsen, J. T. (2001). Floral scent and nectar production in *Parkia biglobosa* Jacq. (Leguminosae: Mimosoideae). *Bot. J. Linn. Soc.* 135, 97–106.

Read, S. M. & Northcote, D. H. (1981). Minimization of variation in the response to different proteins of the Coomassie Blue G dye-binding assay for protein. *Anal. Biochem.* 116, 53–64.

Remis, M. J., Dierenfeld, E. S. Mowry, C. B. & Carroll, R. W. (2001). Nutritional aspects of Western lowland gorilla (*Gorilla gorilla gorilla*) diet during seasons of fruit scarcity at Bai Hokou, Central African Republic. *Int. J. Primatol.* 22, 807–86.

Reynolds, V., Plumptre, A. J., Greenham, J. & Harborne, J. (1998). Condensed tannins and sugars in the diet of chimpanzees (*Pan troglodytes schweinfurthii*) in the Budongo Forest, Uganda. *Oecologia* 115, 331–6.

Riba-Hernández, P., Stoner, K. E. & Lucas, P. W. (2005) Sugar concentration of fruits and their detection via color in the Central American spider monkey (*Ateles geoffroyi*). *Am. J. Primatol.* 67, 411–23.

Schauer, C. L., Chen, M. U., Price, R. R., Schoen, P. E. & Ligler, F. S. (2004). Colored thin films for specific metal ion detection. *Environ. Sci. Technol.* 38, 4409–13.

Silver, S. C., Ostro, L. E. T., Yeager, C. P. & Dierenfeld, E. S. (2000). Phytochemical and mineral components of foods consumed by black howler monkeys (*Alouatta pigra*) at two sites in Belize. *Zoo Biol.* 19, 95–109.

Simmen, B. & Sabatier, D. (1996). Diets of some French Guianan primates: food composition and food choices. *Int. J. Primatol.* 17, 661–93.

Simonne, E. H., Harris, C. E. & Mills, H. A. (1998). Does the nitrate fraction account for differences between Dumas-N and Kjeldahl-N values in vegetable leaves? *J. Plant Nutr.* 21, 2527–34.

Smith, A. C. (2000). Composition and proposed nutritional importance of exudates eaten by saddleback (*Saguinus fuscicollis*) and mustached (*Saguinus mystax*) tamarins. *Int. J. Primatol.* 21, 69–83.

Ungar, P. S. (1995). Fruit preferences of four sympatric primate species at Ketambe, northern Sumatra, Indonesia. *Int. J. Primatol.* 16, 221–45.

Waterman, P. G. (1984). Food acquisition and processing as a function of plant chemistry. In *Food Acquisition and Processing in Primates*, ed. D. J. Chivers, B. A. Wood & A. Bilsborough, pp. 177–211. New York: Kluwer Academic.

Waterman, P. G. & Mole, S. (1994). *Analysis of Phenolic Plant Metabolites*. Oxford: Blackwell.

White, D. W. & Stiles, E. W. (1985). The use of refractometry to estimate nutrient rewards in vertebrate-dispersed fruits. *Ecology* 66, 303–7.

Wrangham, R. W., Conklin-Brittain, N. L. & Hunt, K. D. (1998). Dietary response of chimpanzees and cercopithecines to seasonal variation in fruit abundance. I. Antifeedants. *Int. J. Primatol.* 19, 949–69.

USEFUL INTERNET SITES

http://rsb.info.nih.gov/nih-image/ – for free downloads of NIH Image software.
http://www.gwu.edu/~hebdp/fieldtech/ – for information on some chemical tests.
www.users.muohio.edu/hagermae/tannin.pdf – contains the 'Tannin Handbook'.

CLAIRE M. P. OZANNE, JAMES R. BELL
AND DANIEL G. WEAVER

15

Collecting arthropods and arthropod remains for primate studies

INTRODUCTION

Arthropods, along with other invertebrates, make up 95% of the global fauna, with 1.5 million described species and at least a further 10 million that remain undescribed. These organisms, which include insects, arachnids and myriapods, play a significant role in the life histories of a range of primate species, not least in their diet. Body size plays an important part in determining the degree of insectivory (Kay, 1984); however, examples of insect eating can be found within all lineages of the primate phylogeny. Many strepsirrhines have long been described as insectivores, some having morphological adaptations for insect feeding (e.g. the extractive finger of the aye aye, *Daubentonia madagascariensis*) and others adopting behaviours such as aerial snatching (e.g. galagos, *Galago demidovii*), or stalking specific arthropod prey (e.g. angwantibos, *Arctocebus calabarensis*) (Charles-Dominique, 1974). Some lorisids, like the potto (*Perodicticus potto*), have evolved specialized diets, feeding on arthropods that emit distasteful chemicals (e.g. formic acid from ants: *Crematogaster* spp.) or have the potential to inject venom (e.g. large scolopendrid centipedes: *Spirostreptus* spp.) (Charles-Dominique, 1974). Within the anthropoids, chimpanzees (*Pan troglodytes*) use tools to help eat insects such as bees (*Apis* spp.) and the honey they produce (Tutin & Fernandez, 1992). Even folivores inadvertently eat thousands of insects a day (e.g. mountain gorillas, *Gorilla beringei beringei*, and langurs, *Semnopithecus entellus*). However, apart from the large patas monkey (*Erythrocebus patas*), anthropoid primates are unable to process enough arthropod material for these animals to form a major part of a higher primate's diet (Isbell, 1998).

The time invested in catching some agile insects can be high compared with picking fruit or foliage off a shrub, for example. However, arthropods are a good source of proteins and lipids, providing an energy

Field and Laboratory Methods in Primatology: A Practical Guide, ed. Joanna M. Setchell and Deborah
J. Curtis. Published by Cambridge University Press. © Cambridge University Press 2011.

level equivalent to that of vertebrate tissue (1.93 Kcal/g – see Kay, 1984), and also have other non-dietary uses. For example, the millipede *Orthoporus dorsovittatus* secretes two insect-repellent chemicals known as benzoquinones and is used by wedge-capped capuchin monkeys (*Cebus olivaceus*) to anoint themselves, keeping mosquitoes and botflies away during the rainy season (Valderrama *et al.*, 2000). Not all interactions are beneficial; and a significant number of arthropods can cause disease either by means of vectoring microorganisms or by actual damage (e.g. bites, stings, and myiasis).

In this chapter we confine ourselves to methods associated with collecting terrestrial arthropods (i.e. invertebrates with jointed legs such as grasshoppers and spiders), as there is only limited evidence in the literature of primates eating other non-arthropod invertebrates (although gastropods (slugs and snails) are an occasional exception). There are also a small number of crab-eating primates (e.g. long-tailed macaques, *Macaca fascicularis*: Payne & Francis, 1985; mandrills, *Mandrillus sphinx*: Hoshino, 1985). Crabs are firmly included within the Arthropoda; whereas some crabs are aquatic, others may be almost entirely land-bound. If you wish to collect crabs, you can refine some methods described in this chapter, for example by adding lures and redesigning traps to make them escape-proof.

Arthropod taxonomy and identification

Arthropods are diverse, numerous and in the tropics often poorly known both taxonomically and ecologically. They are morphologically complex (for example, the difference between two species may be the position of a single hair) but can usefully be identified to the ordinal level (e.g. Coleoptera – beetles) or placed in a morpho-taxon group (e.g. beetle-like; see Oliver & Beattie, 1996). If you are new to arthropod studies you can familiarize yourself with general body forms by using colour identification guides (e.g. Chinery, 1993), moving on to more descriptive taxonomic works later when you need more detailed anatomical analyses (e.g. McGavin, 2001). Technical nomenclature is widely used in arthropod taxonomy and an entomological dictionary, such as that by Gordh and Headrick (2003), is useful, although many entomology textbooks contain a glossary sufficient for basic identification work. If you are uncertain of the phylogenetic relationships of the Arthropoda, then visit the Tree of Life website, entering at the level of Metazoa – i.e. multicellular organisms that develop from embryos (see 'Useful Internet sites'). In basic terms, arthropods lack backbones and are thus split from the Chordata, which includes primates, on the basis of embry-ogeny and initial development. They are then split from other animals without backbones (e.g. worms), because of their jointed legs.

You can now use Polymerase Chain Reaction (PCR) and other molecular approaches to identify most organisms (Chapter 21). PCR is a useful tool that amplifies target DNA, offering an alternative to detailed taxonomic description. This can speed up the identification of difficult taxonomic groups, and make it more reliable, but PCR is not without its own problems and caveats. One major sticking point, par-ticularly in tropical systems, is that you need a 'primer', specific to the group of interest, to amplify DNA. If a primer has been developed, then its sequence will be available online through the database Genbank (see 'Useful Internet sites'). There are large gaps in our knowledge, however, and you will need to design a primer for the vast majority of species. This process can be lengthy, and you will need to demon-strate that the primer does not cross-react with other species in the community and that the final product is specific to the taxonomic level for which it has been designed. King *et al.* (2008) gives an excellent account of the best choice of DNA extraction, primer selection and general processes you need to consider.

The types of arthropod selected by primates include the larger-bodied animals such as Orthoptera (grasshoppers, crickets and katy-dids), Dictyoptera (cockroaches, praying mantises and termites), Phasmida (stick insects), Odonata (dragonflies), Lepidoptera (moths and butterflies), Coleoptera (beetles), Hymenoptera (bees, ants and wasps), Hemiptera (cicadas, shield bugs), Arachnida (spiders and scor-pions) and Myriapoda (centipedes and millipedes). For an example of a detailed list of insectivory, refer to Niemitz (1984), who gives an insight into the catholic diet of the genus *Tarsius*.

GETTING STARTED: EQUIPMENT LIST

Two books that have been specifically written for expedition work, being compact, ring-bound and full of information on arthropods, are McGavin (1997) and HMSO (1996). Tuck both into your rucksack before you leave.

Essentials

Collecting net; paint brushes (for delicate specimens); paper and pen-cil; forceps; plastic storage jars and vials; binoculars; plastic bags for foliage samples; hand lens (10 ×–30 ×); white sorting tray (to separate organisms from debris); pooter or aspirator (to suck up arthropods without swallowing them; ideally, take both a manual and a battery-powered aspirator); alcohol (methanol/ethanol for long-term storage, or isopropyl alcohol (2-propanol or 'rubbing alcohol') for field use, but change to a long-term storage solution on your return. All concentrations must be greater than 70% in an aqueous solution to prevent rotting).

Ideally

Binocular microscope (70 ×–100 ×); calico bags; branch clippers; Winkler bags (see 'Vegetation and ground sampling', below); Petri dishes; cold packs; good-quality Global Positioning System (GPS, Chapter 4).

With lots of thought, preparation and good funding

Knockdown apparatus (e.g. misters: Hurricane-Major™ and Stihl™; foggers: Swing-Fog™ and Dyna-Fog™); chemical (natural pyrethrum or synthetic pyrethroids); collecting trays; rope/twine; petrol and 2-stroke oil; refrigerator.

KILLING, SETTING AND STORAGE OF SPECIMENS

If you don't collect arthropods into alcohol you can kill them rapidly by using a killing jar: a polyethylene or glass container (500 ml) labelled 'POISON', with a few drops of ethyl acetate (the chemical in nail varnish remover) as the killing agent. In the long term, most arthropod specimens are best stored in alcohol (ideally 70%–100% solution) in tubes or vials with tight lids. Dry primate faecal samples and store them cool in any air tight container, with silica gel if needed. It is essential to label samples and specimens, either with a paper and pencil or printed on acid-free paper (Fig. 15.1).

You need to pin some insects, such as butterflies and moths, but a short-term solution is to place them flat in small paper envelopes. You can find further curatorial information in McGavin (1997).

METHODS FOR DIRECT OBSERVATION

Where primates are visible, you can often observe their insectivory using binoculars or a camera, although the prey may be concealed and

GPS or Map reference	17°36.945S 145°47.748E
Site Name	Palmerston Nat. Pk. QLD
Capture technique e.g. malaise trap	Pyrethrum mist
Sample code e.g. q1 (quadrat 1)	Planted Trans. 1, 0m
Date	2 Dec 2000
Name of collector	C.Ozanne

Fig. 15.1 Example label for samples and specimens

difficult to identify. Some authors have determined insects to species level by comparing the remains dropped during or after feeding to reference collections, e.g. Orthoptera preyed upon by saddle-back (*Saguinus fuscicollis nigrifrons*) and moustached tamarins (*Saguinus mystax mystax*) (Nickle & Heymann, 1996). You could apply this method even when primates are feeding high in the canopy.

If primates are feeding on small arthropods on ground vegetation or in the leaf litter, you can note the location and collect samples with forceps, fine paintbrushes or aspirators. If they are nest-dwelling insects, collect samples by making a small opening in the side of the nest, then use an aspirator to collect the insects. Ants may leave a bitter taste in your mouth due to the release of formic acid by defensive soldiers, so use a battery-powered aspirator if possible. Social insects can be dangerous (they may sting or bite), so take appropriate precautions when in close proximity to a nest. Always carry medical supplies when in the field if you are allergic to insect bites and stings (antihistamines or an EpiPen™ for extreme reactions).

Determination of primate-insect feeding to a taxon-specific level seems a viable, but largely uncharted, means of gaining dietary information. You can make certain assumptions about the type of insects eaten by using primate behavioural information. For example, field notes that read 'primate seen sitting and picking over leaf litter for extended periods with little investment in searching effort. Cannot see insects, but held between thumb and forefinger' would suggest that the insects are abundant, small, probably social and likely to be either ant- or termite-like. 'Primate seen stalking along tree trunk and moving slowly. Caught insect by snatching action' indicates that the target prey has a good sense of its environment and is a fast flyer, jumper or walker. The likely candidates here are cicadas, dragonflies, grasshoppers or crickets – generally animals with larger eyes or other acute sensory organs. Other ethologies could be related to venomous, leaf-eating or nest-building arthropods, for example.

METHODS FOR INDIRECT OBSERVATION

Where you cannot observe feeding directly, you can use a number of indirect techniques to give detailed information about the arthropods available at the feeding site.

Arthropod by-products: secondary signs

Arthropods often betray their presence, for example by leaving webs (spiders and caterpillars), frass (insect excreta), nests (e.g. wasps and bees), mounds (ants and termites), holes (e.g. solitary wasps and antlions), galleries (e.g. wood-boring beetles) or chewed leaves (e.g. caterpillars and leaf-cutter bees), or by making sounds (e.g. cicadas, grasshoppers and crickets). You can use these signs to detect the type of arthropod selected for consumption or use by a primate. For example, if you see a primate feeding at a hole in a tree trunk, take a picture of the hole and try to extract the concealed arthropod, but never put your hand where you can't see. If the arthropod is determined not to expose itself, carefully collect some small ants in a pooter and blow them into the hole. All arthropods have strong reactions to ants and will take the shortest exit to escape their presence. Likewise, take recordings of audible songs of crickets or grasshoppers that are selected by a primate (Chapter 16). These recordings will help you to identify the animals later (Sueur *et al.*, 2008).

Faecal samples

Faecal sampling provides a measure of the type of arthropod food that has been eaten, and is particularly useful if you are studying the feeding habits of a primate species that is difficult to observe directly. Arthropods are often encased in chitin (sclerotized epidermis), and this passes through the primate gut, later appearing as fragments in the faeces. However, it can be difficult to identify the remains of arthropods: whereas adult beetles are often covered in chitin, the larvae (grubs) have a spartan covering, often in the region around the head and legs, and fly larvae (maggots) may be almost without any identifiable chitin at all. Furthermore, within the faeces, arthropod body parts may be jumbled up, incomplete and obscured by plant debris (but see McGrew *et al.*, 2009 for extraction protocols).

Despite these constraints, many studies have successfully investigated the arthropod diet of primates, restricting the identification of the arthropod fragments to the ordinal (e.g. beetles: Coleoptera) or subordinal level (e.g. true bugs: Heteroptera and Homoptera) but rarely choosing to key to family or species unless the body part is readily identifiable (e.g. the pedicel or waist of an ant (Formicidae) (see Bolton, 1994). Identification of arthropod fragments is made easier with Gentry *et al.* (1991), a two-volume clinical manual for food quality control containing colour photographs of invertebrate fragments. An alternative budget option is to use a set of invertebrate guides, which will introduce the arthropod form to the ordinal level (e.g. Tilling, 1987; Chinery, 1993), and combine these with McAney *et al.* (1997), which gives a detailed account of arthropod fragments in bat droppings. This should be US$300 cheaper and sufficient for most ordinal studies. Make a reference collection of 'whole' arthropods collected from the strata of vegetation used by a primate (see below for the optimum sampling protocol) for comparison with fragments found in the faeces.

Counting arthropod fragments can lead to problems in analysis if you require anything other than presence/absence data. Estimating density or mass is fraught with problems of pseudoreplication and independence: an individual's body parts become fragmented in the gut, and these parts (e.g. legs) can be mistaken for separate individuals, yielding overestimates. Counting only heads can dampen these problems, but arthropods may be decapitated before consumption (e.g. locusts, mantids). Additionally, larval stages do not often have separate head and body segments (e.g. maggots) and you may miss these altogether. In both scenarios, it is simple to demonstrate that data generated from faecal samples can only be considered 'estimates of consumption' that are far from absolute. Lastly, your choice of statistical tests may also help. If you think there is a large, but unquantifiable, error term, then you can make estimates coarser by choosing a nonparametric test (e.g. Spearman rank correlation) over parametric ones (e.g. Pearson product moment correlation).

For projects with a slightly higher budget you can now use molecular approaches to identify prey items from faecal samples (see King *et al.*, 2008). Although there have been no specific studies on primate faecal samples using such techniques, they have been used in other vertebrate species (e.g. bats, Claire *et al.*, 2009). Problems associated with this methodology are the production of false negatives from scavenged material and false positives from secondary predation (for example, if the primate consumes an arthropod predator containing other arthropod DNA).

Sampling arthropods from the subcanopy and canopy

Primates may feed actively on arthropods in the canopy of under-storey and over-storey trees. The two most effective quantitative sampling methods for plant canopies are branch clipping and pyrethrum knockdown (for more detail see Ozanne, 2004a; Basset *et al.*, 1997).

Branch clipping involves cutting off a section of foliage into a bag that you then draw shut to prevent the escape of mobile arthropods. You can shake arthropods off the foliage and identify them live on site, or fill the bag with a chemical such as CO_2 or ethyl acetate (Basset, 1992). In the latter case, pick or brush arthropods carefully off the foliage and take them to the laboratory for storage and sorting. Attach the bag (e.g. plastic, muslin or mesh) and clippers to fixed or telescopic poles (e.g. canes, branches, tent poles), to allow you to push them up into the canopy from the ground. Draw the bag over the foliage, then cut the branch and draw the bag shut with a cord that can be operated from the ground (for construction see Johnson, 2000). Standardize the amount of foliage clipped where possible. This method would be particularly useful for studies of primates such as the patas monkey, which feeds on arthropods in the swollen thorns of *Acacia* trees (Isbell, 1998) or Japanese macaques (*Macaca fuscata*), which bite into woody plant galls to extract aphids (Hill *et al.*, 1995) and primates that feed on insects rolled in dead leaves, such as saddle-back tamarins (Nickle & Heymann, 1996).

Branch clipping is a fairly comprehensive sampling method for most arthropods, particularly sedentary animals, which you can standardize to plant biomass or area units (Schowalter *et al.*, 1981). However, clipping undersamples large mobile (e.g. dragonflies) and some small aerial insects (e.g. midge clouds) (Johnson, 2000) and over-estimates inhabitants of branch tips.

Pyrethum knockdown involves delivering pyrethrum or synthetic equivalents into the canopy in droplet form by using one of two techniques, fogging or misting. In both cases pyrethum and synthetics are shortlived and non-toxic to vertebrates (excluding fish): they cause loss of motor function to arthropods, which then fall towards the ground into collecting trays. Foggers produce a thermal cloud of pyrethrin that rises upwards and outwards, filling the canopy space. This method is useful for high canopies (15 m upward), but is more difficult to operate and control than misting. Fogging must be carried out in still dry air and is typically done for 5–10 min at a time. In contrast, misters use a fan mounted in a backpack to produce a powerful air current that shears off droplets of chemical, forming a fine mist. The mist reaches a more targeted area of foliage than a fog, but you need to hoist the machine into the canopy to reach foliage above 12–15 m. Misting is typically carried out for 1–5 min, depending on the machine, which may be set up for low volume (20–300 l/ha) or ultra-low volume delivery (5–20 l/ha). Suspend cone-shaped collecting trays, constructed from waterproof material (e.g. vinyl or tent material), with rope below the canopy. Allow arthropods to drop into these for about 2 h after application and standardize samples to the area of the tray, usually 1 m^2.

These techniques are particularly applicable where primates feed on a variety of insects in the mid canopy (e.g. moustached tamarins, Nickle & Heymann, 1996) and the high canopy and for those species that feed by collecting insects from the leaf and bark surface, e.g. pied bare-faced tamarin (*Saguinus bicolor bicolor*; Egler, 1992). Fogging and misting give a more accurate representation of arthropod communities than any other foliage sampling method. However, both techniques are expensive, unless you can borrow equipment, and are limited to calm, dry weather conditions. Canopy fogging also does not efficiently sample all arthropods in canopies where you have a large abundance of epiphytes and suspended soil (in a cloud forest for example).

Vegetation and ground sampling

Low vegetation and the litter layer are rich sources of arthropods that are exploited by several primate groups. For example, saddle-back tamarins feed on arthropods in the lower parts of the forest canopy, and lowland gorillas (*Gorilla gorilla gorilla*) and bonobos (*Pan paniscus*) collect ants and termites from the ground (Tutin & Fernandez, 1992; Bermejo *et al.*, 1994). The most effective way to sample low vegetation is by branch clipping or pyrethrum knockdown (see above), but you can also collect arthropods by beating the foliage with a stick so that the animals drop onto a collecting sheet. This only provides relative count data (catch per unit effort), but mimics the technique used by primates such as saddle-back tamarins, which are particularly adept at capturing prey items they have flushed out by shaking the foliage (Peres, 1992). This method is also very effective at sampling lepidopteran larvae (Basset *et al.*, 1997), which some primates (e.g. olive baboons, *Papio anubis*) gorge on during periods of outbreak (Kunz & Linsenmair, 2008). To sample ground vegetation up to 25 cm in height (and even under-storey shrubs), the most effective technique is to use a suction device such as a petrol-driven garden leaf-vacuum modified to include a small muslin bag (Ozanne, 2004b). Vacuuming an area of vegetation yields samples that are highly representative of the arthropod community, although they tend to oversample owing to an edge effect caused by high suction power and can become inefficient if you don't empty the net on a regular basis.

You can sample litter layer arthropods such as ants effectively by using Winkler extractors (Bestelmeyer *et al.*, 2000). This involves collecting leaf litter (normally from a $1\,m^2$ area) and sifting it through a grid mounted in a cotton bag: live animals and fine debris fall through the grid. You then place the sifted material in a Winkler extractor, comprising an inner net bag in which the sample is placed and an outer funnel-shaped bag with a collecting vessel at the base. Hang the extractor in an airy location; as the sample slowly dries out, animals seeking moisture are driven into the second bag and fall or roll down into the jar filled with preservative fluid. This technique would be most appropriate for sampling arthropods that are fed on by non-tool-using primates such as lowland gorillas and vervet monkeys (then *Cercopithecus aethiops*) (Yamagiwa *et al.*, 1994; Harrison & Byrne, 2000). Where primates scrape the soil and feed on soft-bodied invertebrate prey such as worms, slugs and beetle larvae, mustard powder extraction may be a cheap and easy way of sampling (Paulson & Bowers 2002). The active ingredient in mustard, allyl isothiocyanate, is a skin irritant, which drives soft-bodied invertebrates out of the soil and has been shown to be more representative of worm abundance than all other commonly used extraction techniques (Paulson & Bowers, 2002). Bell *et al.* (2010) successfully used the Paulson and Bowers (2002) method to sample worms and slugs within an enclosed high-sided quadrat ($0.5 \times 0.5\,m^2$) over a 15 min period.

A range of traps for collecting arthropods from vegetation and the ground is available or can be constructed from simple materials (Table 15.1). The advantage of using traps to collect arthropods is that they make use of arthropod activity and can be left to sample the habitat passively while you are absent. In doing so, traps estimate the 'active trappability density', but strictly not absolute 'density' (i.e. per m^2). However, you can use this collection technique to gain an understanding of the potential range of prey organisms available to a foraging primate, and arthropods caught in the traps can also be used to build a PCR primer library that could later facilitate the testing of DNA positives in primate faecal samples.

Table 15.1 *Arthropod trap types, their advantages and disadvantages*
For references, see Southwood & Henderson (2000).

Trap type	Dominant arthropod groups collected	Pros	Cons
Malaise	Flies, termites (winged), parasitic wasps	Inexpensive	Set in flight lines; very large catches
Interception	Beetles	Easily constructed; very effective for beetles	Large catches; group specific
Colour and pan traps	Flies, Hymenoptera (bees, wasps), Hemiptera (bugs)	Cheap, simple effective	Location dependent
Baited traps	Bait-dependent: spiders, beetles, ants, moths and butterflies, flies	Bait can be designed to capture particular sections of the fauna	Draw insects from a wide area
Emergence traps	Flies, beetles	Only method of collecting emerging adults	Life cycle stage-specific (adult)
Pitfall traps	Spiders, ants, beetles, millipedes, centipedes, springtails, crustaceans	Inexpensive	Must be carefully set
Light traps	Moths, beetles	Very effective	Night activity only

DISCUSSION AND CONCLUSION

There is a wide range of techniques for collecting arthropods and we can only include a small but representative sample here. For a more detailed treatment, please refer to Southwood and Henderson (2000) and Wheater *et al.* (2011). Whatever arthropod sampling technique you use, there are a number of factors that you should take into account when designing a study. First, quantitative sampling methods yield the most flexible data. You can quantify arthropod densities per unit area or volume and per unit plant biomass, allowing you to apply many analytical tools to your datasets. Second, arthropod spatial distribution is often highly variable, making decisions about the number and location of samples difficult. Conduct a pilot phase to determine population distributions with careful collection of samples in habitats where the primates are feeding to pre-empt major problems. Third, the timing of collection is crucial. Arthropod communities exhibit diurnal and seasonal variation, thus you need to time sampling to coincide with primate feeding periods. Primates may also vary in their selection of arthropods diurnally and seasonally. For example, Demidoff's galago (then *Galago demidovii*) becomes more insectivorous as the night progresses, while reducing gum and fruit intake (Charles-Dominique, 1974), and spectral tarsiers (*Tarsius spectrum*) change their arthropod diet between seasons, taking Orthoptera and Lepidoptera in the dry season and adding a more varied fauna, which includes Coleoptera and Hymenoptera, during the wet season (Gursky, 2000). Primates will disturb arthropods, causing them to disperse or hide, so do not conduct sampling immediately after a feeding event. Instead, mark sites with a GPS and revisit them at the equivalent time on another day.

Finally, not all arthropods on which primates feed can be collected from the habitat in which they live. Commensal insectivory (e.g. among moustached and saddle-back tamarins, Peres, 1992) may be a useful way of supplementing diet as well as reducing parasite loads, forming an important link between diet and social behaviour. Studies of primates that seek to include interactions with arthropods could also consider ways of monitoring insect and arachnid disease vectors (e.g. ticks and mosquitoes) and could investigate symbiotic relationships such as those between dung beetles (Scarabaeidae) and primates (Estrada & Coates-Estrada, 1991).

REFERENCES

Basset, Y. (1992). Host specificity of arboreal and free living insect herbivores in rain forest. *Biol. J. Linn. Soc.* 47, 115–33.

Basset, Y., Springate, N. D. Aberlanc, H. P. & Delvare. G. (1997). A review of methods for sampling arthropods in tree canopies. In *Canopy Arthropods*, ed. N. Stork, J. Adis & R. Didham, pp. 27–52. London: Chapman and Hall.

Bell, J. R., King, R. A., Bohan, D. A. & Symondson, W. O. C. (2010). Spatial co-occurrence networks predict the feeding histories of polyphagous arthropod predators at field scales. *Ecography* 33, 64–72.

Bermejo, M., Illera, G. & Pi, J. S. (1994). Animals and mushrooms consumed by bonobos (*Pan paniscus*) – new records from Lilungu (Ikela), Zaire. *Int. J. Primatol.* 15, 879–98.

Bestelmeyer, B. T. D., Agosti, D, Alonso, L. E. *et al.* (2000). Field techniques for the study of ground dwelling ants: an overview, description, and evaluation. In *Ants: Standard Methods for Measuring and Monitoring Biodiversity*, ed. D. Agosti, J. Majer, L. E. Alonso & T. Schultz, pp. 122–44. Washington, D.C.: Smithsonian Institution Press.

Bolton, B. (1994). *Identification Guide to the Ant Genera of the World*. Harvard, MA: Harvard University Press.

Charles-Dominique, P. (1974). Ecology and feeding behaviour of five sympatric lorisids in Gabon. In *Prosimian Biology*, ed. R. D. Martin, G. A. Doyle & A. C. Walker, pp. 131–50 London: Gerald Duckworth and Co.

Chinery, M. (1993). *Insects of Britain and Northern Europe*. London: HarperCollins.

Claire, E. L., Fraser, E. E., Braid, H. E., Fenton, M. B., Hebert, P. D. N. (2009). Species on the menu of a generalist predator, the eastern red bat (*Lasiurus borealis*): using a molecular approach to detect arthropod prey. *Mol. Ecol.* 18, 2532–42.

Egler, S. G. (1992). Feeding ecology of *Saguinus bicolor bicolor* (Callitrichidae, Primates) in a relict forest in Manaus, Brazilian Amazonia. *Folia Primatol.* 59, 61–76.

Estrada, A. & Coates-Estrada, R. (1991). Howler monkeys (*Alouatta palliata*), dung beetles (Scarabaeidae) and seed dispersal – ecological interactions in the tropical rain-forests of Los-Tuxtlas, Mexico. *J. Trop. Ecol.* 7, 459–74.

Gentry, J. W., Harris, K. L. & Gentry, Jr. J. W. (1991). *Microanalytical Entomology for Food Sanitation Control*. Vols. I & II. Melbourne: Association of Official Analytical Chemists (AOAC) International.

Gordh, G. & Headrick D. H. (2003). *A Dictionary of Entomology*. London: CABI.

Gursky, S. (2000). Effect of seasonality on the behaviour of an insectivorous primate, *Tarsius spectrum*. *Int. J. Primatol.* 21, 477–95.

Harrison, K. E. & Byrne R. W. (2000). Hand preferences in unimanual and bi-manual feeding by wild vervet monkeys (*Cercopithecus aethiops*). *J. Comp. Psychol.* 114, 13–21.

Hill, D. A., Lucas, P. W. & Cheng, P. Y. (1995). Bite forces used by Japanese mac-aques (*Macaca fuscata yakui*) on Yakushima island, Japan to open aphid-induced galls on *Distylium racemosum* (Hamamelidaceae). *J. Zool.* 237, 57–63.

HMSO (1996). *Biodiversity Assessment: A Guide to Good Practice*. London: HMSO.

Hoshino, J. (1985). Feeding ecology of mandrills (*Mandrillus sphinx*) in the Campo Animal Reserve, Cameroon. *Primates* 26, 248–71.

Isbell, L. A. (1998). Diet for a small primate: insectivory and gummivory in the large patas monkey (*Erthrocebus patas pyrrhonotus*). *Am. J. Primatol.* 45, 381–98.

Johnson, M. D. (2000). Evaluation of an arthropod sampling technique for measuring food availability for forest insectivorous birds. *J. Field Ornithol.* 71, 88–109.

Kay, R. F. (1984). On the use of anatomical features to infer foraging behaviour in extinct primates. In *Adaptations for Foraging in Nonhuman Primates*, ed. P. S. Rodman & J. G. H. Cant, pp. 21–53. New York: Columbia University Press.

King, R. A, Read, D. S., Traugott, M. & Symondson, W. O. C. (2008). Molecular analysis of predation: a review of best practice for DNA-based approaches. *Mol. Ecol.* 17, 947–63.

Kunz, B. K. & Linsenmair K. E. (2008). The disregarded west: diet and behavioural ecology of olive baboons in the Ivory Coast. *Folia Primatol*, 79, 31–51.

McAney, C., Shiel, C., Sullivan, C. & Fairley, J. (1997). *Identification of Arthropod Fragments in Bat Droppings*. London: Mammal Society Publication 17.

McGavin, G. C. (1997) (2007 Reprint). *Expedition Field Techniques: Insects and other Terrestrial Arthropods*. London: Royal Geographic Society.

McGavin, G. C. (2001). *Essential Entomology: An Order by Order Introduction*. Oxford: Oxford University Press.

McGrew, W. C., Marchant, L. F. & Phillips, C. A. (2009). Standardised protocol for primate faecal analysis. *Primates* 50, 363–6.

Nickle, D. A. & Heymann, E. W. (1996). Predation on Orthoptera and other orders of insects by tamarin monkeys, *Saguinus mystax mystax* and *Saguinus fuscicollis nigrifrons* (Primates: Callitrichidae), in north-eastern Peru. *J. Zool.* 239, 799–819.

Niemitz, C. (1984). Synecological relationships and feeding behaviour of the genus *Tarsius*. In *Biology of Tarsiers*, ed. C. Niemitz, pp. 59–76. Stuttgart: Gustav Fischer Verlag.

Oliver, I. & Beattie, A. J. (1996). Designing a cost-effective invertebrate survey: a test of methods for rapid assessment of biodiversity. *Ecol. Appl.* 6, 594–607.

Ozanne, C. M. P. (2004a). Sampling methods for forest understory vegetation. In *Insect Sampling in Forest Ecosystems*, ed. S. R. Leather, pp. 58–76. Oxford: Wiley-Blackwell.

Ozanne, C. M. P. (2004b). Techniques and methods for sampling canopy insects. In *Insect Sampling in Forest Ecosystems*, ed. S. R. Leather, pp. 146–67. Oxford: Wiley-Blackwell.

Paulson, L. A. & Bowers, M. A. (2002). A test of the hot mustard extraction method of sampling earthworms. *Soil Biol. Biochem.* 34, 549–52.

Payne, J. & Francis, C. M. (1985). *A Field Guide to the Mammals of Borneo.* Sabah, Malaysia: Sabah Society.

Peres, C. A. (1992). Prey-capture benefits in a mixed-species group of Amazonian tamarins, *Saguinus fuscicollis* and *Saguinus mystax*. *Behav. Ecol. Sociobiol.* 31, 339–47.

Schowalter, T. D., Webb, W. J. & Crossley, D. A. Jr. (1981). Community structure and nutrient content of canopy arthropods in clearcut and uncut forest ecosystems. *Ecology* 62, 1010–19.

Southwood, T. R. E. & Henderson, P. A. (2000). *Ecological Methods.* Oxford: Blackwell Sciences Ltd.

Sueur, J., Pavoine, S., Hamerlynck, O. & Duvail, S. (2008). Rapid acoustic survey for biodiversity appraisal. *PLoS ONE* 3: e4065.

Tilling, S. M. (1987). *A Key to the Major Groups of British Invertebrates.* UK: AIDGAP, Field Studies Council.

Tutin, C. E. G. & Fernandez, M. (1992). Insect-eating by sympatric lowland gorillas (*Gorilla gorilla gorilla*) and chimpanzees (*Pan t. troglodytes*) in the Lopé Reserve Gabon. *Am. J. Primatol.* 28, 29–40.

Valderrama, X., Robinson, J. G., Attygalle, A. B. & Eisner, T. (2000). Seasonal anointment with millipedes in a wild primate: a chemical defense against insects? *J. Chem. Ecol.* 26, 2781–90.

Wheater, C. P., Bell, J. R. & Cook, P. A. (2010). *Practical Field Ecology: A Project Guide.* New York: Wiley-Blackwell.

Yamagiwa, J., Mwanza, N., Yumoto, T. & Maruhashi, T. (1994). Seasonal change in the composition of the diet of Eastern lowland gorillas. *Primates* 35, 1–14.

USEFUL INTERNET SITES

http://phylogeny.arizona.edu/eukaryotes/animals/animals.html – for the Tree of Life (enter at the level of Metazoa).

http://www.ncbi.nlm.nih.gov/Genbank/ – for the Genbank database. See the taxonomy browser for primers to identify organisms.

SUPPLIERS

www.watdon.co.uk/the-naturalists – for Watkins and Doncaster (UK).
www.entomology.org.uk: for nets, rearing cages, etc. (UK).
www.entosupplies.com.au – for entomological supplies (Australia).
www.bioquip.com – for entomological and botanical equipment, books and
    software (USA).

THOMAS GEISSMANN AND STUART PARSONS

16

Recording primate vocalizations

INTRODUCTION

Ornithologists have been exploring the possibilities and the methodology of recording and archiving animal sounds for many decades.
Primatologists, however, have only relatively recently become aware
that recordings of primate sound may be just as valuable as traditional
scientific specimens such as skins or skeletons, and should be preserved for posterity (Fig. 16.1). Audio recordings should be fully documented, archived and curated to ensure proper care and accessibility.
As natural populations disappear, sound archives will become increasingly important (Bradbury *et al.*, 1999).

Studying animal vocal communication is also relevant from the
perspective of behavioural ecology. Vocal communication plays a central role in animal societies. Calls are believed to provide various types
and amounts of information. These may include, among other things:
(1) information about the sender's identity (e.g. species, sex, age class,
group membership or individual identity); (2) information about the
sender's status and mood (e.g. dominance, fear or aggressive motivation,
fitness); and (3) information about relevant events or discoveries in the
sender's environment (e.g. predators, food location). When studying
acoustic communication, sound recordings are usually required to analyse the spectral and temporal structure of vocalizations or to perform
playback experiments (Chapter 11).

This chapter describes how to record non-human primate vocalizations. We begin with definitions of some technical terms, then provide information on the advantages and disadvantages of various types
of equipment (sound recorders, microphones and other equipment).
Next we discuss methods for recording ultrasound, before ending with
tips for better recordings of primate vocalizations, relevant to both the

Field and Laboratory Methods in Primatology: A Practical Guide, ed. Joanna M. Setchell and Deborah
J. Curtis. Published by Cambridge University Press. © Cambridge University Press 2011.

Fig. 16.1. Clarence R. Carpenter and his parabolic reflector microphone in Chiang Dao, Chiang Mai, Thailand, 1937. This installation was used to record the calls of wild gibbons on disk. Photo from Carpenter (1940, p. 26; Johns Hopkins Press, Baltimore)

field and captivity. Ornithologists and chiropterologists have to deal with very similar problems, and we recommend their introductory texts for further study (e.g. Budney & Grotke, 1997; Kroodsma *et al.*, 1996; Parsons & Szewczak, 2009; see also 'Useful Internet sites').

We do not provide information on the prices for sound equipment. As a rule, these prices change markedly over the years, and retailers' prices differ widely. We recommend that you compare prices on the Internet before buying. The specific models of sound equipment that are produced also change rapidly. Some of the recorders and microphones recommended in the first edition of this book (Geissmann, 2003) went out of production even before the book was published. The same will apply to the products discussed in the present chapter. Digital devices tend to be particularly short-lived and products are frequently replaced by follow-up models or completely discontinued. We recommend that you check the product palette on the Internet before deciding on a particular model. We do not describe methods and equipment required for sound analysis, but you can find information on that topic elsewhere (Charif *et al.*, 1995, 2004; Hopp *et al.*, 1998; Pavan, 2006b; Parsons & Szewczak, 2009).

TECHNICAL TERMS

Sound is a wave phenomenon. Sound waves are oscillations in atmospheric pressure.

The **amplitude** of a sound wave is proportional to the change in pressure during one oscillation. The greater the amplitude of the wave, the more energy it transmits. In sound waves, the amplitude relates to volume, so the greater the amplitude, the higher the energy and the louder the sound.

Frequency f in wave motion is the number of waves that pass through a given point per second. It is the reciprocal of the time T taken to complete one cycle (the period), or $1/T$. Frequency is usually expressed in units called hertz (Hz). One hertz is equal to one cycle per second; one kilohertz (kHz) is 1000 Hz.

Ultrasound is sound with a frequency greater than the upper limit of human hearing. Human hearing is normally limited to frequencies between about 12 and 20 000 Hz.

SOUND RECORDERS

An ideal recorder records a signal without alterations, by matching its dynamic and frequency range and by preserving all its features. Traditional analogue tape-recorders are not perfect. They degrade the signals they record by adding hiss, distortion, frequency response alterations, speed variations, print-through effects and drop-outs. Digital recorders do not have these problems. Within the dynamic range and the frequency limits (owing to the number of bits and sampling frequency they use), they record and reproduce signals with great accuracy, low noise, flat frequency response and no speed variations. Nevertheless, the quality of the signal recorded by good analogue tape-recorders is sufficient for most purposes in bioacoustic studies.

The majority of primate species live in tropical rainforests. Any lengthy stay in this kind of environment puts considerable stress on electronic equipment (see also Chapter 17). If making sound recordings is a crucial component of your study, choose sturdy and humidity-resistant recording machines over highly sophisticated but delicate equipment. Also consider acquiring two complete sets of sound-recording equipment, because the equivalent replacement of a broken machine may not be available anywhere in the country where you are working, and a repair may require replacement parts that may be equally unavailable. Similarly, some digital audio recorders are capable of saving two types of media simultaneously (e.g. hard disk and compact flash) thus providing an immediate backup of all recordings.

We discuss several sound recording formats that are no longer supported by the industry (e.g. analogue, DAT) or that are not an option for bioacoustic analyses because they use sound compression (e.g. MP3 recorders) below. They are included here because you may have access to some of this equipment and may want to know whether it may still be useful (and some of it is). If you plan to buy new equipment anyway then proceed directly to the section on 'Hard-disk and solid-state recorders'.

Analogue tape-recorders (audio)

The analogue tape-recorders that were most suited for field studies are not produced any more, but as they include some of the most durable field equipment ever produced for sound recording, we still recommend them, if you can find them. We mention tape-recorders only briefly, but you can find more detailed comments on these machines in Geissmann (2003).

Avoid the use of noise reduction features such as Dolby or DBX during recording, as they add distortion and limit the high frequencies of many sounds. If possible, use a tape-recorder with adjustable sensitivity for signal intensity. The automatic level control (ALC) that is found on many cheaper portable cassette recorders is often ineffective for recording sounds in the field. Unfortunately, there always were very few portable machines with adjustable sensitivity. The use of ALC may also render recordings unusable for playback studies or for analysis of source levels.

The best machines included the small Sony Walkman WM-D6, the TCM-5000EV, and the larger Sony TC-D5M and TC-D5 Pro II. The Walkman's advantage is its small size, whereas the larger machines come equipped with speakers. Several portable cassette recorders with adjustable sensitivity were also produced by Marantz (e.g. PMD-201, PMD-222 and CP-430) and resembled their Sony counterparts in most respects.

The decision between stereo and mono depends on what your recording will be used for. For most field recordings, stereo is an unnecessary complication. However, if you are interested in recording vocal interactions between several individuals or groups, then a stereo machine is preferable. A second audio channel can also be useful for recording researcher comments via a lapel microphone or similar.

Making effective written notes while recording can often be difficult, and audio comments can be invaluable in these situations. Be sure you transcribe all audio comments soon after making recordings.

Digital Audio Tape (DAT)

This is another technology that has virtually disappeared. The DAT recording system stores the sound information in a binary code, thereby making it immune to speed errors, tape noise (hiss) and non-linear frequency response problems. The DAT recorder delivers a sound quality slightly better than compact disc (CD; 48 kHz versus 44 kHz, often with a greater dynamic range) in a small, easy to use and easy to store, long-duration, tape-based format. Recordings made to DAT can be transferred directly to computer, without the need to redigitize. This minimizes any potential degradation of the signal. Both the machines and the recording tapes are very difficult to obtain now. Unfortunately, DATs are generally very sensitive to high humidity and, therefore, not reliable in some recording environments such as rainforest. Most DAT recorders have a 'dew' or humidity sensor built in that shuts the machine down whenever high humidity is detected. Most DAT machines will operate for only two hours per battery charge, and only accept rechargeable cadmium batteries, a problem when you are working in areas with no or unreliable access to line power. Because of these limitations, we do not recommend the use of DAT recorders.

MP3, MiniDisc (MD), etc.

Avoid working with technologies that remove information from the sound. Of these, MP3 (Moving Picture Experts Group Layer 3) is currently the most popular format, but the same reservations apply to formats like MiniDisc, Digital Compact Cassette, MP4 (Moving Picture Experts Group Layer 4), or Liquid Audio and VQF (also called TwinVQ, transform-domain weighted interleave vector quantization). All these technologies use sound compression algorithms that discard sound details that appear to be non-audible to humans, but may be of biological relevance to primates. For this reason, no compressed or filtered sounds should be used for playback experiments. Input filtering and data compression strategies distort the signal in subtle ways that make it more difficult to analyse and render accurate frequency-related measurements impossible. It may also lead to perceptible distortion if you make multiple copies. Any such technology will change (i.e. degrade) the sound you are recording and make your recordings unsuitable for sound analysis.

Hard-disk and solid-state recorders

Direct-to-hard-disk recording using a laptop computer would be an alternative with great advantages, but normal portable computers are not usually designed to survive long in a tropical climate or under the rough treatment that is unavoidable when following primates through the rainforest. We do not, therefore, discuss this method further here, but see Pavan (2006a) for more information on the topic.

A better option for field use is offered by small, compact recorders that record on internal hard disks or on memory cards such as the compact flash memories also used in digital cameras (Chapter 17). A pivotal criterion for choosing one of these digital recorders is whether they offer the option to store sound not only in compressed format such as MP3 (which is not recommended, see above), but also in uncompressed formats such as WAV (Waveform Audio File Format) or AIFF (Audio Interchange File Format). Macs and most PCs can read these files and edit them (with appropriate software). These recorders range from pocket-sized to book-sized. Some models record up to 4 channels simultaneously and some can sample at rates of up to 192 kHz. Some are not designed for scientific recording, but are more music-oriented.

The most promising technology for carrying out sound recordings in the field are solid-state recorders that record sound on a flashcard and thus use no moving mechanical parts. Examples are Flash Recorders of the PMD series from Marantz (e.g. PMD 620, 660, PMD 670, PMD 671), but it appears that they are no longer produced. Recorders made by Sound Devices, such as the 722 and 744, are capable of recording to compact flash, hard disk, or both. Other, very small, digital recorders that use internal flash memory include the Sony PCM series (PCM-D1, PCM-M10/B, PCM-D50) or the ZOOM H4n. These come equipped with electret condenser microphones. The built-in microphones may not be sufficient for field recordings of primates, unless the primates are calling close to the recorder, but the recorders have input jacks for external directional microphones.

Other recorders suited for field recordings include the following: Edirol R1, R-4/R-4Pro and R-09HR; M-Audio MicroTrack II; Fostex FR-2 and FR-2LE; Tascam HD-P2. Each device has its own features, advantages and disadvantages that you will need to evaluate according to research needs and budget. Moreover, new and interesting devices are likely to appear in the near future.

Specifications given by manufacturers may be confusing, especially those related to the noise of the microphone pre-amplifiers. Most are suitable for recording loud concerts, but the self-noise of the microphone and of the recorder may add an annoying hiss that limits the possibility of capturing low-level sounds in quiet environments. Among the recorders mentioned above, the Sound Devices 7xx series 'has the best reputation for reliability, flexibility and overall sound quality, in particular for the low noise mic preamplifiers' (Pavan, 2006a).

MICROPHONES AND PARABOLIC REFLECTORS

If possible, use a directional microphone or a microphone with a parabolic reflector (also called a parabola). Directional microphones improve sound collection in nature considerably by reducing unwanted ambient noise and sounds from other directions, and are the most useful in bioacoustic field recordings. Various names are in use for microphones with various degrees of directionality. Here, we differentiate between shotgun or ultra-directional microphones and semi-directional microphones. Parabolic reflectors improve sound collection by giving emphasis to the sounds coming frontally.

Shotgun or ultra-directional microphones

A shotgun microphone is a cardioid microphone fitted with an interference tube on its frontal face. It is characterized by a flat frequency response and is less sensitive to wind and handling noise, but offers a lower sensitivity, than a microphone mounted in a parabola. The interference tube cancels off-axis signals, while the on-axis signals reach the microphone's diaphragm without attenuation or gain. As a rule, shotgun microphones work in mono only, so you would need two of them if you wanted to make a stereo recording.

There are many brands of high-quality microphone available, but they are relatively expensive. As a simple rule, the degree of directionality in shotgun microphones correlates with their length (and, usually, with their price). Unfortunately, longer microphones are not very handy in dense tropical forests – they get in the way and are more easily damaged than shorter microphones. We have tested various models of the Sennheiser line of microphones. They are expensive but they have proven to be very durable, exhibit low noise figures, and are relatively immune to high humidity.

Sennheiser's combination of the ME 66 microphone capsule and the K6 pre-amp/power unit using an AA battery is particularly useful in the field. This system has a total length of 32 cm. For better directionality, you can plug the longer ME 67 into the K6, for a combined length of 44 cm. The top-of-the-range Sennheiser shotgun microphones (e.g. MKH 416-P48U3 [25 cm], MKH 418-S [28 mm], MKH 60 [28 mm], MKH 70-1 [41 cm]) have particularly flat frequency responses but tend to be more expensive. Note that you also need special microphone cables with appropriate connectors for the selected recorder for some Sennheiser directional microphones (available separately).

Directional microphones are also produced by a number of other companies including Audio-Technica (e.g. AT 8015, AT 8035, AT 875R, AT 897), Røde (NTG-1, NTG-2, NTG-3) and Edirol (CS-50). Some are less expensive than the Sennheisers but still exhibit good directionality.

On windy days, you may need to add a rubber foam windscreen to your directional microphone. This functions to keep the wind turbulence as far away from the microphone surface as possible. Some microphones come equipped with a windscreen, in others it is an extra. In some cases, it is relatively expensive (about US$30), although it is basically just a tight-fitting glove of rubber foam for your microphone. A similar windscreen of your own manufacture should work reasonably well and costs almost nothing. Be careful that your windscreen does not selectively filter (remove) certain frequencies from your recordings.

Semi-directional microphones

Semi-directional microphones produced for video cameras (Chapter 17) are typically less directional than the shotgun microphones, but are cheaper and smaller. They may suffice if you are recording 'loud-calls' of primates or if you are studying captive animals, where background noise is frequently less of a problem and where the microphone can usually be positioned closer to the vocalizing animals. However, the degree of directionality and other quality characteristics vary strongly (Wölfel & Schoppmann, 1994). One of us (TG) has used semi-directional microphones to record loud vocalizations of various primates, including gibbons (Hylobatidae), leaf monkeys (Presbytini), macaques (Macaca) and indris (Indri indri), in the field with good results. Usually, these video-microphones can be plugged directly into the sound recorders with no special connecting cable needed (unlike some of the directional micro-phones from Sennheiser), but we recommend buying an extension cable (1.5 m) for field use. Example semi-directional microphones are Hama's RMZ series (RMZ-10, RMZ-12, and RMZ-14), Røde VideoMic Directional Video Condenser Microphone, the video microphone from Sennheiser (MKE 400), the DM series from Canon (DM-50, DM-100), and Audio-Technica's ATR 6550. Be careful to select a microphone with a frequency range suitable for your needs. The frequency range of some of the cheaper video microphones does not exceed 10 kHz, which makes them less suited to tape-recording primates with high-pitched vocaliza-tions (e.g. callitrichids).

If you are working with analogue recorders, the microphone should never be too close to the cassette recorder while recording; otherwise you might have the noise of the machine on your sound-recordings.

Parabolic reflectors

A parabolic reflector focusses incoming sound waves that are parallel to its axis onto a single point, the focus, where a microphone is placed. Whereas a shotgun microphone simply screens off sound from direc-tions other than the sound source, a parabolic reflector can actually amplify the sound from the target direction. It also acts as a highpass filter, especially for low-frequency environmental sounds. The effect-iveness is determined by the diameter of the reflector in relation to the wavelength of the sound. Gain and directivity increase propor-tionally as the diameter : wavelength ratio increases. For wavelengths larger than the diameter of the parabola, the response is predominantly that of the microphone itself. As the wavelengths become smaller than the parabola diameter, gain and directivity increase with the frequency. Common diameters are 45 cm, 60 cm and 90 cm, with directionality starting at about 750, 550 and 375 Hz, respectively. We only recommend the larger dishes for primate studies: they are able to cover the fre-quency range of calls produced by the smaller primate species, at least. Many larger species, however, produce calls that go well below 375 Hz, which would require a parabola diameter above 90 cm. Obviously, this would be difficult to use in tropical forests, even if some plastic parab-olas can be rolled up for travelling (Telinga). Parabolas may be of better use in stationary recording sites or in more open habitat.

A variety of parabolic reflectors are available, made of metal, fibre-glass or clear plastic. Some manufacturers, such as Telinga and Saul

Mineroff Electronics, sell complete systems that include a parabolic reflector with a microphone and shock-mount system. If you are working on a tight budget, you might consider building a parabolic reflector yourself. You can add some directionality to any non-directional microphone with a large bowl- or umbrella-shaped dish. Position the microphone in the focal point of the reflector, which, fortunately, is not very narrowly defined. The sound is concentrated in this area, and increases sensitivity.

OTHER EQUIPMENT

Many ornithologists recommend the use of good-quality headphones that allow you to listen while recording, making it easier to aim the microphone, as well as giving a clear idea of the quality of the recording being made. We find it inconvenient, however, to monitor the vocal activities of primates, often calling from several directions at the same time, with headphones on.

If you are making analogue recordings, we recommend using quality audio-cassettes (e.g. TDK, Maxell). Much cheaper cassettes are available, but they may yield fuzzy recordings and some frequency levels may be under-represented. Of the available cassettes (types I, II and IV), type II tape offers the best high-frequency response and the lowest signal-to-noise ratio. Interesting vocalizations tend to occur exactly when the cassette is full, and you will miss them while swapping cassettes. We therefore prefer cassettes that allow you to record for 90 or 100 minutes (instead of 60 minutes). Shorter tapes, on the other hand, are more time accurate, because the tape is thicker.

Always carry a set of new, non-rechargeable batteries with you, for emergencies. Animals often decide to vocalize after you have spent hours recording background noise and all your rechargeable batteries have expired. In such cases, an emergency set of batteries will come in handy.

ULTRASOUND

A number of small primates, e.g. pygmy marmosets (*Cebuella*), galagos (*Galago*), mouse lemurs (*Microcebus*) and lorises (*Nycticebus*), have been shown to emit ultrasound vocalizations; others, e.g. the greater bamboo lemur (*Prolemur*), have been reported to produce sounds with harmonics that can extend into the ultrasound range (Bergey & Patel, 2008; Cherry *et al.*, 1987; Glatston, 1979; Pariente, 1974; Pola and Snowdon,

1975; Zietemann, 2000; Zimmermann, 1981). Fundamental frequencies of some note types of the grey mouse lemur (*Microcebus murinus*) reach maximum values of about 45 kHz (Zietemann, 2000). Because detecting, recording and analysing ultrasound vocalizations requires special equipment, such calls may be more widespread among small primates (and other small mammals) than has been documented so far. Fortunately, much research is being carried out on ultrasound calls of bats, and scientists wishing to study ultrasound vocalizations of primates can greatly benefit from exploring the expertise of, and the technologies being used by, bat researchers (see Parsons & Szewczak, 2009 for a recent review).

Bat detectors

Bat detectors or ultrasound detectors were developed to provide researchers with instruments to study bat echolocation, but are also used for research on other small mammals and grasshoppers. Bat detectors are based on both analogue and digital techniques to detect and record ultrasounds and transform them into sounds audible to the human ear. The detectors available on the market use three main systems: heterodyne frequency shifting, frequency division and time expansion.

Heterodyne detectors and frequency division (or count-down) detectors are real time methods (i.e. you hear the sound from the detector at the same time as it is emitted by the bat). Heterodyne detectors remove all spectral and temporal information from the sounds they transform and are thus of limited use. However, they are small and light and so useful for determining whether an animal is producing ultrasound or not. Use caution in interpreting the output of such detectors, as they often have a wide listening bandwidth (up to 16 kHz). This means that vocalizations may appear to be ultrasonic when they are not. Heterodyne detectors are also only capable of monitoring a narrow bandwidth at one time, so may miss the vocalizations of animals calling outside the monitored frequencies.

Frequency-division detectors lower the frequency of a vocalization by reducing the number of cycles that it contains (usually by a factor of 10). At the same time, many also remove amplitude information. This inevitably reduces the amount of information contained in the vocalization, making such detectors of limited use experimentally. However, the detectors' ability to work in real-time, their retention of the spectral content of the vocalization, albeit at a reduced resolution, and their ability to monitor a wide range of frequencies simultaneously makes them an excellent monitoring tool.

If you want to study the frequency content of ultrasonic vocalizations in detail, you will need a time-expansion detector (e.g. Pettersson D240x, D980, D1000x; Ultra Sound Advice: U30 + PUSP). Time expansion relies on the inverse relation between time and frequency: if time is doubled, frequency is halved. Time expansion detectors function by sampling a sound at a high rate, then playing it back at a lower rate, thus effectively stretching it in time. The result is a sound lowered in frequency, but that retains all the temporal, spectral, and intensity information of the original sound. The method is similar to making a high-speed tape recording of the sound and then playing it back at a lower speed. Clearly, this is not a real-time conversion method, but it does offer a number of benefits over and above heterodyne and frequency-division methods. Since the signal is stretched out in time, it is possible to hear details of the sound not audible with other types of detector (for example, you can actually hear frequency differences in single short pulses).

Digital data acquisition

It has recently become possible to record ultrasound in real time, without the need for transformation. This is ideal if you wish to make high-quality recordings suitable for detailed analysis in the lab, and for playback in the field. Ultrasound is acquired by a suitable microphone, digitized at a very high rate, and then either held in RAM or saved to hard disk or some other storage medium (e.g. compact flash). The simplest and cheapest option for acquiring ultrasound directly is a digital bat detector. These possess microphones with relatively flat frequency responses that extend beyond 150 kHz. As mentioned previously, many are capable of transforming the ultrasound, but several also give the researcher access to the untransformed ultrasound (e.g. Pettersson D980, D1000x; UltraSound Advice U30). Detectors such as the D1000x will also save the ultrasound to compact flash, meaning that you do not need immediate access to a laptop computer. However, all others require the ultrasound output to be sent to a specialist recorder capable of sampling at rates sufficiently high to capture the ultrasound.

Some solid-state recorders (e.g. those made by Sound Devices) are capable of sampling at rates up to 192 kHz and so can acquire most vocalizations and their harmonics. Newer laptop computers are also equipped with built-in sound cards capable of sampling at similar rates.

If this is not the case, then you will need separate USB or PCMCIA sound acquisition cards (e.g. the DAQCard 6062E made by National Instruments). If you use a laptop computer to record the ultrasound, you will need software to control the process. Software programs that can interface with built-in sound cards as well as USB and PCMCIA-based systems include Raven (Cornell University Laboratory of Ornithology), BatSound Pro (Pettersson Elektronik AB) and AviSoft Recorder (Avisoft Bioacoustics).

Specialist solutions are also available that allow you to acquire multiple channels directly to laptop computer in real-time, and are field-portable (e.g. those made by AviSoft Bioacoustics). These come with microphones, amplifiers, digital acquisition systems and the software to run them.

Acquisition of ultrasound in real time has two primary disadvantages: cost and storage. Ultrasound recording systems are significantly more expensive than those suited to lower frequencies and, owing to the high sampling rates used, file sizes are often large. If your budget allows for the purchase of such a system, ensure that you also purchase sufficient storage to hold, and back up, all your recordings.

RECORDING PROCEDURE

Get into the habit of making frequent comments while recording your animals. This will be very helpful later when you analyse your recordings. For instance, if several individuals are vocalizing together, it is often impossible to determine who made which vocalization later on. Record information about the identity, sex and age of a vocalizing animal, and the context in which each vocalization is produced. It is usually impossible to describe a long vocalization bout in detail from memory after the recording has been made. Therefore, do it as the vocal bout goes on but try to speak during intervals between vocalizations. Of course, it is best not to voice input at the same time as your target animal is calling.

Always narrate 'stop' or some other indication that the recorder is turned off before stopping. The lack of critical voiced information on tape, especially where one recording ends and another begins, is one of the greatest failings of the beginner. If you interrupt a tape-recording ('cut'), say so on the tape. Otherwise, it is often difficult to detect that the recorder was paused, and confusion can occur during later analysis. If you continue recording the same or new target subjects following the cut, add some narration to say so.

Make long, uninterrupted recordings. Vocalizations tend to occur when you think they won't. If you are waiting for a vocalization to occur, it may be worth while recording continuously, even while your animals are not vocalizing. By doing so, you will catch complete vocalizations or vocal bouts, of which you would otherwise miss the beginning.

Hold the microphone very firmly and steadily. Even moving your fingers may interfere with recordings. If possible, use a tripod to hold the microphones. Small field-portable models are cheap, light and durable. Alternatively, if the calling animal is stationary and known to produce long calling bouts, orient the microphone in the correct direction on a branch (this usually requires an extension cable). This avoids the risk of disturbing the sound recordings when your fingers and arms get tired during a long vocal bout. Moreover, it frees your hands and allows you to take notes, consult your watch, or take compass bearings during the recording.

Record as close to the subject as possible. Directional microphones and parabolic reflectors are no substitute for proximity. Very soft, quiet vocalizations are lost in ambient noise if you are too far away from the animal. Occasionally, though, closer may not be better. If you encounter a loud sound source, such as calling cicadas, somewhere between you and your target animals, going closer would bring you closer to the cicadas as well. In this case, it may be better to change position.

Two special recommendations if you are carrying out analogue audio-recordings are:

1. If no vocalizations occur, rewind your cassette from time to time. If you already have vocalizations on the same side of the cassette on which you intend to record, set the rotation-counter to zero before starting to record again. This will allow you to rewind to the end of your last recording when you have been tape-recording for some time with no vocalizations.

2. When the side of an audio-cassette is filled with recordings, always break out the record-enabling chip on the upper left, thin side of the cassette, to ensure that you do not accidentally record on this side again and delete important material. You may decide to give provisional field numbers to your tapes and give them final inventory numbers when you are back in the laboratory. In any case, it is important that tapes can be identified at any time and referred to in your field notes.

Be consistent in your methods and the way that you record data. That way, someone else listening to your recordings later can learn your pattern, in your absence, and retrieve important information.

Recordists frequently overlook the importance of adequate documentation of recordings, often obtained with painstaking effort and sometimes at great expense. Lack of documentation seriously weakens the scientific value of your sound recordings. After making the recording, narrate the following information on tape:

1. Recordist and equipment configuration: name of the recordist, type of recorder, microphone and parabolic reflector.
2. Identification of target subject and the degree of certainty: name of species, type of contact (heard only, seen), number of target subjects on record, sex and age class of target animal(s) or group composition (if known), approximate distance of subject from microphone, cross-references if same individual can be heard on previous recordings.
3. Time and location: date, time at the beginning and the end of recording, time at important occasions (swapping cassettes, calls of new individuals, etc.), location (accurate locality, district, province, country).
4. Other information: type of habitat, weather conditions, behaviour of subject, identification of other sounds in the background.

Make a written edit of your tape as soon as possible after recordings are completed. Keep a journal of your recordings detailing their contents. You can also add information that was not narrated onto the sound recording. Editing recordings takes a lot of time but makes your collection much more usable. It is a good idea to listen to your recordings while the day's work is still fresh in your memory, and make a written record of what you have recorded.

Keep your personal collection of wildlife sounds well documented and in one place. Label both your sound media (e.g. files, cassettes, DVDs) and your storage boxes clearly.

Review recordings as frequently as possible, preferably at the end of each day. Often, recordings that appear perfect in the field are less so when reviewed later.

Finally, digitize recordings to computer as soon as possible. This provides a necessary backup against degradation of analogue recording media and an additional copy of digital media. We use the 3-2-1 protocol for backing up our recordings: three backups, on two different media, one of which is off-site. Few things are more tragic than the loss of a life-time's work due to theft of a computer or loss of media.

ACKNOWLEDGEMENTS

We are grateful to Deborah J. Curtis, Joanna M. Setchell, Sabine Schmidt, Robert Dallmann and Marina Davila Ross for reading and commenting on earlier versions of this manuscript.

REFERENCES

Bergey, C. & Patel, E. R. (2008). A preliminary vocal repertoire of the greater bamboo lemur (*Prolemur simus*): classification and contexts. *Nexus* 1, 69–84.

Bradbury, J., Budney, G. F., Stemple, D. W. & Kroodsma, D. E. (1999). Organizing and archiving private collections of tape recordings. *Anim. Behav.* 57, 1343–4.

Budney, G. F. & Grotke, R. W. (1997). Techniques for audio recording vocalizations of tropical birds. *Ornithol. Monogr.* 48, 147–63.

Carpenter, C. R. (1940). A field study in Siam of the behavior and social relations of the gibbon (*Hylobates lar*). *Comp. Psychol. Monogr.* 16, 1–212.

Charif, R. A., Mitchell, S. & Clark, C. W. (1995). *Canary 1.2 User's Manual.* Ithaca, NY: Cornell Laboratory of Ornithology.

Charif, R. A., Clark, C. W. & Fristrup, K. M. (2004). *Raven 1.2.1 User's Manual.* Ithaca, NY: Cornell Laboratory of Ornithology.

Cherry, J. A., Izard, M. K. & Simons, E. (1987). Description of ultrasonic vocalizations of the mouse lemur (*Microcebus murinus*) and the fat-tailed dwarf lemur (*Cheirogaleus medius*). *Am. J. Primatol.* 13, 181–5.

Geissmann, T. (2003). Tape-recording primate vocalisations. In *Field and Laboratory Methods in Primatology: A Practical Guide*, ed. J. M. Setchell & D. J. Curtis, pp. 228–38. Cambridge: Cambridge University Press.

Glatston, A. R. (1979). *Reproduction and behaviour of the lesser mouse lemur (Microcebus murinus, Miller 1777)* in captivity. PhD thesis, University College, University of London.

Hopp, S. L., Owren, M. J. & Evans, C. S. (eds.) (1998). *Animal Acoustic Communication – Sound Analysis and Research Methods.* Berlin: Springer Verlag.

Kroodsma, D. E., Budney, G. F., Grotke, R. W. *et al.* (1996). Natural sound archives: guidance for recordists and a request for cooperation. In: *Ecology and Evolution of Acoustic Communication in Birds*, ed. D. E. Kroodsma & E. H. Miller, pp. 474–86. Ithaca, NY: Cornell University Press.

Pariente, G. F. (1974). Importance respective du reperage visuel et auditif (absence d'écholocation) chez *Microcebus murinus*. *Mammalia* 38, 1–6.

Parsons, S. & Szewczak, J. M. (2009). Detecting, recording, and analyzing the vocalizations of bats. In *Ecological and Behavioral Methods for the Study of Bats*, ed. T. H. Kunz & S. Parsons, pp. 91–111. Baltimore, MD: Johns Hopkins University Press.

Pavan, G. (2006a). *Instruments and Techniques for Bioacoustics.* Università degli Studi di Pavia. Centro Interdisciplinare di Bioacustica e Ricerche Ambientali (CIBRA). http://www.unipv.it/cibra/edu_equipment_uk.html#reco, accessed 28 Oct. 2009.

Pavan, G. (2006b). *Software for Sound Analysis*. Università degli Studi di Pavia. Centro Interdisciplinare di Bioacustica e Ricerche Ambientali (CIBRA). http://www-3.unipv.it/cibra/res_software_uk.html, accessed 28 Oct. 2009.

Pola, Y.V. & Snowdon, C.T. (1975). The vocalizations of pygmy marmosets (*Cebuella pygmaea*). *Anim. Behav.* 23, 825–42.

Wölfel, M. & Schoppmann, J. (1994). Vergleichstest: Richtmikrophone – Lasst Mikros sprechen. *VIDEOaktiv* 1994, 4–7.

Zietemann, V. (2000). *Artdiversität bei Mausmakis: Die Bedeutung der Akustischen Kommunikation*. PhD thesis, Institut für Zoologie, Tierärztliche Hochschule Hannover.

Zimmermann, E. (1981). First record of ultrasound in two prosimian species. *Naturwissenschaften* 68, 531–2.

USEFUL INTERNET SITES

Internet sites with review texts on recording and analysing nature sounds:
http://blb.biosci.ohio-state.edu/
http://zeeman.ehc.edu/envs/Hopp/sound.html
http://www.birds.cornell.edu/brp
http://www-3.unipv.it/cibra/ (see 'Bioacoustic Equipment')

17

Photography and video for field researchers

INTRODUCTION

Photography and video are valuable for research, presentations, teaching, or to interest more general audiences. Good imaging conveys the quality of light, emotion, and the subject's essence to the viewer and comes only with practice. Photography and video have much in common but each requires a different set of skills and thought processes. Here we provide general guidelines for capturing images in the field using photography and video, discuss features common to both still and video equipment, and then move on to specific advice concerning digital still and video cameras.

GENERAL GUIDELINES

Both still and video cameras have dozens of controls and large instruction manuals. Some features are used only occasionally but are nevertheless essential. There will be no time to experiment in the field, so master your cameras beforehand.

Protect your equipment and media

Electrical devices do not like environmental extremes or rapid changes in humidity or temperature. Binoculars, cameras and lenses corrode and are a substrate for fungus in tropical conditions. Store them in waterproof bags with silica gel. Dry the silica gel often in a frying pan or an oven. Use a big waterproof bag or plastic garbage bag to protect all valuables during boat trips and stream crossings and from rain and leaky roofs. If electricity is available, it may be worth building a dry box with small holes and a 50 W incandescent bulb to keep equipment dry.

Field and Laboratory Methods in Primatology: A Practical Guide, ed. Joanna M. Setchell and Deborah J. Curtis. Published by Cambridge University Press. © Cambridge University Press 2011.

If a camera gets wet with freshwater, dry it out in a low-temperature oven (50 °C) or with a hair dryer. Saltwater will probably render it worthless. Where humidity and exposure to water are ongoing hazards, a water-resistant housing may be necessary. You can use an umbrella for shooting in the rain.

Do not expose recordable media to direct sun, high humidity or water; keep it in zip-top bags wherever it is coolest. Carry only the media you need for the day (300 still images, 180 minutes of video).

Travel wisely with your equipment

If travelling abroad, make a list of your equipment in advance to expedite paperwork on arrival at customs. For each item, include description, manufacturer, serial and model numbers, quantity, weight, and cost; if possible, attach photocopies of purchase receipts. Customs usually ignore small video and digital still cameras if you have a tourist visa. Pack photographic equipment in carry-on luggage and prepare to be searched. Have the battery charged and media in the camera; security officials may ask you to demonstrate that it is, in fact, a working camera. Videotape and digital media are not normally affected by airport X-ray equipment.

An equipment case should protect its contents both in transit and in the field. It should be water-resistant and small enough to qualify as carry-on, but large enough for the camera, charger, lights, batteries, recording media, lens tissue, a small torch, headphones (for video), zip-top plastic bags, desiccant, duct tape and indelible markers to label media. Padded backpacks may be suitable, but it is often hard to grab cameras quickly from them.

Consider security and the local government and culture. Avoid unnecessary displays of gear, particularly in urban areas and in crowds. Know the government regulations – recording prohibited images can lead to detention, fines, and confiscation of equipment and media. The most likely problems involve border areas, sites of civil disturbance, police-patrolled locations, military installations, industrial structures (including harbour, rail, and airport facilities) and aerial pictures.

The most important aspect of photographing and videotaping is to treat people and their places with respect and to err on the side of caution. Always ask permission of human subjects before taking a photograph or video. Do they expect payment? Are certain places or things, particularly of a cultural or religious nature, subject to photographic or filming restrictions? Providing pictures of themselves to village heads and other leaders may enhance relations between you and the community, and carrying photos of yourself, your family, and your home to show to local people may help them understand you better too.

Combining research and imaging

For the most part you cannot collect good data and take good photos or video at the same time but you may be able to arrange to collect data on some days and take photos on others. If you lack the time, money or ambition to take pictures but need photos or video, one solution might be to have a professional come to the study site. Choose a professional who has good equipment, likes fieldwork and can get good pictures. Make sure that it is understood that you need your own set of good photos of subjects you want and that the photographer must pay her or his own way, plus contribute financially to the fieldwork.

Post-processing

Digital post-processing is beyond the scope of this book, but be aware that images captured are only the basic material of the final presentation. Saved media may be acceptable as is but its presentation can probably be improved by post-processing and editing software. That said, in some circumstances manipulating images may be regarded as tampering with them rather than enhancing them. Follow the *Digital Manipulation Code of Ethics* of the (US) National Press Photographers Association, which states in part, 'We believe the guiding principle of our profession is accuracy; therefore, we believe it is wrong to alter the content of a photograph in any way that deceives the public.'

Maintain files systematically

Devise a straightforward organization method in advance; you will be able to locate items later only if a system is in place from the start. Have a robust backup strategy. Conserve your original material; a common mistake is to lose the original high-quality media after manipulating them to meet your immediate needs, or to save space, or after a hard-drive crash. When you have copies that have been converted to another file type, cropped or otherwise modified, develop versioning conventions to indicate what each file contains and how it relates to the original.

FEATURES COMMON TO BOTH STILL AND VIDEO EQUIPMENT

In digital photography and video, an image is captured by a sensor consisting of light-sensitive picture elements (pixels). The camera's microprocessor transforms the image into digital bits that are then stored (Johnson, 2001). The greater the number of unduplicated pixels, the higher (sharper), the picture's resolution. The larger the physical size of the sensor, the larger the amount of light that may land on it. Larger chips are therefore more accurate and more sensitive. High pixel density on a smaller sensor means less light is available to each pixel. Smaller sensors are thus noisier and provide grainier images despite their higher resolution.

The dimensions of an image are relevant to the final product. To print a photo, translate pixels into dots per inch (dpi), by dividing the number of pixels by the physical print size. Good-quality prints need a resolution of 200–300 dpi, depending on the printer (300 dpi for publication). One megapixel (1280 × 960 pixels) is enough for a photographic-quality 3 × 5 inch (8 × 13 cm) print. The more megapixels, the more the photo can be enlarged – and cropped as needed – while remaining sharp. In video, a variety of digital high-definition (HD) standards have overtaken standard-definition (SD) video cameras. 'High-definition' is a relative term. SD, originally an analogue format mostly recorded on videotape, typically has a 4 × 3 aspect ratio and a resolution of roughly 640–480 pixels, and is captured at 24–30 frames per second (fps). HD has an aspect ratio of 16 × 9 and a resolution of 1100–1920 × 480–1080 pixels and is captured at up to 60 fps.

All digital cameras come with a liquid crystal display (LCD) to view the image. LCDs are difficult to see in bright light, so a viewfinder is useful, but not all cameras have one. The major advantage of digital technology is that you can view the result immediately on the LCD.

White balance

White balance – compensating for lighting conditions – is important for colour accuracy in both still and video. Most cameras have multiple settings for white balance. The automatic setting may be okay outdoors, but indoors in artificial light adjust the setting to the type of light or manually set a white balance using a white card or T-shirt. You can shoot stills in RAW format (all metadata associated with an image are saved and bundled with *raw* data, hence the name) and balance the colours later.

Tripods

A tripod is essential for video and telephoto still photos. Few people can hand-hold a camera at maximum zoom. With still cameras, a tripod allows slow shutter speeds to get sharp pictures and more depth of field. Use a cable release or remote control to trigger the camera without touching and possibly shaking it. Weight adds stability, so most good tripods are between 2 and 4 kg. Graphite tripods are lighter yet stable, but more expensive. For stills of animals, a ball head is the most rapidly adjustable type; most have a quick-release plate for changing cameras or lenses faster. Specialized heads for video cameras can be levelled and stabilize the camera when it is panned or tilted, keeping the picture level. Even a light-weight tripod is better than none. Monopods are easier to carry but not as steady and do not allow hands-free shooting. When observing primates from a car, a window clamp or a bean bag support cushion is useful to hold the camera.

Storage media

Buy good-quality tape or non-volatile recording media (cards) from a well-known brand. If using tape, also buy a head-cleaning tape. Make sure you purchase the type of card required for your camera. Price varies with storage capacity and data transfer speed. With fast-moving primates and a good high-megapixel camera, card speed is important; for most video, only the fastest cards are acceptable. Check online for comparisons of speeds. Media drops continuously in price, and the best practice is to have as many cards as you think you will need but never fewer than two. Transfer files to a media recorder, laptop computer or DVD periodically.

Batteries and chargers

The number of batteries you will need depends on how often you can charge them and how many hours you spend shooting per day. Most rechargeable batteries are rated in milliamp, amp or watt hours. A higher number is better. Extensive use of flash reduces battery life. Some cameras operate on standard-sized batteries but most operate with proprietary batteries sold by the manufacturer. Many video cameras act as their own charging stations, but you can't charge the batteries when you are carrying the camera, so consider purchasing a separate charging station. Many cameras have a second small battery to power the camera's internal settings, so carry a spare battery of this type for each camera. You need a battery charger that works with different voltages (120–240 V). In the field you may need a charger that uses 12 V from a vehicle or solar panels.

Filters, extension lenses, and other lens add-ons

A polarizing filter can improve images of habitats and landscapes by minimizing reflections and consequently enhancing the recorded colours. This filter requires up to 1.5 stops extra exposure (Shaw, 2000) (see 'Aperture (f-stop)'). A graduated neutral density filter, which balances the contrast between sky and foreground, is also useful. You can attach additional elements to the lens to increase the angle of view on cameras with fixed lenses. Wide angle and telephoto adapters may help extend the range.

Working with flash and at night

Flash or strobe adds light, showing more colours of subjects in low light, and effectively freezes motion. Many consumer video cameras and almost all digital cameras come with a small built-in flash, often with a limited range of 2–3 m. For stills, a separate flash unit compatible with the camera's through-the-lens (TTL) metering system will be much more powerful, up to 10 m. A flash cord enables you to remove the flash from the camera to light close-up subjects such as insects and plants. You can often use flash to fill in the detail of primates in shade or backlit in daylight, if they are habituated enough not to flee the flash.

Night photos and video present special challenges. A flash of white light is likely to disturb nocturnal animals, but their eyes are usually less sensitive to red light. A helper holding a torch covered with red cellophane is often necessary to spot the subject. Most nocturnal primates have a *tapetum lucidum* that reflects red, orange, or yellow when flash is used. Don't use LED (light-emitting diode) torches to study nocturnal primates because of the sensitivity of nocturnal animals to light (Chapter 22). For stills, you need a sophisticated autofocus camera and a powerful TTL flash for shooting anything that isn't very close and cooperative (Shaw, 2000). Camera-mounted lights that operate on a separate camera battery are helpful for relatively close shooting.

DIGITAL STILL PHOTOGRAPHY

Digital cameras are currently the best way to record photographic images. You can select file format and compression options on many cameras. The most common file formats for storing camera images are JPG (JPEG = Joint Photographic Experts Group) and RAW. JPG has better storage efficiency and is easier to use after acquiring the image, but JPG file compression may subtly alter colours and introduce other artefacts. Bigger files contain more information but require more storage space; always record the highest-quality JPG feasible. A RAW file provides more information and more accurate colour but is much larger, slower to transfer to other media, and must be processed later. Some cameras take RAW and JPG photos at the same time.

ISO (derived from the Greek 'isos', meaning 'equal') is the standard rating of the camera sensor's sensitivity to light. You can change the ISO setting as conditions dictate. In good light the best setting for sharpness and resolution is 100–200. The higher the ISO, the faster the shutter speed and the less likely it is that the image will blur from camera shake. Images taken with high ISO settings (1600–6400), which may be necessary in low light, will show 'noise', making them less likely to be publishable but fine for research.

Types of digital still cameras

Five types of digital camera are used for still images: mobile phone, point-and-shoot (PSCs), camera traps, single-lens reflex (SLRs), and video (see 'Video', below). Phone cameras are ubiquitous and cheap. They are easy to use, and the lens and sensor are usually small, with little or no zoom. Although the image quality is not very high, yet, it is sufficient for Web presentations and quick communication with others.

PSCs are useful for documenting fairly close subjects such as habitat, camp life, local people, and primates. They are small and light enough to be carried at all times, but their disadvantages are a smaller sensor, more noise, and delayed opening of the shutter after pushing the button. PSCs come in an enormous array. They weigh 86–220 g; have autofocus, zoom, and built-in flash; and cost US\$50–600. Zooms are commonly 3–6 ×, extending from about 38 mm (mild wide angle) to 200 mm (mild telephoto) in the equivalent of a 35 mm film camera. Some DPSCs have 10 × and up to 24 × (26–624 mm) lenses. The zoom should have an 'optical zoom' rating covering the desired ranges. 'Digital zoom' enlarges the central portion of the image digitally, so disregard this specification because you can do this later (Johnson, 2001). Weatherproof models that withstand splashes and waterproof models (to a depth of 3 m) are practical in tropical forest and when travelling on water.

You can use camera traps (also called game cameras) to determine the presence and abundance of elusive mammals and to record otherwise unobservable behaviours in primates (Morgan *et al.*, 2008). Set up along trails or at nest or feeding sites, camera traps acquire still or video images when triggered by motion sensors. Some use a flash; others use infrared light to make them less detectable by the subject. They record date and time on the image and sometimes moon phase and temperature. Make sure you choose a model appropriate for your study animals (game cameras are built for deer, for example, so make sure they react to the motion of smaller animals).

Digital SLR cameras are the professional standard. Their advantage is that the image is recorded when the shutter button is pushed. They have fast interchangeable lenses, from super-wide-angle (10 mm) to long telephoto (800 mm); larger, faster sensors with less noise at a high ISO; and cost US$600–5000. However, they are bulky and heavy (0.3–1.3 kg without a lens), and a long telephoto lens is expensive (US$1000–10 000) and heavy (up to 2.7 kg), requiring a tripod (2.3–3.6 kg). The best option is a brand-name camera body (Canon, Minolta, Nikon, Olympus, Pentax, or Sony) with manual focus (necessary in the field), manual exposure, and exposure compensation (EV), which enables more or less exposure than that set automatically. The latest innovation in high-end SLR cameras is the ability to take high-definition video, but the audio quality is not reliably good. Conversely, digital video cameras usually can take stills, but image quality often is lower than with a still camera.

Zooms and macros

A 6 × zoom lens (35–200 mm equivalent) will cover many subjects and a 10 × zoom (18–200 mm) will cover most; the wide angle (18 mm) is useful in forest and indoors. Some DSLR cameras increase lens magnification by 1.3–1.6 times because of sensor size. This can be good for more telephoto, as 200 mm becomes 300 mm, but 35 mm becomes 52 mm, which is not a wide angle. A serious photographer would want three lenses: wide angle (12–24 mm), 10 × zoom, and telephoto. Sigma and Tamron lenses are less expensive than other brands, but make sure they are compatible with the camera and that the metering works correctly. As a rule, they are comparable to big-name brands in sharpness but are not built for hard use.

A longer telephoto will be useful if you have the money and are willing to carry it. Canon and Nikon make image-stabilized 100–400 mm zooms that allow an exposure of 1/60 s without a tripod (approx. US$1500). As a rule, shutter speed should approximately equal the camera's focal length to obtain a sharp picture (for example, with a 200 mm lens, shoot at 1/250 s). Most people cannot hand-hold any camera below 1/30 s without shaking.

Most zooms have macro capability, but usually only 4:1 magnification (a 4-inch subject will be 1 inch on the sensor). To take close-ups at 1:1, buy a 105 mm or 200 mm macro lens, which allows you to be farther away from the subject than a 50 mm macro does.

Aperture (f-stop)

The aperture, or f-stop, of a lens (usually f/4–f/22 on a zoom) determines the depth of field. The greatest depth of field is provided by the highest number (f/22), with almost everything in focus, whereas with f/2.8 perhaps only a primate's eyes will be in focus. Telephoto lenses generally have less depth of field, even at f/22; the largest aperture available on such lenses is often f/4. DPSCs have much more depth of field because of their small lenses. Shutter speed works inversely with f-stop, so for any given amount of light, as one goes up, the other must go down to obtain a good exposure. For instance, to expose correctly at 100 ISO on a sunny day, you can take the picture at f/16 and an exposure of 1/60 s for great depth of field, or at f/4 and 1/1000 s with very little depth of field. The reality of photographing primates is that you usually have little choice of f-stop and shoot with the lens aperture wide open at the lowest f-stop to employ the fastest possible shutter speed for the sharpest picture.

Remember that the camera's meter expects to read an average of a normal scene with the sun behind the photographer, so if you point it at a dark primate against a bright sky, it will expose for the sky. To get a good picture of the primate, increase the exposure compensation by 1.5 to 2 f-stops. If your camera has a spot meter, it may help to obtain a better exposure if pointed at the primate with little or no sky included. If the picture is important, bracket exposures: take three (or more) successive exposures at the same shutter speed – one as measured by the camera and one each over- and under-exposed by one f-stop – to ensure that one exposure will be good. Some cameras have an automatic bracket function. Although a bad exposure can often be fixed with software, it is always better and less work to have a good exposure. Most cameras can display a histogram (barred graph) of the exposure in photo review mode. Correct exposures show a bell curve in the middle of the graph. If most of the information is to the left, the image is underexposed; to correct this, lower the shutter speed, raise the ISO, or put the EV in the plus range. Information mostly to the right means over-exposure, so raise the shutter speed, lower the ISO, or put the EV in the minus range.

VIDEO

With the digital revolution, video equipment has become cheaper, better and more diverse, leading to professional results with amateur equipment. There is still a relationship between quality and performance but this is far more dependent on the outcomes required. Someone shooting video for a television documentary will undoubtedly work with different devices from someone who needs supplementary documentation for a preliminary field study.

Planning ahead

Videotaping technique is as much about storytelling as it is about image and sound recording. The final products will depend on the recording choices in the field, so make an inventory of your needs and intentions in advance. A list of outcomes might include personal mementos; science-related aids to taxonomic identification and behavioural analysis; educational or public relations products (e.g. for project documentation, fundraising, or funding or sponsor requirements); commercial purposes (e.g. for stock libraries, magazines, textbooks, educational products, or entertainment); and online uses (e.g. file sharing, social networking, or a website).

Video production techniques

Make a habit of narrating annotation into the microphone. This might include date, time, location, behaviours observed, the number of individuals in the group, and so on. Also record a few minutes of 'wild' sound – ambient sound with no distractions. You can edit this in later to replace human voices and other inappropriate sounds. When recording people important to your story, make sure you record the spelling of their full name and their title at the beginning of the recording. Never interview without using headphones to monitor the audio.

Follow these guidelines for video you would like other people to watch. The camera should be square and level on its tripod before recording. Compose the frame so that even if no primate were present, there would still be an attractive, balanced picture. Keep the camera still. With small cameras this usually means composing the shot, hitting 'Record', and then taking your hands off the camera. Pan and zoom as little as possible. When you do pan or zoom, move from one well-composed picture at the beginning to another at the end. Try several pans or zooms so that you can choose among them later. If possible, shoot both close-ups and shots wide enough to show the animals' habitat and social relationships, and change your position several times so that each subject has a variety of backgrounds. To end a shot, let the animal move out of frame. Always shoot lots of material from many angles so you have a choice of edits.

With primates, it is often necessary to point the camera almost straight up. If the sky fills the background, the animal will typically be a silhouette without detail, so try the settings for 'backlight' or 'spotlight', or adjust the iris to overexpose. It is usually better to reposition the camera so that some of the background is leaves or at least the sun's angle creates less contrast. Note that many tripods are not designed to shoot straight up. The solution is to mount the camera backwards on the tripod.

Always label both recording media and their jackets as soon as the media are spent, and if you keep a journal, reference the media in it. Playing back videotape and deleting files in the field is risky because you may forget having done so and shoot over something important.

Types of video cameras

Video cameras can be distinguished in part by price. Cameras costing US$200–1300 are typically consumer-oriented. Some take surprisingly good pictures, and a few have features such as integrated GPS (Global Positioning System) and night capabilities valuable to field researchers and not available in more expensive cameras. Between US$1300 and US$6000 are the 'prosumer' models, with professional features such as better lenses, professional audio connections, more image controls and larger video sensors. Cameras costing more than US$6000 are typically used by professionals.

Video formats

NTSC (National Television Standards Committee), PAL (Phase Alternate Lines) and SECAM (Séquentiel Couleur avec Mémoire) are regional standards for broadcast and distribution of video. Digital conversion among them is fairly simple if you have the time, software and equipment. If your camera shoots in a format other than the local standard, it may be hard to locate equipment for playback in the field. Computers can manage most regional standards, so playback on a laptop with proper connections is often the easiest method.

Common recording media and file types

Analogue tape includes VHS (Video Home System), Hi8, S-VHS and other older SD recording formats. These are older formats, which have low definition and poor colour rendition.

DV tape (Digital Video tape) has been the standard for low-cost field recording since the late 1990s. Designed to be digitally recorded on 6 mm tape in standard definition 4 × 3, it can provide outstanding video and can be edited with inexpensive editing software on lower-powered computers. However, we do not recommend DV for new purchasers because it is now being replaced by high-definition formats.

HDV tape shares the same cassettes as DV but digitally records High Definition (1440 × 720) in a 16 × 9 aspect ratio at a single fixed rate of compression. It is the most commonly available format in tape-based cameras. Good cameras can be found in this format, the medium is inexpensive and editing can be performed on most modern computers.

DVD (Digital Video/Versatile Disc) cameras have a special niche. They record video to DVD ±R (DVD recordable) or DVD ±RW (DVD rewritable) disks in SD only. It is hard to edit and there are no quality cameras available. DVD players are ubiquitous throughout the world, so a DVD camera may be useful if easy playback is important.

AVCHD (Advanced Video Codec High Definition) is more properly a family of recordable video file types. It may be encountered under other names and is used in video devices other than video cameras including phones, computers and digital still cameras. Written to non-volatile recording media and in-camera hard drives, it can be recorded at a variety of compression rates up to full HD (1920 × 1080). The quality of the video relative to the size of the files is superior to all other formats. Compression and decompression is computationally intense so editing AVCHD challenges even the most modern computers at present. Non-volatile recording media and internal hard drives are relatively expensive storage so you will need to move your files to external hard drives routinely.

Many inexpensive still cameras record SD video to AVI (Audio Video Interleaved), MOV (QuickTime movie) and other computer file formats. Always record the largest file sizes feasible in the field.

Useful (and not useful) video features and a few additional necessities for field videographers

Select a video camera based on budget, image quality, size, weight, ease of use, battery options, and connectivity with other recording equipment.

Some prosumer video cameras have **three image sensors (3 chips)**, devoting a separate sensor to each primary colour. At present, '3-chip' cameras are best, but the advent of new cameras with very large image sensors with high pixel counts will likely soon supplant 3 chips.

Manual focus is essential. Avoid cameras where the position of the adjustment is so poor that the feature is not usable or where you can only access the manual mode through internal menus.

Even expensive cameras may have inferior internal or attached microphones, and a **microphone input jack** allows you to attach an external microphone and cable. Microphones are better the closer they are to the subject. Make sure you have the correct cables, connections, and power sources to attach a specific microphone to a specific camera. When ordering microphone cables, specify the two devices you intend to connect.

When audio is important, an **earphone jack and headphones** are essential for monitoring audio. People will accept very poor images but will stop watching if the audio is unpleasant.

Zoom considerations are similar to those for still cameras, but video zooms are motorized to operate easily and smoothly while the camera is recording. A good camera will have variable or multiple zoom speeds. Inexpensive cameras have only one speed.

Image stabilization is never as good as using a tripod but may help. It is intended to dampen small movement in handheld cameras. Mechanical stabilization is better than electronic stabilization.

There's often not much light in the forest. If you are choosing between two cameras, choose the one with the lowest **low-light sensitivity**. A few cameras (mostly Sony) offer **night acquisition** through increased infrared sensitivity. With an infrared spotlight, a camera can double as a nightscope. Such cameras are valuable for studying nocturnal animals.

Most cameras record **time code** that uniquely numbers each frame. You can elect to see the time code during playback, and editing software reads it. Some cameras embed camera make and model and the time of day of the recording. A few cameras have built-in GPS.

Most cameras have some internal **electronic effects**, including fades to black, date stamp, insertion of titles and colour effects. Never use these in-camera effects and turn off the date stamp if not required. You can replicate any effect later with an editing program, but if you add it in-camera, it can be impossible to remove.

Video cameras have a variety of **connections** to transfer media, view video, and remotely control the camera. Make sure you have all the cables and controllers necessary to transfer video or remotely control the camera.

REFERENCES

Johnson, D. (2001). *How to Do Everything with Your Digital Camera.* Berkley, CA: Osborne/McGraw-Hill.
Morgan D., Sanz, C., Eyana, C. & Ndolo, S. (2008). *Practical Guide for Remote Video Monitoring of Wildlife in the Congo Basin.* Unpublished report to the Wildlife Conservation Society's Global Health Program. Republic of Congo.
Shaw, J. (2000). *John Shaw's Nature Photography Field Guide.* New York: Amphoto Books, an imprint of Watson Guptill Publications.

OTHER USEFUL RESOURCES

Compesi, R. & Sherriffs, R. (1997). *Video Field Production and Editing.* Boston, MA: Allyn and Bacon.
Zettl, H. (1999). *Television Production Handbook.* (7th edn.) San Francisco, CA: Wadsworth.
www.camcorderinfo.com – for reviews of current video cameras
www.dpreview.com – for independent camera and lens evaluation
www.photoxels.com/tutorial_raw.html – for digital photography fundamentals
http://picasa.google.com – for free editing software
www.reconyx.com – for game cameras
www.electrophysics.com – for night acquisition
www.advanced-intelligence.com/video.html – for specialty cameras

18

Chronobiological aspects of primate research

INTRODUCTION

Terrestrial animals live in an environment that undergoes regular variations, ultimately induced by the geophysical conditions prevailing in our solar/Earth system. Solar radiation and gravity, in combination with the Earth's rotation around its inclined axis and its orbit around the sun, and the moon's revolution around Earth, produce marked diurnal, seasonal and lunar as well as tidal periodicities in important physical environmental factors, such as light intensity, ambient temperature, humidity, precipitation, day and night length, and duration of twilight. As a consequence of the superimposition of these periodicities on one another, many relevant biotic environmental factors, such as food availability and predator pressure, as well as social contact, communication and competition with conspecifics, and intra- and interspecific competition for food may also vary diurnally, seasonally, lunar periodically or tidally. In this way, each animal's environment has a highly complex time structure, is highly repetitive in time and thus highly predictable. Reliable predictability provides a good substrate for genetically fixed adaptations. Hence, in addition to other general or specific physiological, ecological and/or behavioural adaptations, animals have also evolved endogenous diurnal (circadian), annual (circannual), lunar (circalunar) and/or tidal (circatidal) rhythms.

In non-human primates (hereafter 'primates'), adaptation to the time structure of the physical and biotic environment is restricted mainly to the development of a circadian timing system, which is involved in the regulation of the pronounced daily (circadian) organization of physiology and behaviour. Although many primate species show a seasonal pattern of reproduction, strong evidence for the existence of a circannual rhythm underlying this rhythmicity has so far

Field and Laboratory Methods in Primatology: A Practical Guide, ed. Joanna M. Setchell and Deborah J. Curtis. Published by Cambridge University Press. © Cambridge University Press 2011.

been found only in rhesus macaques (*Macaca mulatta*; Michael & Bonsall, 1977); weak clues to an endogenous annual rhythm have been obtained in squirrel monkeys (*Saimiri sciureus*; DuMond, 1968), ring-tailed lemurs (*Lemur catta*; van Horn, 1975) and mouse lemurs (*Microcebus murinus*; Petter-Rousseaux, 1975). Circatidal rhythms can be ruled out in primates, and the unique report on an endogenous lunar periodicity in *Saimiri* (Richter, 1968) is not conclusive owing to methodological shortcomings.

Against this background of current knowledge about endogenous timing processes that have evolved in primates, this chapter focusses on diel (24 hour) and circadian rhythmicity. I give special emphasis to fundamentals of comparative chronobiology that you should consider when carrying out primatological research in the field. Ultradian rhythmicity (period length significantly less than 24 h), which may be found in sleeping behaviour, thermoregulatory processes (Chapter 19) or hormone secretion (Chapter 20) is not dealt with here and neither is ovarian cyclicity – although they are also of interest in chronobiology.

DAILY ORGANIZATION OF PHYSIOLOGY AND BEHAVIOUR

Many behavioural activities and most physiological functions of the organism vary regularly over the 24 h day. Several hundred diel rhythms have been established so far in mammals, including primates. The time course of the different parameters may vary considerably over the 24 h day. Some functions show a more or less pronounced unimodal pattern, whereas in others a more bimodal, trimodal or even multi modal pattern may dominate. Furthermore, the amplitude and the phase of the numerous rhythms, as indicated by the time relation of peaks and/or troughs to certain phases of the external day, may differ. However, the mutual phase relationship of these rhythms is usually relatively constant. Thus the whole system has a relatively stable internal time structure (for review see Moore-Ede *et al.*, 1982) that may be essential for the well-being of the individual and for the organism's adaptation to the challenges of its temporally changing environment.

Depending on the part of the solar day in which a given species exhibits most behavioural activities, primates are usually assigned to one of the two main ecotypes of diurnal or nocturnal animals. Almost all haplorrhines – with the exception of the mostly nocturnal South

American owl monkeys (*Aotus* spp., Martin, 1990) – are strictly diurnal. Most strepsirrhines – except a few diurnal Madagascan lemurs such as the indris and sifakas – show a predominantly nocturnal lifestyle (Martin, 1990). True crepuscular species, which limit their daily activity period to dawn and dusk, have not been described in primates. However, some species in the family Lemuridae (Tattersall, 1982; Overdorff & Rasmussen, 1995; Colquhoun, 1998; Donati et al., 2001) and the Argentinian owl monkey *Aotus azarai* (Wright, 1989; Fernandez-Duque, 2003) have been shown to be active during day and night. This kind of activity has been called cathemeral (Tattersall, 2006) and there is ongoing discussion as to whether it represents an ecological adaptation (Tattersall, 1982; Engqvist & Richard, 1991; Pereira et al., 1999), is ancestral to the Lemuridae (Tattersall, 1982; Curtis & Rasmussen, 2006), or represents a transitional stage between a primarily nocturnal and diurnal lifestyle (Kappeler & Erkert, 2003).

THE CIRCADIAN TIMING SYSTEM

Under constant laboratory conditions that do not provide any external 24 h day time cues, i.e. in an environment with constant lighting (LL) of given illuminance or constant darkness (DD), constant ambient temperature and relative humidity, and without daily social contact with conspecifics or the keeper, representatives of all primate species studied so far continue their diel rhythms of activity, core temperature, feeding and drinking for some time. The period of such free-running rhythms usually deviates systematically from the 24 h of the solar day and the approximate 24.83 hours of the lunar day. This reveals an endogenous origin and led to the designation 'circadian rhythms'. According to this definition the term 'circadian' always implies an endogenous origin for the rhythm, and you should only use it if the rhythm has been measured while free-running under constant conditions. Otherwise you should speak of a diel rhythm.

Table 18.1 summarizes the range of the spontaneous periods of the free-running circadian rhythms in the few primate species studied under constant environmental conditions. The circadian period varies from about 21 to 26 h in primates and there is no clear relationship either between circadian period length and systematic affiliation or between period and general ecotype.

Lesion experiments in squirrel monkeys have shown that in primates, as in rodents and all other mammals, the main pacemaker regulating circadian rhythmicity is located in the hypothalamic

Table 18.1 *Circadian periods (τ) in nonhuman primates*

Abbreviations: di, diurnal; no, nocturnal; cat, cathemeral; loc., locomotor activity; feed., feeding; temp., core temperature; PD, physiological darkness, ≤ 10⁻⁶ lux.

	Activity	n	Parameter	Range of luminance (lux)	Period length τ (h)		Source
					$\tau \pm SD$	$\tau_{min} - \tau_{max}$	
Strepsirrhini							
Eulemur fulvus albifrons	cat	5	loc.	0.1 – 240	25.3 ± 0.4	24.1 – 26.5	Erkert & Cramer, 2006
Microcebus murinus	no	30	loc.	DD	22.5 ± 0.6	20.7 – 23.3	Schilling et al., 1999
Galago senegalensis	no	5	loc.	PD – 0.1	23.6 ± 0.6	23.4 – 24.0	Schanz & Erkert, 1987
Otolemur garnettii	no	9	loc.	PD – 0.1	22.6 ± 0.7	21.1 – 23.9	Erkert et al., 2006
Haplorrhini							
Aotus lemurinus griseimembra	no	6	loc.	0.02 – 360	25.1 ± 0.9	24.2 – 26.2	Thiemann-Jäger, 1986
		10	loc.	0.2	24.4 ± 0.4	23.9 – 25.6	Rappold & Erkert, 1994
		12	loc.	0.5	24.3 ± 0.1		Rauth-Widmann et al., 1991
Callithrix jacchus	di	14	loc.	0.1 – 400	23.3 ± 0.3	22.7 – 24.0	Erkert, 1989
		8	loc.	<0.5	23.3 ± 0.4		Glass et al., 2001
Saimiri sciureus	di	8	loc.	0.1 – 400	25.2 ± 0.4	24.5 – 26.2	Aschoff & Tokura, 1986
		16	feed.	1 – 600	25.0 ± 0.5	24.3 – 26.3	Sulzman et al., 1979
			temp.		24.9 ± 0.6	23.3 – 25.8	
Macaca mulatta	di	4	loc.	270	24.0 ± 0.3	23.8 – 24.4	Yellin & Hauty, 1971
		4	loc.	0/0.4 – 300	24.1 ± 0.3	23.8 – 24.9	Martinez, 1972
Macaca nemestrina	di	3	loc.	0.003 – 100	23.1 ± 0.4	22.3 – 23.8	Tokura & Aschoff, 1978
Macaca irus	di	2	loc.	0/50	24.2	23.7 – 24.6	Hawking & Lobban, 1970
Pan troglodytes	di	1	feed.	1 – 85	24.4	23.7 – 25.1	Farrer & Ternes, 1969

suprachiasmatic nuclei (SCN). These paired nuclei are situated on the base of the third ventricle, just above the optic chiasma (Albers *et al.*, 1984; Moore-Ede *et al.*, 1982). Details of the cellular structure of the brain's master clock in the SCN, its transmitter and neuropeptide content, afferent and efferent connections, the molecular mechanisms involved in its rhythm generation and the transmission to the various effector systems can be found in Klein *et al.* (1991) and Moore (1999).

ENTRAINMENT OF CIRCADIAN RHYTHMS BY ZEITGEBERS

Since their spontaneous period usually deviates from the 24 h solar day, circadian systems must be synchronized (entrained) to it by environmental periodicities called zeitgebers. As in other mammals, the most potent zeitgeber cycle that phase-sets a primate's circadian periodicity to the external 24 h rhythmicity is always the natural (or artificial) light–dark cycle (LD). In some species, large-amplitude temperature cycles may also function as a (relatively weak) zeitgeber and entrain circadian rhythmicity under otherwise constant conditions. This has been shown in pig-tailed macaques (*Macaca nemestrina*), squirrel monkeys and common marmosets (*Callithrix jacchus*). Social entrainment, i.e. synchronization of circadian rhythmicity by periodic contact with a conspecific (mate), has been demonstrated in *Callithrix*, but not in *Saimiri*. In *Saimiri*, entrainment was also achieved by feeding cycles (for review see Erkert, 2008).

To become effective as a zeitgeber, LD cycles must have a minimum amplitude and the level of luminance prevailing during the light time must exceed a specific threshold. While, for example, circadian activity rhythms in owl monkeys and mouse lemurs can be entrained by 12:12 h LDs with only about 0.1 lx (corresponding to full moon luminance) during light time and 0.001 lx, or physiological darkness, during dark time, these LDs failed to entrain the free-running circadian rhythm in Senegalese and Garnet's galagos (*Galago senegalensis* and *Otolemur garnettii*) (Erkert *et al.*, 2006; Erkert, 2008). In *Otolemur*, the threshold for photic entrainment was between 3–5 and 30–50 lx. Such a high threshold for photic entrainment in strictly nocturnal species may be of adaptive value in that it protects their circadian system and its mechanism of photic entrainment from perturbations by moonlight.

According to current ideas, entrainment of circadian rhythms by a zeitgeber cycle is achieved by a phase-setting mechanism (detailed by Moore-Ede *et al.*, 1982). This mechanism compensates exactly for the time difference between the circadian period and the 24 h zeitgeber period in such a way that the endogenous rhythmicity adopts a certain phase relationship to the outer day. It is described as the phase angle difference between arbitrarily defined corresponding phases of both rhythms, such as activity onset or end, and sunrise and sunset in diurnal species, or sunset and sunrise in nocturnal species. Changes in zeitgeber parameters such as amplitude (strength) and form (e.g. light-time : dark-time ratio, steepness and duration of rise and fall of light intensity – all of which vary greatly with latitude and season), as well as in certain characteristics of the circadian system, may lead to changes in the phase position of the entrained biological rhythm to its main zeitgeber cycle.

MASKING: DIRECT EFFECTS OF ENVIRONMENTAL FACTORS

Many environmental factors, such as light, ambient temperature, social contact, noise and periodic availability of food, may have a dual effect on primate circadian rhythms in both physiology and behaviour. Besides their phase-setting (entraining) zeitgeber effect, they often lower or raise, inhibit or enhance the basic level of a parameter as determined by the circadian system. Such rhythm-modulating direct effects of external (and internal) factors have been called 'masking effects' (Aschoff et al., 1982). They represent a further control mechanism, in addition to the circadian mechanisms, which is of equal importance for the organism's adaptation to varying environmental factors (Erkert, 2008).

More or less pronounced masking direct effects of light, ambient temperature, relative humidity or social factors may be observed in primates both in their natural environment and under controlled conditions in the laboratory. However, you can only separate the two effects clearly under certain laboratory conditions. As an example, Figure 18.1 shows the activity rhythm of a male Otolemur garnettii in which a large amplitude LD 12:12 of 330:0.3 lx failed to synchronize the free-running circadian rhythm. It did, however, induce pronounced masking, owing to a strong activity-inhibiting direct effect of the high luminance prevailing during the light time of the LD (see also Erkert et al., 2006).

Under natural conditions, strong rhythm-masking direct effects occur in several nocturnal and cathemeral primates, such as Aotus lemurinus, A. azarai and Eulemur fulvus. In all three species, activity

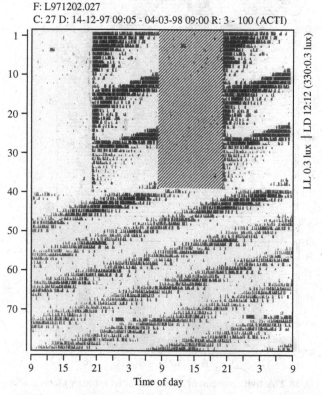

Fig. 18.1 Free-running circadian activity rhythm of a male Garnet's galago (*Otolemur garnettii*) kept in an artificial light–dark cycle (LD 12:12 h) of 330:0.3 lx (days 1–40; L: 08:00–20:00, hatched area) and subsequently (days 41–80) under constant dim light (LL) of 0.3 lx (25 ±1 °C; 60 ±5 %). Owing to the individual's very short spontaneous period, *τ*, of only 21:20 h in LL and 22:20 h in the LD, the zeitgeber period T of 24 h was outside the circadian system's range of entrainment. The LD therefore failed to entrain its free-running endogenous rhythm. However, due to a strong activity-inhibiting direct effect of the high luminance during the light time, it produced a marked diurnal masking of the circadian rhythm, which immediately disappeared after the LD was switched off on day 40.

inhibition by light intensities below full moon luminance (about 0.1–0.5 lx) leads to a lunar periodic modulation of the activity pattern and presumably also of the patterns of other circadian functions (Fig. 18.2; Erkert, 1974; 2008; Fernandez-Duque & Erkert, 2006; Kappeler & Erkert, 2003). Based on the limited data available, nocturnal activity in *Tarsius*, some *Galago* species and some cathemeral lemurs

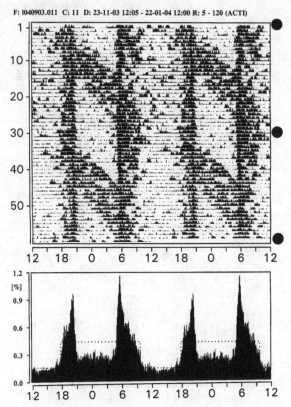

F: I040903.011   C: 11   D: 23-11-03 12:05 - 22-01-04 12:00 R: 5 - 120 (ACTI)

Fig. 18.2 Activity rhythm of a wild male owl monkey (*Aotus a. azarai*) living in its natural habitat, a subtropical gallery forest in the North Argentinean Gran Chaco near Formosa (58° 11' W, 25° 58' S). *Above* Double-plotted original data (5 min values) of a 60 d section (23 Nov. 2003 to 22 Jan. 2004) of a 6 month continuous recording (15 Sept. 2003 to 16 March 2004) carried out with an Actiwatch-AW4® accelerometer motion logger. The spacing between the histogram lines corresponds to 5–120 counts per 5 min interval. Black circles on the right margin indicate new moon nights. Note that intense locomotor activity occurs mainly during dusk and dawn and that there is clear lunar periodic variation in the animal's nocturnal activity which may be attributed to a pronounced activity inhibiting (masking) direct effect of the low light intensities prevailing during the night hours without moonlight. Daytime activity in the morning (06:00–09:00) is more pronounced around new moon, which may be interpreted as a compensatory reaction to the concurrently depressed night-time activity. *Below* Averaging over the entire 60-day record reveals a pronounced bimodal mean activity pattern characterized by two prominent activity peaks during dusk and dawn and a considerably higher activity level at night than during the day. Ordinate values are given in % of average total activity per day. The dotted line indicates the average activity and resting time as calculated by best-fitting a rectangular curve to the averaged activity pattern.

mainly parallels the moonlit night hours, as in the owl monkeys, whereas in other cathemeral lemurs and nocturnal strepsirrhines, such as *Microcebus*, moonlight may be expected to have a less activity-enhancing or even a more inhibitory direct effect on the activity rhythm (Schanz & Erkert, 1987; Erkert, 2008).

Finally, from a chronobiological point of view, almost all variation in environmental factors and changes in the organism's physiological status, such as its hormonal, nutritional, reproductive and/or developmental state, may induce masking effects on circadian or diel rhythmicity, independent of whether they recur regularly or occur stochastically.

ADVICE FOR FIELDWORK

General considerations

All field and laboratory primatological studies should bear in mind that most physiological functions and behavioural activities are organized and structured on a circadian basis. Hence physiologists, behaviourists and ecologists should first of all try to establish the diel pattern(s) of the function(s) of interest, rather than only making observations, taking measurements or collecting samples at one particular time of day (i.e. at only one fixed circadian phase) or at randomly varying circadian phases. Reliable diel patterns can be obtained only when sampling is evenly distributed over the 24 h cycle and sampling intervals are not too large. Avoid mere two-point (e.g. day–night or morning–evening) measurements/samples. Although 'the shorter the better' may be a valid statement in general, choosing the most promising sampling interval for the determination of the diel pattern in a given function is always an optimization task. Spread sampling periods evenly over the animals' entire activity time, even when you are establishing time budgets for the various behavioural activities. In long-term studies that require only one trial or measurement per day, week or month, keep the circadian phase of sampling or at least the time of day (and if necessary also the lunar phase) constant.

Bearing in mind that certain environmental factors may act as entraining and/or masking agents for circadian rhythmicity, and that such factors may vary systematically with latitude, season, lunar phase, and/or time of day, it is obvious that both laboratory and field studies should provide detailed information on all environmental factors that may produce circadian effects or directly influence (mask) diel and circadian rhythms. Among these are, for instance, climate factors such as the time course of ambient temperature and relative humidity (Chapter 5), as well as luminance and all other relevant ecological and social factors.

It is important to determine and describe the exact geographic coordinates (latitude and longitude) of your study area (Chapter 4) and to record the dates of the beginning and end of the study as well as day and night length, duration of twilight and the times of sunrise and sunset and their variation during the study period. Whereas day length, often referred to as 'photoperiod', is clearly defined as the span between the true local times of sunrise and sunset, the duration of night is a matter of defining twilight. According to the differentiation between civil, nautical and astronomical twilight, in addition to the 'solar night' which lasts from sunset to sunrise, we differentiate between civil, nautical and astronomical night. Civil, nautical and astronomical twilight in the evening correspond to the spans lasting from the true local times when the sun's upper margin 'touches' the horizon (about $0°$) to when its centre is $6°$, $12°$ or $18°$ below the horizon, respectively (dusk). In the morning, the span lasts from the true local times at which the centre of the sun crosses $18°$, $12°$ and $6°$ below the horizon, respectively, to when its upper margin crosses the horizon (dawn). Correspondingly, the astronomical, nautical, civil or solar night and day extend between the times of sun positions of -18 to $-18°$, of -12 to $-12°$, of -6 to $-6°$ (centre of disk) and of 0 to $0°$ (upper margin), respectively. When classifying a primate species as diurnal, nocturnal, crepuscular or cathemeral, include the definition of night/nighttime, day/daytime and twilight used. For ease of comparison, it is best to refer to solar day and night in most cases.

When studying primates in the field and evaluating data from such studies, we should be aware that our anthropocentric measurement of (solar) time and the use of certain phases of the 'solar cycle' as reference points (e.g. position of the sun in zenith or at its culmination point corresponds to 12:00 h local time) may lead to problematic conclusions. Also bear in mind that we are diurnal animals, and as such have a diurnally biased view when studying animals with activity rhythms different from our own. Times of day at which light intensity rapidly changes, i.e. the periods of dawn and dusk, are much more important for the phase-setting mechanism of an animals' biological clock or circadian system than the reference points we use. On cloudless new moon days, the luminance prevailing on open areas is about 10^5 lx around midday, 10^3 lx at sunrise and sunset, and during the astronomical twilight it falls to and rises from a relatively constant level of about 5×10^{-4} lx throughout the astronomical night. On clear full moon nights the luminance of the moonlit sky amounts to about 3×10^{-1} lx. Thus, according to phase and position of the moon, luminance during the night may vary by a factor of about 1:1000, which is more than the relative change in light intensity between sunrise and midday or midday and sunset, when luminance varies only by a factor of about 1:100. Cloud cover can produce a reduction in light intensity of up to 90% or even 99%, and vegetation, depending on its density, can lead to a reduction in luminance by factors of up to 10^3–10^4. As an example, Curtis *et al.* (1999) give data from luminance measurements in a Madagascan seasonally dry, secondary forest, where the amount of light transmitted through the canopy varied 0.4%–1.5% during the wet season and 1%–16 % during the dry season. These relationships show that the best reference phases of the environmental LD-zeitgeber pattern to which the onset and end of the primates' activity times and activity patterns should be related are the times of sunrise and sunset, and the beginning and end of astronomical twilight.

Give all times as true local time, i.e. corrected according to latitude and longitude of the study site. Never use summer time when carrying out studies in the field or under semi-natural conditions. This quite artificial and annually changing mode of time measurement may lead to considerable confusion and misinterpretation. Also ensure that the clock on any computer used for data acquisition, programming and reading out data loggers is not set to summer time. You can request local times of sunrise and sunset, beginning and end of civil, nautical and astronomical twilight, as well as the dates of the four phases of the moon and the times of moonrise and moonset of the year and period of the study, from national institutions and authorities for astronomy, or obtain them from tables in *The Astronomical Almanac* issued every year by the Nautical Almanac Office of the United States, Washington, D.C., and Her Majesty's Nautical Almanac Office, London. Alternatively, you can obtain these data more easily from the web site of the US Naval Observatory data services or by using a software package (New Moon V1.0) designed for this purpose (Thomas & Curtis, 2001).

In nocturnal and cathemeral and probably also some diurnal primate species, the level of locomotor activity and other behaviours, as well as the body temperature, may depend to different degrees on the luminance prevailing during the night hours. Therefore, you may need to consider the lunar cycle, i.e. the lunar phase and the hours of moonlight during sampling nights, in field studies. In all cases in which you observe strong direct effects of moonlight on behaviour and/or the recorded physiological functions, evaluate (i.e. average) data separately for the four phases of the moon or, at least, for new moon and full moon days. When doing this, use both the data from the day and night in which the respective phase of the moon (new moon, waxing moon, full moon, waning moon) occurs, and data for one or two days before and after it. This compensates to some degree for the effects of accidental variations in luminance, due to varying degrees of cloud cover or unfavourable weather conditions.

Measuring luminance

From a chronobiological point of view, the lighting conditions prevailing throughout a study are one of the most important environmental factors. Continuous recordings of light intensity during fieldwork are thus necessary. For luminance measurements, choose only devices with a sensor that has a spectral sensitivity that corresponds approximately to the spectral sensitivity of your study species. Since most primate eyes have a spectral sensitivity similar to that of the human eye (Kremers *et al.*, 1999), it seems reasonable to measure light intensities in lux or footcandles (1 fc corresponds to about 10.71 lx). Many photometers (luxmeters) have sensors adapted to human spectral sensitivity; some others, such as the Optometer P9710, allow you to choose between sensors with a spectral sensitivity corresponding to photopic (day) or to scotopic (night) vision. Only use the latter when studying strictly nocturnal species that possess an almost pure rod retina and that start activity well after sunset and end it long before sunrise.

For measurement, use a cosine-corrected cell and always direct it vertically towards the sky in an open area. You can calculate reduction factors to estimate the light intensities to which animals are really subjected during their activity by taking simultaneous reference measurements at sites that they use preferentially. Depending on habitat structure, primates may be subjected to a very broad range of illumination intensities ($<10^{-4}$–10^5 lx). Their eyes code light intensity logarithmically over a broad dynamic range. Consider this when measuring and describing the luminance prevailing at certain sites and times in a study area. Report the data rounded up or down on a logarithmic scale to one decimal place (e.g. $2.3 \pm 0.2 \times 10^{-1}$ lx). Use luxmeters that are as sensitive as possible (minimum threshold of at least 10^{-3} lx) and that allow measurements over a broad range of illumination intensities (10^4/10^5 lx) for field studies of nocturnal or cathemeral species. Less sensitive devices covering a range from about $10^{-1}/10^0$ to $10^4/10^5$ lx may be sufficient when studying strictly diurnal species. Digital devices and data-loggers are very useful (e.g. Optometer P9710; or similar devices, see 'Useful Internet sites'), but only use those that have a sensor with an adequate spectral sensitivity and allow measurements over a broad range of light intensities. STOWAWAY® and HOBO® light intensity loggers, for instance, are biased because the spectral sensitivity of their sensors is strongly shifted to the near-infrared (about 930 nm), which is invisible to primate eyes and hence ineffective for the circadian phase-setting mechanism and for eliciting masking effects. Light intensity loggers designed for chronobiological research in humans, such as Actiwatch-L®, Actiwatch-L plus®, Actiwatch Spectrum®, Actilight®, and Micro Light Sensor LS®, are more suitable because they measure illumination intensity in lux. However, the range of measurement is limited to 1–4.000 lx (resolution 1 lx) in the Micro LS®, 0.5–40 000 lx (resolution 0.1–200 lx, depending on the luminance level) in the Actiwatch-L® and Actilight®, and 0.01–50 000 lx (0.01 lx intervals) in the Actiwatch-L plus®. Depending on the sampling interval you use, these loggers allow you to record light intensity automatically over periods of up to 35–45 days. Unfortunately, these devices and the appropriate software are very expensive and need to be specially adapted for use in primatological fieldwork.

Recording activity

The best and most common, although most time-consuming and strenuous, way to record primate activities in the field is through observation. You can do this either directly with a good pair of binoculars (7–10 ×) in diurnal species and/or with a sensitive night vision device (starlight scope) in nocturnal species, or indirectly by normal or infrared videotaping (Chapter 17) and later evaluation on screen. Both options require you to follow individuals or groups of interest throughout several entire activity periods and record the behaviours of focal animals for fixed successive time intervals (e.g. 5 or 10 minutes). The great advantage of recording primate activities by observation is that you can establish the diel patterns of various behavioural activities like travelling, foraging, grooming, play or other social interactions and resting, and determine individual and/or group time budgets simultaneously (cf. Alonso & Langguth, 1989). Continuous long-term recordings of locomotor activity, core temperature, heart rate and/or haemodynamic parameters in freely moving primates are best carried out with automated radiotelemetric recording devices (Chapters 10 and 19). A reasonably good, economical method for recording primate locomotor activity in the laboratory or in outdoor enclosures is to use passive infrared motion detectors (PIDs), available from suppliers of electronic and/or security equipment. Combined with a program-controlled multiplexer and counter device (i.e. an adequately programmed computer), PIDs provide a simple and efficient device for automatic long-term activity recordings, which can be adapted for almost all primate species (Lerchl et al., 1988). However, this method detects only passages of infrared sources in a given area or space. Hence, it allows only activity registrations of singly kept individuals or of the entire group. For some research questions, you may need to use several PIDs and focus the sensors on different sites in the enclosure, such as feeding and resting sites or the entrance to the sleeping box. This yields a more differentiated picture of space use over time and/or the onset and end of daily activity. If you use PIDs for automatic activity records, you will need to validate the method initially by direct observation or videotaping to find out which kinds of behaviour and locomotion contribute to the activity recorded.

In the field, primate locomotor activity and activity patterns are best recorded continuously over long time-spans by radio-tracking (Chapter 10). Use either the distance covered or the time moving around per time unit (usually per hour) as a measure of locomotor activity. Alternatively, record changes in the transmitter signal intensity and use these as an indicator of motor activity.

An alternative method for automatically recording the (loco)motor activity of individual wild-living or captive primates over weeks, months or even years is to use small motion loggers (accelerometers) designed for chronobiological research in humans. At present only a few devices can be adapted for use in medium-sized to larger primates (> 1 kg): Actiwatches® AW4, AW 7, AWM (Mini) and Actical®, the Micro Mini-Motionlogger® and the Micro SleepWatch Motionlogger®. Depending on memory capacity, the sampling interval chosen (5 minutes is a suitable standard for field studies) and battery life, Actiwatch devices allow continuous activity records of up to about 210 days for the AW4 (with a lithium-ion battery CR2032, RENATA®, Switzerland) or about 450 days for the AWM®. Although the AW4 has already been used successfully in primatological field studies (see below), the other devices have not yet been tested. However, the Mini Actiwatch® and Actical® devices have been used quite successfully in laboratory studies of Callithrix jacchus (E. Fuchs,

Fig. 18.3 Common marmoset (*Callithrix j. jacchus*) carrying a coated globechain collar (left) with an Actiwatch Mini® accelerometer motion logger (right) protected against physical damage and seeping in of water and urine by several layers of adhesive tape. Photos kindly provided by Dr Christina Schlumbohm, German Primate Center, Göttingen, Germany.

personal communication; Fig. 18.3, left) and *Macaca mulatta* (Hunnell *et al.*, 2007), respectively. The Actical device only allows you to sample epoch lengths of up to 1 minute, meaning that its recording time is limited to 44 days. Hence I recommend it for use in captive settings but not for activity recordings in wild-living primates, for ánimal welfare reasons. In contrast, based on a very long recording time of up to 450 days at a programmable sampling interval of 5 minutes, the small and lightweight Actiwatch Mini (Fig. 18.3 right; 7.5 g, depth 7 mm, diameter 24 mm) should be tested for suitability in primatological field research.

All devices require a computer with an interface and dedicated software to program (initialize) and read the loggers and evaluate the data recorded. Protect the activity loggers against mechanical damage, humidity and water by wrapping them in thin, waterproof polythene sheeting (available in supermarket household departments) and fitting them into a small, completely closed, hard cover (e.g. aluminium) box, which you can fix to a collar and put on the animal's neck. This collar should not be too broad and should be as flexible as possible. For larger primate species, such as macaques, you can obtain aluminium housings for AW4 and Actical devices and special collars commercially. Alternatively, you can protect devices such as the AW4 by using several layers of strong adhesive tape (e.g. Tesaband®), fix them to a plastic-coated, flexible double-globechain collar and waterproof them using acrylic spray. This is especially recommended for species with a relatively short neck, such as the owl monkey, in which a broader (>8mm) rigid collar may cause abrasions dorsally.

As an example of long-term actimetry with an Actiwatch® AW 4 motion logger, Fig. 18.2 (upper diagram) shows a two month sector of the activity rhythm of a wild-living owl monkey (*Aotus a. azarai*). Erkert (2008) gives further demonstration of the potential of long-term actimetry with AW 4® devices in various wild-living diurnal, nocturnal and cathemeral primate species, as well as of some methods of data evaluation.

SUMMARY AND CONCLUSIONS

Adequate timing of behavioural activities and physiological functions is an essential prerequisite for survival and reproduction. Thus, you should consider the fundamentals of chronobiology when planning and carrying out primatological field studies or laboratory research.

1.   The mammalian organism shows a distinct circadian organization, and most of its physiological functions and behavioural activities undergo more or less pronounced daily (circadian) variations regulated by a genetically fixed endogenous timing system.
2.   The spontaneous period produced by the circadian system usually deviates from 24 h and must therefore be phase-set to the outer solar day by certain environmental periodicities called zeitgebers, the most important of which is the light–dark cycle (LD).
3.   The entrained and free-running circadian rhythms may be modulated (masked) more or less intensely by inhibitory or enhancing direct effects of certain environmental and/or internal factors.

These facts lead to the following conclusions for primate research.

1.   Whenever possible you should establish the average diel patterns of the behavioural or physiological function(s) of interest, or at least carry out sampling out at an invariable circadian phase.
2.   Adequate assessment of the prevailing zeitgeber and masking conditions requires detailed information on the geographical

coordinates of the study site, the season or vegetation period of data collection, the lighting conditions or times of sunrise, sunset and twilight duration, as well as the ambient temperature, precipitation, food abundance, predator pressure, social system, etc.

3. Since moonlight may exert more or less pronounced masking effects on the entrained circadian rhythms of locomotor activity and other parameters in most nocturnal strepsirrhines and owl monkeys, you should note moon phases and check for moonlight effects in field studies on such species.

4. Only use devices (luxmeters) that have a broad sensitivity range and a spectral sensitivity corresponding to that of the study species for luminance measurements.

5. Out of general chronobiological interest, field studies of a given primate species and/or population should record and report the timing of onset and end of the activity and/or the sleeping period in relation to sunrise or sunset.

6. You can adapt accelerometer data-loggers for automated quantitative locomotor activity recordings in the field. However, for animal welfare reasons, only use devices that allow continuous recording over at least 5-6 months. You can make more economical recordings in outdoor enclosures or in laboratory set-ups for medium-sized species using computer-based passive infrared sensors. However, all automatic recordings first require calibration by comparing the recordings from several days with simultaneous behavioural observations.

REFERENCES

Albers, H. E., Lydic, R., Gander, P. H. & Moore-Ede, M. C. (1984). Role of suprachiasmatic nuclei in the circadian timing system of the squirrel monkey. I. The generation of rhythmicity. *Brain Res.* 300, 275–84.

Alonso, C & Langguth, A. (1989). Ecologia e comportamento de *Callithrix jacchus* (Primates: Callitrichidae) numa ilha de floresta atlantica. *Rev. Nordest. Biol.* 6, 105–37.

Aschoff, J., Daan, S. & Honma, K. I. (1982). Zeitgebers, entrainment, and masking: some unsettled questions. In *Vertebrate Circadian Systems. Structure and Physiology*, ed. J. Aschoff, S. Daan & G. A. Groos, pp. 13–24. Berlin: Springer-Verlag.

Aschoff, J. & Tokura, H. (1986). Circadian activity rhythms in squirrel monkeys: entrainment by temperature cycles *J. Biol. Rhythms* 1, 91–9.

Colquhoun, I. C. (1998). Cathemeral behaviour of *Eulemur macaco macaco* at Ambato Massif, Madagascar. *Folia Primatol.* 69, 22–34.

Curtis, D. J. & Rasmussen, M. A. (2006). The evolution of cathemerality in primates and other mammals: a comparative chronoecological approach. *Folia Primatol.* 77, 178–93.

Curtis, D. J., Zaramody, A. & Martin, R. D. (1999). Cathemeral activity in the mongoose lemur, *Eulemur mongoz. Am. J. Primatol.* 47, 279–98.

Donati, G., Lunardi, A., Kappeler, P. M. & Borgognini Tarli, S. M. (2001). Nocturnal activity in the cathemeral red-fronted lemur (*Eulemur fulvus rufus*), with observations during a lunar eclipse. *Am. J. Primatol.* 53, 69–78.

DuMond, F. V. (1968). The squirrel monkey in a seminatural environment. In *The Squirrel Monkey*, ed. L. A. Rosenblum & R. W. Cooper, pp. 87–145. New York: Academic Press.

Engqvist, A. & Richard, A. (1991). Diet as a possible determinant of cathemeral activity patterns in primates. *Folia Primatol.* 57, 169–72.

Erkert, H. G. (1974). Der Einfluss des Mondlichtes auf die Aktivitätsperiodik nachtaktiver Säugetiere. *Oecologia* 14, 269–87.

Erkert, H. G. (1989). Characteristics of the circadian activity rhythm in common marmosets (*Callithrix j. jacchus*). *Am. J. Primatol.* 17, 271–86.

Erkert, H. G. (2008). Diurnality and nocturnality in non-human primates: comparative chronobiological studies in laboratory and nature. *Biol. Rhythm Res.* 39, 229–67.

Erkert, H. G. & Cramer, B. (2006). Chronobiological background to cathemerality: circadian rhythms in *Eulemur fulvus albifrons* (Prosimii) and *Aotus azarai boliviensis* (Anthropoidea). *Folia Primatol.* 77, 87–103.

Erkert, H. G., Gburek, V. & Scheideler, A. (2006). Photic entrainment and masking of prosimian circadian rhythms (*Otolemur garnettii*, Primates). *Physiol. Behav.* 88, 39–46.

Farrer, D. N. & Ternes, J. W. (1969). Illumination intensity and behavioural circadian rhythms. In *Circadian Rhythms in Nonhuman Primates*, ed. F. H. Rohles, *Bibl. Primat.* 9, 1–7. Basel: S. Karger.

Fernandez-Duque. E. F. (2003). Influences of moonlight, ambient temperature and food availability in the diurnal and nocturnal activity rhythms of owl monkeys (*Aotus azarai*). *Behav. Ecol. Sociobiol.* 54, 431–40.

Fernandez-Duque, E. F. & Erkert, H. G. (2006). Cathemerality and lunar periodicity of activity rhythms in owl monkeys of the Argentinean Chaco. *Folia Primatol.* 77, 123–38.

Glass, J. D., Tardiff, S. D., Clements, R. & Mrosowsky, N. (2001). Photic and non-photic circadian phase resetting in a diurnal primate, the common marmoset. *Am. J. Physiol.* 280, R191–7.

Hawking, F. & Lobban, M. C. (1970). Circadian rhythms in macaca monkeys (physical activity, temperature, urine and microfilarial levels). *J. Interdisc. Cycle Res.* 1, 267–90.

Hunnell, N. A., Rockcastle, N. J., McCormick, K. N. *et al.* (2007). Physical activity of adult female rhesus monkeys (*Macaca mulatta*) across the menstrual cycle. *Am. J. Physiol.* 292, E1520–5.

Kappeler, P. M. & Erkert, H. G. (2003). On the move around the clock: correlates and determinants of cathemeral activity in wild redfronted lemurs (*Eulemur f. fulvus*). *Behav. Ecol. Sociobiol.* 54, 359–69.

Klein D. C., Moore, R. Y. & Reppert, S. M. (eds.) (1991). *Suprachiasmatic Nucleus. The Mind's Clock.* New York, Oxford: Oxford University Press.

Kremers, J., Silveira, L. C. L., Yamada, E. S. & Lee, B. B. (1999). The ecology and evolution of primate color vision. In *Color Vision. From Genes to Perception*, ed. K. R. Gegenfurter & L. T. Sharpe, pp. 123–42. Cambridge: Cambridge University Press.

Lerchl, A., Küderling, I., Kurre, J. & Fuchs, E. (1988). Locomotor activity registration by passive infrared detection in saddle back tamarins and tree shrews. *Physiol. Behav.* 44, 281–4.

Martin, R. D. (1990). *Primate Origins and Evolution. A Phylogenetic Reconstruction.* London: Chapman & Hall.

Martinez, J. L. (1972). Effects of selected illumination levels on circadian periodicity in the rhesus monkey (*Macaca mulatta*). *J. Interdisc. Cycle Res.* 3, 47–59.

Michael, R. P. & Bonsall, R. W. (1977). A 3-year study of an annual rhythm in plasma androgen levels in male rhesus monkeys (*Macaca mulatta*) in a constant laboratory environment. *J. Reprod. Fert.* 49, 129–31.

Moore, R. Y. (1999). Circadian timing. In *Fundamental Neuroscience*, ed. M. J. Zigmond, F. E. Bloom, S. C. Landis, J. L. Roberts & L. R. Squire, pp. 1189–206. San Diego, CA: Academic Press.

Moore-Ede, M. C., Sulzman, F. M. & Fuller, C. A. (1982). *The Clocks that Time Us. Physiology of the Circadian Timing System.* Cambridge, MA, London: Harvard University Press.

Overdorff, D. J. & Rasmussen, M. A. (1995). Determinants of nighttime activity in "diurnal" lemurid primates. In *Creatures of the Dark: the Nocturnal Prosimians*, ed. L. G. Alterman, G. A. Doyle & K. Izard, pp. 61–74. New York: Plenum Press.

Pereira, M. E., Strohecker, R. A., Cavigelli, S. A. Hughes, C. L. & Pearson, D. D. (1999). Metabolic strategy and social behavior in Lemuridae. In *New Directions in Lemur Studies*, ed. B. Rakotosamimanana, H. Rasamimanana, J. U. Ganzhorn & S. M. Goodman, pp. 93–118. New York: Kluwer Academic/ Plenum Publishers.

Petter-Rousseaux, A. (1975). Activité sexuelle de *Microcebus murinus* (Miller 1777) soumis à des régimes photopériodiques experimenteaux. *Ann. Biol. Anim. Biochem. Biophys.* 15, 503–8.

Rappold, I. & Erkert, H. G. (1994). Re-entrainment, phase-response and range of entrainment of circadian rhythms in owl monkeys (*Aotus lemurinus g.*) of different age. *Biol. Rhythm Res.* 25, 133–52.

Rauth-Widmann, B., Thiemann-Jäger, A. & Erkert, H. G. (1991). Significance of nonparametric light effects in entrainment of circadian rhythms in owl monkeys (*Aotus lemurinus griseimembra*) by light-dark cycles. *Chronobiol. Int.* 8, 251–66.

Richter, C. P. (1968). Inherent twenty-four and lunar clocks of a primate – the squirrel monkey. *Comm. Behav. Biol.* 1, 305–32.

Schanz, F. & Erkert, H. G. (1987). Resynchronisationsverhalten der Aktivitätsperiodik von Galagos (*Galago senegalensis, Galago crassicaudatus garnettii*). *Z. Säugetierk.* 52, 218–26.

Schilling, A., Richard, J. P. & Servière, J. (1999). Duration of activity and period of circadian activity-rest rhythm in a photoperiod-dependent primate, *Microcebus murinus*. *C. R. Acad. Sci.* 322, 759–70.

Sulzman, F. M., Fuller, C. A. & Moore-Ede, M. C. (1977). Environmental synchronizers of squirrel monkey circadian rhythms. *J. Appl. Physiol.* 43, 795–800.

Sulzman, F. M., Fuller, C. A. & Moore-Ede, M. C. (1979). Tonic effects of light on the circadian system of the squirrel monkey. *J. Comp. Physiol.* 129, 43–50.

Tattersall, I. (1982). *The Primates of Madagascar.* New York: Columbia University Press.

Tattersall, I. (2006). The concept of cathemerality: history and definition. *Folia Primatol.* 77, 7–14.

The *Astronomical Almanac*. Issued each year by The Nautical Almanac Office, USA and Her Majesty's Nautical Almanac Office, Royal Greenwich Observatory. Washington, D.C.: US Government Printing Office; London: HMSO.

Thiemann-Jäger, A. (1986). *Charakteristika der circadianen Aktivitätsperiodik von Nachtaffen* (Aotus trivirgatus Humboldt 1811). Unpublished dissertation, University of Tübingen.

Thomas, R. M. & Curtis, D. J. (2001). A novel software application for the study of photoperiodic cueing mechanisms underlying circadian and circannual rhythms and lunar-periodic modulations. *Folia Primatol.* 72, 187.

Tokura, H. & Aschoff, J. (1978). Circadian activity rhythms of the pig-tailed macaques *Macaca nemestrina*, under constant illumination. *Pflügers Arch.* 376, 241–3.

van Horn, R. N. (1975). Primate breeding season: photoperiodic regulation in captive *Lemur catta*. *Folia Primatol.* 24, 203–20.

Wright, P. C. (1989). The nocturnal primate niche in the New World. *J. Hum. Evol.* 18, 635–46.

Yellin, A. M. & Hauty, G. T. (1971). Activity cycles of the rhesus monkey (*Macaca mulatta*) under several experimental conditions, both in isolation and in a group situation. *J. Interdisc. Cycle Res.* 2, 475–90.

USEFUL INTERNET SITES

http://www.camntech.com – for Actiwatch AW4, AW7, AWM, Actilight, Actiwatch-L/L plus®.

http://www.minimitter.com; http://www.actiwatch.respironics.com – for AW4, Actical, Actiwatch Spectrum, aluminium housings and collars.

http://aa.usno.navy.mil/USNO/astronomical-applications – Astronomical Applications Department of the US Naval Observatory.

http://www.ambulatory-monitoring.com – for MicroMini-Motion logger and Micro SleepWatch®.

www.psgb.org/Software/index.html – for New Moon V1.0 software

http://www.gigahertz-optik.de – for Optometer P9710

www.onsetcomp.com – for STOWAWAY® and HOBO® light intensity loggers.

19

Thermoregulation and energetics

INTRODUCTION

The study of how animals apportion time and energy (energetics) can provide much insight into physiology, ecology and evolution (Bartholomew, 1982; Geiser, 2004; Schmidt-Nielsen, 1997). Body temperature has a profound effect on the ability of animals to function effectively. Since all animals generate heat internally to some extent, energetics is closely linked to the problem of heat management and thermoregulation. For example, homeothermic or 'warm-blooded' animals (birds and mammals) must produce a great amount of heat in order to maintain a high and constant body temperature in cold as well as in warm surroundings (Schmidt-Nielsen, 1997). Moreover, natural environments can be extremely variable in their thermal attributes and consequently animals show behavioural and physiological adaptations that enable them to cope with these external gradients.

Over the past decade, there has been a tremendous increase in methodologies and techniques applicable to studies of energy expenditure and thermoregulation. This is particularly true for the study of daily energy requirements and body temperature rhythms of wild animals behaving normally in their natural habitats. In this contribution, I therefore review methods for the study of energetics and mechanisms of temperature regulation in primates (although these can generally be applied to almost all mammal species). I also briefly mention other possibilities for physiological measurements. All procedures described here require capture of the study animals, and some require invasive surgical intervention. They therefore raise ethical questions when dealing with wild animals and/or endangered species, and require governmental permits and authorization.

Field and Laboratory Methods in Primatology: A Practical Guide, ed. Joanna M. Setchell and Deborah J. Curtis. Published by Cambridge University Press. © Cambridge University Press 2011.

SCIENTIFIC TERMS AND DEFINITIONS

BMR (basal metabolic rate): The energy required by an animal when it is resting and not performing any metabolic work to digest food or maintain its body temperature within its **thermoneutral zone** (the temperature range within which the metabolic heat production is unaffected by ambient temperature). BMR can only be measured under controlled conditions in the lab.

RMR (resting metabolic rate): The energy required by a calm resting animal. This can be measured in the field, and refers to the metabolic rates of animals that are measured when they are resting without any further information of what else they may be doing (e.g. digesting).

DEE (daily energy expenditure): The total amount of energy an animal expends during an entire day (including the BMR, costs of thermoregulation, locomotion, digestion, reproduction, etc.).

FMR (field metabolic rate): DEE of a wild animal.

RQ (respiratory quotient): The ratio of carbon dioxide produced to oxygen consumed (gives information about the fuel used in metabolism).

T_b**core (core body temperature)**: The internal temperature of an animal's body.

T_b**skin (skin body temperature)**: The external skin temperature of an animal's body.

T_a **(ambient temperature)**: The temperature of the surrounding environment.

STUDYING ENERGY EXPENDITURE

Indirect calorimetry (respirometry)

Basal and maximal energy expenditures, and energy demands while engaged in specific activities, can be measured routinely using indirect calorimetry. This standard procedure involves measuring the heat production of an animal indirectly from its gas exchange. The metabolic rate of an animal is measured as the rate of oxygen consumption or the rate of CO_2 production and then converted to energy expenditure (for the caloric equivalent, see Schmidt-Nielsen, 1997). For respirometry measurements, subjects are restricted to a relatively small chamber. In general, the size of the animal you can study is limited by the power of the gas analysis system, and the size of the metabolic chamber available. To conduct respirometry studies you will need to collaborate with specialists who have a fully equipped laboratory at their disposal.

Respirometry systems involve extensive and sensitive technical equipment; until approximately two decades ago, measurements could only be carried out in the laboratory. Nowadays, metabolic rate can be measured for animals in the field using portable O_2- and CO_2-analyser assemblies and portable electrical supplies (see, for example, Schmid et al., 2000). These are available from various manufacturers (e.g. FoxBox from Sable Systems; see 'List of suppliers'). They generally measure O_2 and CO_2 simultaneously and include a variable pump and a mass flow meter. This portable gas analysis system is ideal, because free-living animals in their natural environment have an extensive behavioural repertoire and thus physiological data for wild animals can contribute a great deal to our understanding of rates of energy utilization. Another advantage of this technique is that if your study animal naturally uses tree holes or burrows as sleeping places, you can use these sleeping hollows as metabolic chambers, which significantly reduces disturbance to the animal. Alternatively, you can keep individuals in outdoor enclosures surrounded by the species' habitat under conditions of natural photoperiod and temperature. Equip the enclosures with sufficient vegetation (depending on the species) and one or several nest-boxes, which you can provide as sleeping quarters as well as use as metabolic chambers for the measurement of metabolic rate. The advantage of the latter set-up is that the animals under investigation remain in the enclosure, and thus continuous measurements are possible without moving the experimental equipment. The estimates of O_2 consumption and CO_2 production made using such field systems are, however, not as reliable and accurate as those from laboratory systems.

A general problem of all standard respirometry techniques is that animals cannot behave naturally in the small area of a metabolic chamber (whether you use a natural chamber in the field or an artificial chamber in the laboratory).

Time–energy budgets

An alternative approach to estimating the daily energy demands of wild animals is to prepare time–energy budgets. In theory, you should watch the animal carefully throughout the day, and accumulate records of the time spent in each activity. In primates, this is generally done by radio-tracking and direct observation (Chapters 10 and 11). To convert this time budget to an energy budget, you need knowledge of the energy expenditure associated with each behaviour logged. Various approaches have been employed to do this (see, for example, Schartz & Zimmermann, 1971; Weathers *et al.*, 1984). However, time and energy budgeting has general problems as a method for estimating daily energy demands. It is often difficult to record the behaviours of the subject animals in any great detail; assigning realistic values for the energy expenditure associated with the different behavioural classes of the time budget can be problematic; and finally, obtaining a detailed time budget is very time-consuming.

Doubly labelled water

Theory

Energy expenditure and water metabolism can be measured routinely in both the field and the laboratory by means of doubly labelled water $(D_2{}^{18}O)$ (Lifson & McClintock, 1966; Speakman, 1997). In principle, this measures the CO_2 production of the animal in a quantitative manner, and this is then converted to energy demands. The technique involves labelling an animal's body water with two isotopes, deuterium (^2H) and heavy oxygen (^{18}O), and determining the washout rates of both isotopes. The ^2H is incorporated into water molecules in the animal and traces water flux through the animal. The ^{18}O equilibrates rapidly between water and CO_2 dissolved in body water, and so washes out faster than ^2H, because it leaves the animals as $C^{18}O_2$ as well as $H_2{}^{18}O$. Thus, the difference between the washout rates of deuterium and ^{18}O represents CO_2 production alone and is a measure of metabolic rate. Doubly labelled water probably provides the most powerful method for the measurement of CO_2 production, and hence energy expenditure, of free-living animals over protracted periods of one to several days, and can be applied just as well to animals kept in captivity.

The doubly labelled water method gives an integrated estimate of energy demands over relatively long periods (days). By combining several different measurements for animals with varying time budgets, you can factor out the costs of major activity components. Although the doubly labelled water method is less accurate than respirometry, the advantages are less disturbance for the animal, no need for electrical supplies, and that the resultant estimate includes passive and active phases of metabolism. Furthermore, it is easy to make simultaneous measurements on several individuals, and to measure individuals in their natural habitat and social situation. There are three major disadvantages to the doubly labelled water approach: first, a complete and successful measurement requires recapture of the injected individual within a certain period of time; second, it is very expensive; and finally, the need for capture and manipulation of animals means that it may not be ethical to conduct such research on endangered species and you may also have problems obtaining permission from the local authorities to carry out such research.

Practice

There are several important factors to be considered during the execution of a study using doubly labelled water. The technique is relatively complex and difficult to perform and it is therefore indispensable that you contact a specialist and consult the relevant literature for a precise description. Details concerning the efficacy of the technique in its application to various animal species can be found in Speakman (1997). Here, I outline the general procedure and important aspects that need to be taken into account when planning a study.

The practical limit on the minimum size of animals that can be studied by using doubly labelled water is a body mass of about 0.9–1.2 g (Speakman, 1997). The upper size limit relates only to the costs and availability of sufficient isotopes to label the animal. Inject the animals selected for the experiment with doubly labelled water ($D_2{}^{18}O$). The quantity of isotopes injected depends on the body mass of the animal, the duration of the experimental measurement, and the enrichment of the injectate. The minimum duration of an experimental measurement is defined by the time required for sufficient divergence of the labels, and for sufficient isotopes to have been washed out of the body. It is important you know the exact quantity of stable isotopes administered and the exact body mass of the animal. Use precise electronic balances (in the laboratory) or less accurate spring balances (in the field) for small animals. After injection, hold the animals captive to allow complete equilibration of the isotopes in the body water. Estimates of time taken for isotopes to reach equilibrium vary from about 15–60 min (body mass less than 100 g) to several hours (humans) (Speakman, 1997). Then take an initial sample. The most desirable sample sources are blood and saliva, since both are direct measures of the instantaneous state of the body water pool. Urine samples are not suitable for two reasons. First, the bladder acts as a water reservoir, which is not in complete exchange with body fluids, and bladder content therefore represents an integrated sample over an unknown time period prior to the sample collection (Speakman, 1997). Second, you are unlikely to be able to obtain a urine sample at a specific time of day as you cannot control the timing of urination!

The best approach for blood sampling small animals is some form of peripheral venipuncture (Chapter 8). Flame seal samples for isotope analyses into capillary tubes. Capillaries are best protected in sample tubes containing some cotton wool to avoid breakage. You can store them at room temperature (around 10–20 °C) but do not leave them in the cold because they burst if frozen.

Take a second or final sample 24 h to several days after the first sample (depending on the experimental measurement). You therefore need to recapture the injected animals. The problem is, however, that recapture cannot be guaranteed for any released animal in the field, particularly not within a certain time frame (before the isotopes have been washed out of the body).

Before using doubly labelled water, consider the following. Is one measurement for a given individual sufficient to provide reliable data? If not, what are the ethical implications of multiple captures and recaptures on one animal in terms of disturbing the animals and possibly also their place in the hierarchy within their group (Chapters 7 and 8)?

Which technique to use

To decide which is the most appropriate technique to measure the energy expenditure of animals you must weigh up the advantages and disadvantages of the various alternative procedures. Furthermore, there are several important factors and practical aspects to be considered when choosing a particular methodology:

- Are the experiments to be carried out in the laboratory or in the field?
- Does the chosen technique provide a sufficiently reliable answer to the specific question asked?
- What animal do you wish to study and does its natural activity or anatomy rule out the possibility of using a particular technique?
- What training is required for the application of a particular technique? How and where will you fulfil these training requirements and how much time is needed?

- What are the licensing requirements of the proposed procedures and will they be acceptable to the legal authorities of the host country?

STUDYING BODY TEMPERATURE

Core body temperature

You can measure the ($T_b$core) of an animal routinely by means of implanted temperature-sensitive transmitters. There are two types of transmitter: temperature-sensitive radio transmitters (Mini Logger®, accuracy of ±0.1 °C) or data-loggers (Thermochron ibutton®, temperature range from –40 °C to +85 °C; accuracy of ±0.5 °C). The major difference between the two types is that data-loggers have internal memories and store data automatically. In contrast, you need to record signals from radio transmitters manually or automatically by using a telemetry assembly consisting of receiver and antenna (Chapter 10). When using data-loggers, recorded data are downloaded to a computer via its serial or parallel interface. Setting and resetting of the data-logger, retrieving the data and sampling methods definition are performed by custom-made computer programs. The disadvantage of using data-loggers is that you must recapture the same individual to download the data, or you lose the data. If you use radio transmitters, you need to calibrate each transmitter against a mercury thermometer in a water bath before implantation. Seal both data-loggers and transmitters with a paraffin-wax coating before implantation into the animal to minimize the risk of infections or allergic reactions.

Weigh animals to the nearest 0.5 g before surgery. You can implant transmitters either into the peritoneal cavity or under the skin on the animal's back. The use of subdermal implantation depends on the size and anatomy of the species under investigation. Carry out all operations under deep anaesthesia (Chapter 8). Allow animals to recover from the surgery for several hours before release. Repeat surgery to remove the transmitters when the experiment is finished (the duration of measurement depends on the experimental design and your time-table).

A non-invasive technique is to measure $T_b$core rectally by using a probe inserted 2–4 cm (depth depends on the species) into the rectum. This is easy to carry out and is frequently used in the laboratory and in the field (Chapter 8).

A further approach to measuring $T_b$core is to use temperature-sensitive transponders. Transponders are especially designed for animal identification and are suitable for use even in the smallest species owing to their small size (length: 10 mm). The system contains two basic elements: a passive transponder (ID tag) and a reader. The reader emits a flow-frequency magnetic field via its antenna and its ID code is received when a transponder passes within range. Transponders are individually packaged in disposable syringes and are injected subdermally (usually on the animal's back) without anaesthesia. Temperature-sensitive transponders give you both a reading of the ID of the animal and its body temperature. You don't need to calibrate the transponders before usage. One disadvantage of this approach is that the range of the reader is very small and to get a reading you need to be very close to the ID tag (1–2 cm).

Skin body temperature

You can use radio-collars with integrated temperature sensitive sensors (e.g. Biotrack) to monitor $T_b$skin (Chapter 10). Place the radio-collar so that the temperature sensor is in contact with the animal's skin (usually around the neck). This $T_b$skin is a reasonable estimate of core body temperature, particularly when resting or torpid animals are in a curled-up position with the transmitter pressed against the ventral surface when it is placed around the neck (Audet & Thomas, 1996; Dausmann, 2005). Calibrate each radio-collar against a precision mercury thermometer in a water bath before application. Transmitter signals are received automatically or manually using receiver and antenna.

Note: During all body temperature measurements, you should record the corresponding ambient temperature (T_a) continuously by using data-loggers or calibrated thermometers (Chapter 5).

Which technique to use

The main practical aspect of monitoring body temperature of animals is to decide which transmitter type is the most appropriate technique for the problem at hand. Transmitters differ in frequency, signal range and reception, battery lifespan, type of data recorded and mass. There are practical limits on the minimum size of an animal that can be studied using data-loggers or radio-collars. Generally, the mass of the transmitter used should be below the tolerable weight limit of 5% of the animal's body mass (Gursky, 1998; Chapter 10).

The most significant advantages of data-loggers are reduced disturbance of the study animal (no need to follow the animal in the wild), and that you can make measurements continuously for a long period without 'time out' periods. Furthermore, you can measure large numbers of individuals at the same time as well as at different study sites. The resultant measurements are truly simultaneous (repetition), which effectively allows you to control for possible external factors.

In the field, the most obvious technique is to use temperature-sensitive radio-collars and telemetry equipment. The advantage of this approach is that it is possible to make measurements of home range, distance travelled and body temperature simultaneously (Chapter 10). Even when data are not recorded automatically (stationary telemetry set-ups can record data automatically but they are expensive and time-consuming to install), you can still make simultaneous measurements of a number of individuals at one site over a period of time. Such manually collected data will involve more irregular measurements, but still allow you to detect significant effects.

All these methodologies for the measurement of body temperature are applicable to animals both in the field and in the laboratory. However, measurements of the body temperature of free-living animals in their natural environment are more difficult to obtain than for animals in the laboratory. The principal difficulties of using these various techniques to monitor body temperature of free-living animals are the need for surgery and recovery, and the need to remain in close contact with the animal in the field. You can avoid the latter problem, however, by using data-loggers. You can either implant the data-logger (to measure $T_b$core) or attach it to the animal's skin using special skin glue or a neck collar (to measure $T_b$skin). At present, the smallest and lightest temperature loggers that are specially made for tagging small mammals weigh less than 2 grams (e.g. Weetag Collars from Alpha Mach Inc.; see List of suppliers).

Implanting transmitters or data-loggers is theoretically and practically feasible in the field, but it is advisable to contact a veterinarian or someone who is experienced in surgery (Chapter 8). Furthermore, once you have decided to use data-loggers in the field, you need to recapture the same individuals to remove the transmitter and to download the data (Chapter 7). Thus, you should only apply this methodology to species that are generally easy to trap and, particularly, to individuals that have a high trap–recapture rate.

OTHER POSSIBILITIES

Schnell and Wood (1993) have described a telemetry system for the continuous measurement of blood pressure and heart rate in conscious marmosets moving freely in their home cages. In this case, pressure transmitters are implanted into the peritoneal cavity under anaesthesia and aseptic conditions. A fluid-filled sensor catheter connected to the transmitter is placed in the aorta below the renal artery (pointing upstream). Several biotelemetry receivers are distributed around the cage and signals are received, computerized and stored at defined intervals. The application of other techniques to record data on blood pressure and heart rate are provided by Michel *et al.* (1984) and Wood *et al.* (1987). As a cautionary note, all techniques to measure blood pressure and heart rate are invasive and should be carried out by a veterinarian or someone who is experienced in surgery (Chapter 8).

LEGAL ASPECTS

Once you have decided to work with animals in the field or in the laboratory, the first consideration is the legal aspect of applying a specific technique. Applying for permits can take some time, and you need to obtain the authorizations for your study in time to avoid a delay. Unfortunately, there are no global rules and regulations concerning the legal requirements, and legislative requirements are different in different countries if you intend to work abroad. Ethics committees usually review proposals for animal experiments and decide whether the research project is justified and how it should be carried out. If the procedures are classed as animal experiments, then you will also need to comply with the appropriate legislation. Contact the ethics committee in your host country for further details concerning animal experimentation and specific requirements.

You will require permits or licences for both you and your assistants to capture and mark animals at your proposed field study site. You may also need a special permit for procedures such as injecting an animal, using stable isotopes, taking blood samples and implanting data-loggers or transmitters. Generally, you need to submit a project proposal including information on the study animal, study site and a precise description of the techniques and material that will be used. For laboratory studies, you may also need to meet certain requirements concerning the conditions in which animals will be kept in captivity.

Furthermore, you may have to pass special training courses to prove you can perform to the standards of the country where you will be working. If you hold a licence to perform procedures in one country, this does not necessarily mean that other countries will accept your accreditation of competence, since standards differ between countries.

The final factor to consider is transportation of blood samples (and perhaps also isotopes) between countries. You will need a permit to export and import blood samples for the doubly labelled water analyses, and if the blood you have sampled is from an animal that is listed in Appendices 1–3 of the Convention on International Trade in Endangered Species (CITES) regulations, you will need a CITES permit to import your samples.

TIME NEEDED IN THE FIELD AND IN THE LABORATORY

Estimates of time needed in the field and in the laboratory vary from about one week to several weeks, months, or up to a few years. In general, the length of a project depends on the experimental set-up, the questions asked, and the main emphasis of the project. Other time factors are the animal species with which you want to work, its abundance and 'cooperation' (e.g. trapping success, recapture rate; Chapter 7). Furthermore, if the technologies and materials you will be employing are not already established or used routinely, you may need to allow extra time, since establishing and setting up a new methodology is very time-consuming.

COSTS

The costs of a study on thermoregulation and energetics fall into several categories: first, personnel to conduct the work; second, equipment and materials (such as traps, radio-collars, receivers, data-loggers, balances, isotopes); and third, costs of the analysis. In addition to these major costs, there are also travel and subsistence costs (inclusive of daily rates for the study site) for field and laboratory excursions. The costs for each study will depend, principally, on the methods you wish to apply, the sample size you wish to end up with, the country where you will carry out the study, and your personal access to technical equipment and analysis. Prices for specific equipment you may need for your study are best found on the Internet or by consulting suppliers, since they change rapidly. For example, to complete a doubly labelled water study, at present day prices (2010), heavy oxygen costs US$30 per ml and a 50 g bottle of deuterated water (minimum order) costs about US$75. The quantity of isotopes required will depend on the size of the animal and the required final sample size. Isotope analysis costs are currently about US$10 per isotope determination. In total, the costs for a two-sample protocol (initial sample and final sample), analysed in duplicate, come to about US$120 per animal successfully recaptured (i.e. 8 samples at US$10 each plus US$40 for backgrounds and standards).

REFERENCES

Audet, D. & Thomas, D. W. (1996). Evaluation of the accuracy of body temperature measurement using external transmitters. *Can. J. Zool.* 74, 1778–81.

Bartholomew, G. A. (1982). Energy metabolism. In *Animal Physiology: Principles and Adaptations*, ed. M. S. Gordon, pp. 46–93. New York: Macmillan.

Dausmann, K. H. (2005). Measuring body temperature in the field: evaluation of external vs. implanted transmitters in a small mammal. *J. Therm. Biol.* 30, 195–202.

Geiser, F. (2004). Metabolic rate and body temperature reduction during hibernation and daily torpor. *A. Rev. Physiol.* 66, 239–74.

Gursky, S. (1998). Effects of radio transmitter weight on a small nocturnal primate. *Am. J. Primatol.* 46, 145–55.

Lifson, N. & McClintock, R. (1966). Theory of use of the turnover rates of body water for measuring energy and material balance. *J. Theor. Biol.* 12, 46–74.

Michel, J. B., Wood, J. M., Hofbauer, K. G., Corvol, P. & Menard, J. (1984). Blood pressure effects of renin inhibition by human renin antiserum in normotensive marmosets. *Am. J. Physiol.* 246, F309–16.

Schartz, R. L. & Zimmermann, J. L. (1971). The time and energy budget of the male dickcissel (*Spiza americana*). *Condor* 73, 65–76.

Schmid, J., Ruf, T. & Heldmaier, G. (2000). Metabolism and temperature regulation during daily torpor in the smallest primate, the pygmy mouse lemur (*Microcebus myoxinus*) in Madagascar. *J. Comp. Physiol.* B 170, 59–68.

Schmidt-Nielsen, K. (1997). *Animal Physiology: Adaptation and Environment.* Cambridge: Cambridge University Press.

Schnell, C. R. & Wood, J. M. (1993). Measurement of blood pressure and heart rate by telemetry in conscious, unrestrained marmosets. *Am. J. Physiol.* 264, H1509–16.

Speakman, J. R. (1997). *Doubly Labelled Water. Theory and Practice.* London: Chapman and Hall.

Weathers, W. W., Buttemer, W. A., Hayworth, A. M. & Nagy, K. A. (1984). An evaluation of time-budget estimates of daily energy expenditure in birds. *Auk* 101, 459–72.

Wood, J. M., Heusser, C., Gulati, N., Forgiarini, P. & Hofbauer, K. G. (1987). Sustained reduction in blood pressure during chronic administration of renin inhibitor to normotensive marmosets. *J. Cardiovasc. Pharmacol.* 7, Suppl. 10, S96–8.

LIST OF SUPPLIERS AND THEIR INTERNET SITES

Data-loggers for measuring body temperature, ambient temperature and humidity

www.maxim-ic.com, www.ibutton.com – for Thermochron ibutton®
www.alphamach.com – for Weetag Collar
www.geminidataloggers.com – for Gemini Data Loggers (UK) Ltd
www.synotech.de – for synoTECH™ Sensor und Meßtechnik GmbH (in German)

Radio transmitters with attached temperature sensors

www.biotrack.co.uk – for Biotrack Ltd
www.minimitter.com – for Mini-Logger®

Micro transponders

www.euroid.com – for EURO I.D. Identifikationssysteme GmbH & Co.
www.trovan.com – for Trovan™ Electronic Identification Systems, Ltd.

Telemetry receivers

www.telonics.com – for Telonics
www.positioning.televilt.se – for Televilt TVP Positioning AB

Gas analysis system

www.ametek.de – for Ametek Precision Instruments Europe GmbH (in German)
www.draeger.com – for Drägerwerk AG
www.ahlborn.com – for Ahlborn Mess- und Regelungstechnik GmbH

20

Field endocrinology: monitoring hormonal changes in free-ranging primates

INTRODUCTION

Field endocrinology can be considered as the application of non-invasive methodologies to examine behavioural–endocrine interactions in primates living in natural conditions and social settings. In bringing together laboratory and field-based research methods, the discipline provides new and exciting opportunities for developing a more integrated approach to studies of primate behavioural ecology.

Traditionally, field studies have relied mainly on visual measures, such as behaviour and/or morphology. Although this has generated a great deal of essential information, its limitation is that interpretation of the observations is often based on assumptions concerning the physiological context in which they were made. The availability of non-invasive methodologies based on measurement of hormones in either urine or faeces now provides us with quantitative measures of physiological status by which the significance of observational data can be gauged. This greatly facilitates the testing of hypotheses concerning the adaptive significance of behavioural and morphological traits and mating systems and is helping to provide new insights into reproductive processes in an evolutionary context. Field endocrinology also facilitates a better understanding of the impact of anthropogenic activities on primate physiology. In combination with studies of the health status of wild primates (Chapter 8), this can help to elucidate the link between environmental stress, health and reproductive parameters and thus to estimate the viability of threatened populations (Pride, 2005; Chapman et al., 2007).

In this chapter, we review the most important aspects fieldworkers need to be aware of when planning and carrying out

Field and Laboratory Methods in Primatology: A Practical Guide, ed. Joanna M. Setchell and Deborah J. Curtis. Published by Cambridge University Press. © Cambridge University Press 2011.

endocrinological studies on free-ranging primates. You should consult Whitten *et al.* (1998a) for a comprehensive overview of the topic of field endocrinology and its application to studies in primatology.

Urine

Options for collecting urine samples under field conditions include (i) using sheeting, a tray or bowl to collect urine mid-stream (arboreal or semi-arboreal species); (ii) aspiration (using a syringe or pipette) or absorption onto filter paper from foliage; or (iii) recovery (usually using centrifugal force) from samples of earth. Volumes as small as 0.1 ml may be sufficient (provided further reduction during transport and storage does not occur), and more than 1 ml is generally not necessary.

Faeces

Faecal samples are usually easier to identify and locate than urine and you can collect them directly from the ground. A thumb-nail sized amount (roughly 0.5–2 g) provides more than enough material for analysis. In species where the bolus is larger, homogenize the sample on a leaf or other flat surface using your gloved hand or an improvized spatula, and transfer the required amount to the storage container (not more than 50% of total volume). If you store samples and extract them directly in alcohol, don't vary the amount of faeces collected too much, since hormone concentrations related to the mass of faeces may be influenced by the amount of the sample collected (i.e. very small samples lead to proportionately higher hormone concentrations per gram faeces compared to large samples). Avoid the inclusion of large amounts of non-faecal material such as seeds and non-digested foliage.

General points

It is essential that you only collect samples of known origin, in other words only from individuals actually seen to be urinating or defecating. This is particularly important for collection of samples from sleeping sites. It is also essential to avoid sample contamination. Cross-contamination (principally of faeces with urine) is the most likely and can be reduced partially in big samples by collecting an aliquot from within the faecal bolus. Other sources of contamination are dilution of urine with water and sample-to-sample contamination through use of the same syringe, spatula for homogenization, glove, etc. Where you cannot use disposable materials, rinse the implement with water and wipe it dry, and avoid the use of detergents.

You need to control for time of collection. Diurnal patterns of secretion are particularly pronounced for some hormones (e.g. testosterone, cortisol). Although these patterns are likely to be more evident in urine than in faeces, they may still be seen in the faeces of certain, particularly small-bodied, species in which faecal passage rate is high (e.g. callitrichids, *Callithrix*: Sousa & Ziegler, 1998). Therefore, whenever possible, restrict the period of collection to roughly the same time of each day. If you can only collect samples opportunistically, test the potential effect of time of day on hormone concentration (e.g. by comparing hormone levels in morning and afternoon samples from the same individuals). Remember that urine and faeces, like all other body fluids and tissues from primates, are a potential source of pathogens and thus infection. Handle them with care and take appropriate precautions to reduce the risk of infection at all stages of collection, storage, transport and subsequent analysis.

SAMPLE STORAGE

Whilst most steroids (progesterone is one exception) are relatively robust and stable at ambient temperature for several days, conjugated forms (the predominant form in urine; present in variable amounts in faeces) and creatinine (the protein breakdown marker most often used to index urine concentration) are less so. Steroid breakdown due to bacterial metabolism can be a potential problem in faeces, where gastro-intestinal bacteria are abundant. Thus, in general, keep the interval from sample collection to preservation as short as possible. Immediate preservation at the site of collection is preferable, but otherwise keep the sample cool (in an insulated cool-box) until you preserve it properly after return to camp.

Urine

The principal, most commonly applied, and preferable method of long-term storage of urine is freezing (– 10 °C or below). This may limit long-term sample storage at field sites and, since repeated freezing and thawing should be avoided, also requires the use of dry ice (or ice packs) to maintain samples in a frozen condition throughout all stages of transport (this can be expensive and certain airlines impose restrictions on transportation of dry ice). Alternative methods are (i) storage as an ethanolic solution, (ii) use of sodium azide (0.1%) as a preservative or (iii) absorption onto filter paper and dry storage. Ethanol is preferable and enables you to store samples at ambient temperatures without deterioration for several weeks. The final solution should be within the range of 30–50% ethanol, and since this must be kept constant for all samples you need accurately measured volumes of urine. One major disadvantage with the use of sodium azide is that it might interfere with the hormone assay step, if you use enzyme immunoassays. Storage of aqueous or ethanolic solutions on filter paper has been shown to be useful, but can introduce a number of potential analytical errors and needs to be very carefully controlled (cf. Shideler et al., 1995).

Faeces

The most effective way of preserving faeces for prolonged periods of time is undoubtedly freezing at −20 °C. In the past, this has generally not been possible for remote field sites, but solar-powered freezers are now available. The most widely used approach for faecal sample storage under field conditions has been to use alcohol (>80% ethanol in a minimum volume to mass ratio of 2.5 : 1), one of the main advantages being that it can be applied at ambient temperature. Although ethanol has the additional advantage that you can take vials or tubes containing a known volume of ethanol into the field, enabling you to transfer samples to the preservation medium immediately after collection, several studies in primate and non-primate species have now shown clearly that this method can markedly alter faecal hormone levels during both short- and long-term storage, even when samples are frozen (see, for example, Khan et al., 2002). This can potentially affect data interpretation, particularly when comparing absolute hormone levels in samples that have been stored for long and/or variable periods before analysis. This storage effect, however, appears to be both species- and hormone-specific and use of ethanol as a preservation medium may thus be less of a problem in some species and/or for the measurement of certain hormones in the species of interest (Fichtel et al., 2007; Daspre et al., 2009). Furthermore, when you are only interested in relative hormone concentrations within individuals, as for example in the case of day-to-day changes in progestogen levels used for timing of ovulation, a potential storage effect is unlikely to lead to

errors in data interpretation since each animal serves as its own control. For some questions, you can also control for the effect of storage
on faecal hormone levels during data analysis, for example by calculating average concentrations for certain time periods (e.g. months)
or from a cross-section of the study population, thereby generating
residual values that can be compared (Pride, 2005). Nevertheless, whenever you plan to use alcohol as the preservative for faeces, you should
take into consideration the potential effect of storage on hormone
levels.

Additional potential problems with the use of ethanol include
evaporation (acceptable in limited amounts) and spillage or leakage.
Pay particular attention to the type of storage container used. Many are
not as leak-proof as they are claimed to be (particularly when using
alcohol) and sample spillage during transport will seriously affect the
reliability of your results. Where possible, use screw-cap vials (10–20 ml
range) and test them rigorously beforehand with ethanol. As an extra
precaution, ensure that sample vials are kept upright at all times. Glass
vials are more inert, but are fragile and need careful packing before
transport; plastic tubes are safer to transport, although there is a small
risk that steroids might stick to the wall during long storage periods.

Alternative methods to freezing and storing faecal samples in
alcohol are reviewed by Ziegler and Wittwer (2005). Among these are (i)
drying the samples (e.g. using a solar oven) and storing them in
moisture-proof packages with desiccant until shipment to the laboratory, (ii) extracting and purifying the sample in the field using alcoholic
solutions and filtration and (iii) preserving the extract on octade-
cylsilane (C-18) cartridges. Although these methods have been
shown to stabilize steroids for prolonged times and result in 'samples'
that are lighter in weight and meet the requirements for international
transport of potentially infectious material better, they require
more skill and equipment in the field as well as regular access to
organic chemicals. Moreover, as these techniques involve several
processing steps, they may result in steroid loss and higher methodological variability. Whatever method of faecal sample storage you use,
you should validate it for the species in question and samples should
generally be transported to a freezer or the laboratory as soon as
possible.

Label sample containers properly, giving animal name/ID
number and date (and time of collection, if useful). Special sticky
paper labels that resist solvents and freeze–thawing processes are
preferable and you can apply clear waterproof tape over the labels

as an additional precaution. Additionally, write relevant information directly on the tube or cap (preferably both), using an indelible, black, waterproof marker. Keep a separate list of samples for each animal for crosschecking purposes. Inappropriate labelling can result in loss of information on sample identity (usually during transport), rendering analysis meaningless. This continues to be one of the major causes of loss of data in this type of work, and of course you can be sure that the unidentifiable samples will always be the ones of particular importance.

LABORATORY SAMPLE PREPARATION

Urine

Steroids in urine are present predominantly in the conjugated form, either as mono- or multi-conjugated sulphate or glucuronide residues. You can measure these either directly in appropriately diluted urine or by first cleaving the conjugate by hydrolysis, followed by an assay designed specifically for the parent compound (e.g. measurement can be either as oestrone-3-glucuronide or after hydrolysis as oestrone). You can perform urine hydrolysis by incubation with an enzyme preparation (e.g. from *Helix pomatia* or *Escherichia coli*) or via a non-enzymatic procedure using organic solvents (solvolysis, see Ziegler *et al.*, 1996 for details). *Helix pomatia* (*HP*) preparations are most commonly used, although since *HP* juice contains enzymes other than glucuronidase/arylsulphatase, you should view its application with caution, particularly for urinary testosterone measurements where hydrolysis using *HP* preparations can result in steroid transformation (e.g. androst-5-ene-3β,17β-diol into testosterone; Hauser *et al.*, 2008). If you use hydrolysis, it is usual to carry out an extraction step, whereby steroids are removed into an organic phase (e.g. diethyl ether), which you reconstitute in aqueous buffer for assay after evaporation. Steroid conjugates are best extracted with aqueous alcohol or (more usually) assayed directly without an extraction step. You need a separate aliquot (usually 0.02–0.05 ml) of each sample to determine creatinine content, either using a creatinine analyser or a micro-titre plate method (see Bahr *et al.*, 2000). More recently, specific gravity measurement has been used as an alternative to creatinine for estimating urine concentration (Anestis *et al.*, 2008). Apart from avoiding potential problems with creatinine degradation, you can also easily measure specific gravity in the field by using a small battery-powered, handheld refractometer (Anestis *et al.*, 2008).

Faeces

An extraction step is always necessary to measure steroids in faeces. Various methods exist for extracting steroids from frozen–thawed faeces, the more commonly used involving agitating (shaking/vortexing) a known weight of sample with an aqueous solution (40–80%) of methanol or ethanol. Allow the suspension to settle (or centrifuge it) and take a portion of the supernatant either for assay or for further purification steps (such as re-extraction or use of Sep-Pak C-18 mini-columns). The final hormone content is expressed per unit mass (g) of wet or dry faeces after determination of the dry mass of the faecal pellet. Use similar procedures with samples stored in ethanol. In our experience, the most efficient way of dealing with these is to homogenize the sample in the original solvent (e.g. by using a metal spatula), shaking it on a vortexer for 15 min (when the amount of faeces collected is relatively small) or on a shaker overnight (when the amount of faeces is large) and, if necessary (depending on the efficiency of the procedure), re-extracting it with additional (e.g. 10 ml) 80% methanol.

Alternatively (our own preferred method), you can freeze-dry frozen faecal samples directly and then pulverize them (usually using a pestle and mortar) and sieve them through a fine wire mesh before methanol extraction (see Heistermann et al., 1995a). Although the combined process is relatively time consuming, the advantages are that: (1) freeze-drying compensates for differences in faecal water content and allows you to remove non-faecal material (e.g. seeds, stones, undigested fibre) easily; (2) pulverization produces a homogeneous powder which you can pre-weigh, aliquot and store further at room temperature and (3) the extraction step is easier and generally does not require further purification steps before assay. With an efficient freeze-drier, you can process up to 150 samples simultaneously, requiring a total of about 72h.

In certain circumstances, you may need chromatographic procedures (thin layer chromatography, TLC; high performance liquid chromatography, HPLC) to detect and/or aid identification of individual steroid components of sample extracts, usually for assay validation purposes (see Bahr et al., 2000).

ASSAY METHODOLOGY

Measurements of hormones and their metabolites are usually carried out by immunological procedures using hormone- or hormone-group-specific antibodies. Two main types of immunoassays are available:

radioimmunoassays (RIA), which use radioactively labelled hormone as the competitive tracer in the quantification process, and enzyme-immunoassays (EIA) in which either enzyme or biotin labelled preparations are employed. Being non-isotopic, EIAs avoid the problems associated with use and disposal of radioactivity and are also cheaper.

Since all immunoassays are highly sensitive, you need to assess assay performance carefully both during the initial set-up phase and during routine use. There are four main criteria of laboratory validation: **sensitivity** (minimum amount of hormone that can be detected); **precision** (within- and between-assay repeatability), **accuracy** (ability to detect the correct amount of hormone in the sample) and **specificity**. The latter has two components: the degree of specificity of the antibody itself and the possible influence of interfering substances excreted with urine and faeces (matrix effects), which you need to control for and remove if present by incorporating additional sample purification steps. Check carefully for the presence of such matrix effects before any routine use of an assay.

Concerning antibody specificity, specific assays may be useful when the identity of the major metabolite is known (or the hormone of interest is not heavily metabolized, e.g. oestrogens) and when you are interested in species comparisons. Since, however, excreta (especially faeces) usually contain numerous metabolites of the parent hormone, a specific measurement is often difficult and might be less useful in cases where the antibody detects only metabolites of low abundance. Group-specific assays use antibodies that cross-react with several metabolites of related structure. Since knowledge of the relative abundance of individual metabolites is not necessary, these assays have advantages in that they can usually be applied to a wider range of species (Heistermann et al., 1995b, Schwarzenberger et al., 1997), thus helping to overcome the problems of species specificity in hormone metabolism. When measuring faecal androgens to assess testicular function, however, group-specific antibodies may lead to problems due to the co-measurement of androgens of adrenal origin (see below).

Available data show that, in the majority of primates, the direct measurement of oestrone conjugates (E1C) and pregnanediol-3-glucuronide (PdG) is most useful for monitoring ovarian function and pregnancy by using urine analysis (see Heistermann et al., 1995b for review). In macaques, you can also use the non-specific measurement of C19/C21-progesterone metabolites, such as androsterone (structurally an androgen) or 20α-hydroxyprogesterone. Urine also provides a matrix for the measurement of polypeptide and proteohormones (e.g. FSH, LH, oxytocin, prolactin), which you cannot measure in faeces.

Information on the measurement of androgens in primate urine is limited. To date, most studies have measured immunoreactive testosterone, although this is a relatively minor component (Möhle et al., 2002; Hagey & Czekala, 2003). There is some evidence that measurement of 5-reduced androstanes might reflect testicular endocrine activity better (Möhle et al., 2002; Hagey & Czekala, 2003). The measurement of urinary cortisol provides a reliable method for monitoring glucocorticoid output in a variety of primate species (Whitten et al., 1998b; Robbins & Czekala, 1997), but the more abundant 5β-reduced cortisol metabolites with a 3α-hydroxy,11-oxo and 3α,11β-dihydroxy structure might be more suitable, at least in some species (see Bahr et al., 2000).

The most appropriate assays for the measurement of oestrogen and progesterone metabolites in faeces are shown in Table 20.1. Assays for oestradiol-17β or oestrone (or a collective measurement of both) are generally useful for monitoring ovarian function and pregnancy, but oestrogen measurements in faeces appear to be unreliable in some species. The majority of progestogen measurements have been based on the use of non-specific progesterone and pregnanediol assays cross-reacting with a broad range of pregnanediones and hydroxylated pregnanes, which are known to represent abundant progesterone metabolites in the faeces of most mammals, including primates (Schwarzenberger et al., 1997). Which assay reflects female reproductive status best depends on the species of interest (Table 20.1), although faecal progestogen profiles are generally easier to interpret than oestrogen profiles, particularly with respect to monitoring of ovarian cycles and timing of ovulation.

Measurement of adrenal steroids (glucocorticoids) in primate faeces has increased substantially over the past couple of years. In some species (e.g. ring-tailed lemur, Lemur catta: Cavigelli, 1999; chimpanzee, Pan troglodytes: Whitten et al., 1998b), cortisol assays appear to yield useful information, although cortisol itself is often either barely detectable or not present (Bahr et al., 2000; Heistermann et al., 2006). Since it is impossible to predict whether cortisol is present in substantial amounts in the faeces of any given species, group-specific assays, capable of measuring a range of faecal cortisol metabolites, might be generally more suitable, in that they are more likely to detect at least some of the more abundant metabolites present and also have greater

Table 20.1 *Selected studies in which faecal oestrogen and progestogen assays have been used to assess female reproductive status in primates*

Abbreviations: E_t = total oestrogen; E_2 = oestradiol-17β; E_1C = oestrone conjugates; PdG = pregnanediol(glucuronide); 20α-OHP = 20α-hydroxyprogesterone; P_4 = progesterone; 5-P-3OH = 5α-pregnane-3α-ol-20-one.

Taxa	Species	Oestrogen	Progestin	Reference
Lemuridae	Eulemur mongoz	E_t	20α-OHP	Curtis et al., 2000
	Eulemur fulvus rufus		5-P-3OH	Ostner & Heistermann, 2003
	Hapalemur griseus	E_t	5-P-3OH	P. Gerber et al., unpubl.
Indriidae	Propithecus verreauxi	E_2	P_4	Brockman & Whitten, 1996
Lorisidae	Nycticebus coucang	E_1C	—[a]	Jurke et al., 1997
Cebidae	Callithrix jacchus	E_2	Pd, P_4	Heistermann et al., 1993
	Saguinus oedipus	E_2	P_4, Pd	Ziegler et al., 1996
	Saguinus fuscicollis	?[b]	Pd	Heistermann et al., 1993
	Saguinus mystax	E_t	PdG	Löttker et al., 2004
	Leontopithecus rosalia	E_1C	PdG	French et al., 2003
	Callimico goeldii	E_t	?[b]	Pryce et al., 1994
	Cebus apella	—[a]	P_4, Pd	Carosi et al., 1999
	Saimiri sciureus			Moorman et al., 2002
Pitheciidae	Pithecia pithecia	E_1C	PdG	Shideler et al., 1994
Atelidae	Brachyteles arachnoides	E_2	P_4	Strier & Ziegler, 1997
	Ateles geoffroyi	E_1C	PdG	Campbell et al., 2001
	Alouatta pigra	E_2	P_4	Van Belle et al., 2008
Cercopithecinae	Macaca fascicularis	E_1C	PdG	Shideler et al., 1993
	Macaca silenus	E_2	5-P-3OH	Heistermann et al., 2001a
	Macaca fuscata	E_1C	PdG	Fujita et al., 2001

Table 20.1 (*cont.*)

Taxa	Species	Oestrogen	Progestin	Reference
	Macaca sylvanus	E_t	5-P-3OH	Möhle *et al.* 2005
	Cercocebus torquatus	E_2	P_4	Whitten & Russell, 1996
	Papio cynocephalus	E_2	P_4	Wasser *et al.*, 1991
	Papio hamadryas anubis	E_t	PdG	Higham *et al.*, 2008
Colobinae	*Semnopithecus entellus*	E_2	20α-OHP PdG	Heistermann *et al.*, 1995a
	Pygathrix nemaeus	E_t	5-P-3OH	Heistermann *et al.*, 2004
Hylobatidae	*Hylobates lar*	—[b]	5-P-3OH	Barelli *et al.*, 2007
Hominidae	*Pan paniscus*	—[a]	P_4, Pd	Heistermann *et al.*, 1996
	Pan troglodytes	E_2	P_4	Emery & Whitten, 2003
	Gorilla gorilla	E_2	P_4	Miyamoto *et al.*, 2001

[a] Measurement not successful.
[b] No information available.

potential for cross-species application (Heistermann *et al.*, 2006; Fichtel *et al.*, 2007). However, such group-specific assays have the potential to cross-react with structurally related testosterone metabolites (chimpanzee: Heistermann *et al.*, 2006), which can confound the actual glucocorticoid measurement, although the degree to which this occurs may still be acceptable (Heistermann *et al.*, 2006; Fichtel *et al.*, 2007). Nevertheless, take potential co-measurement of metabolites that do not originate from cortisol into account when using faecal glucocorticoid assays.

Information on the metabolism of testosterone and the nature of the metabolites excreted into primate faeces remains surprisingly limited (but see Möhle *et al.*, 2002). As with cortisol, testosterone metabolism is complex and often species-specific, resulting in excretion of a number of metabolites, with native testosterone usually being quantitatively of minor importance or virtually absent in faeces (Möhle et al., 2002). Nevertheless, testosterone assays have been widely used and in the majority of cases this has yielded informative results (Brockman et al., 1998; Lynch et al., 2002; Setchell et al., 2008). The measurement of 5-reduced androstanes has also been successfully applied to monitor androgen status in male primates (Girard-Buttoz et al., 2009). However, irrespective of the type of assay used, reliable assessment of male testicular androgen secretion in Old World monkey species and great apes might generally be difficult owing to the potential co-measurement of metabolites derived from androgens of extra-testicular (e.g. adrenal) origin. For example, dehydroepiandrosterone (DHEA), a weak androgen from the andrenal gland, may potentially confound assessment of male testosterone secretion since it is metabolized to products very similar (if not virtually identical) to those of testosterone itself (Möhle et al., 2002). Specific measurement of testicular endocrine activity (rather than a measurement of overall androgen status) from faecal androgen measurements is therefore likely to remain difficult in these primate taxa.

For a comprehensive overview of references concerning the application of urinary and faecal androgen and glucocorticoid measurements in primates and other vertebrate taxa see Hodges et al. (2010).

GENERAL CONSIDERATIONS

Species variation in metabolism and excretion

Steroids circulating in the bloodstream undergo a series of metabolic changes before finally being eliminated from the body. The nature of these changes can vary considerably between species (even closely related ones), resulting in differences not only in the nature and identity of the metabolites themselves, but also in their preferred route of excretion. This can have important consequences for the selection of an appropriate measurement system and correct interpretation of results obtained (Heistermann et al., 2006).

Time lag

Whereas circulating hormones more or less reflect real-time changes in endocrine activity, hormones in excreta reflect events that have occurred in the past. In urine the time lag is usually only about 4–8 h (see Bahr et al., 2000), but even so, with once-daily sampling, this can be enough to delay detection of an event (e.g. peak, defined rise) by one day. Time lags associated with faecal measurements are longer and more variable, both between and within species, and gut passage times can be affected by a variety of factors including diet, health status and stress level. In most large-bodied species for which data are available, steroids are excreted in faeces 36–48 h after appearance in circulation (see Shideler et al., 1993), although there are exceptions (e.g. 22 h for testosterone and cortisol in the chimpanzee; Möhle et al., 2002; Bahr et al., 2000). In contrast, passage time is quicker in smaller animals (e.g. 4–8 h for testosterone and cortisol in the common marmoset, Möhle et al., 2002; Bahr et al., 2000). The two main consequences of these time lags (especially for faeces) are (i) a delay between the occurrence of a specific event and its detection, which you need to account for when using faecal measurements to determine the timing of events such as ovulation or implantation and (ii) a dampening effect, reducing the amplitude of hormonal changes, and making it more difficult to detect short-lived endocrine responses to acute situations (e.g. stress).

Sampling frequency

Sampling frequency is largely determined by the type of information required, although how often samples can actually be collected is influenced by numerous practical considerations. Information on overall physiological condition/status requires less frequent sampling than information on dynamics or timing of events. As a general rule, to compensate for intra-individual (sample-to-sample) variation (and also for statistical considerations), we recommend that you collect no fewer than 6 and preferably 10 samples per condition (i.e. before vs. after birth; breeding vs. non-breeding season; before and after change in rank; animal 1 vs. animal 2).

Weekly samples should be sufficient to follow the course of pregnancy, although increased sampling (2 per week) may be useful during the period leading up to parturition. To detect the presence of ovarian cycles, twice weekly samples are the absolute minimum, whereas you need more regular samples collected at a higher frequency to define the duration or interval between two events, dependent on the margin of error that is compatible with the objectives of your study. Generally, this will be 1 day at 3–4 samples per week and 2 days at 2 samples per week. To time ovulation, collect daily samples. It is useful to aim for a slightly higher frequency than is necessary, since it is unlikely that you will collect all samples as planned.

In addition to sampling frequency, a variety of other factors can also affect the reliability of the information obtained from hormone assays of field samples. Important among these are assay precision and criteria used to define the events or parameters under investigation. The first refers to the amount of inherent variability in the measurement system and the second is basically a question of how you interpret the results. As no two samples will have identical hormone content, it is essential to differentiate between random variation and patterns of potential significance in relation to the parameters under investigation. One elevated progesterone value doesn't indicate an ovulatory cycle, but how many do, at what intervals and what frequency? What is elevated? Ovulation is best timed retrospectively according to the rise in progesterone that occurs at the onset of the luteal phase. Here, the question is, what is a rise? There are no hard and fixed rules, but in our experience, the increase (and maintenance) above a threshold value determined as the mean plus two standard deviations of preceding (3–5) baseline values provides a statistically significant, useful and informative method for defining a rise in hormone levels (see Heistermann *et al.*, 2001b). Appropriate definitions (providing objective criteria for assessment) are essential to meaningful interpretation of data.

Influence of diet on hormone levels

Many plant species contain phytosteroids, which may potentially interact with an animal's intrinsic physiology and influence reproductive processes. In primates, it has recently been shown that the consumption of certain plants in the genus *Vitex* can confound urinary and faecal measurements of oestrogen and, in particular, progesterone metabolites in baboons and chimpanzees (Higham *et al.*, 2007; Emery Thompson *et al.*, 2008). Effects include increased fluctuations and sustained elevations in hormone levels, making interpretation of results (and assessment of reproductive status) difficult. Since *Vitex* species feature prominently in the diet of many African primates, their effect on female hormone levels may represent a more general phenomenon. To what extent other steroids (glucocorticoids and androgens) may also be affected by consumption of these plants is not really known, but at least for chimpanzees it has been shown that the consumption of *Vitex* does not alter urinary testosterone levels in males (Emery Thompson *et al.*, 2008).

REFERENCES

Anestis, S. F., Breakey, A. A., Beuerlein, M. M. & Bribiescas, R. G. (2008). Specific gravity as an alternative to creatinine for estimating urine concentration in captive and wild chimpanzee (Pan troglodytes) samples. Am. J. Primatol. 70, 1–6.

Bahr, N., Palme, R., Möhle, U., Hodges, J. K. & Heistermann, M. (2000). Comparative aspects of the metabolism and excretion of cortisol in three individual non-human primates. Gen. Comp. Endocrinol. 117, 427–38.

Barelli, C., Heistermann, M., Boesch, C. & Reichard, U. H. (2007). Sexual swellings in wild white-handed gibbon females (Hylobates lar) indicate the probability of ovulation. Horm. Behav. 51, 221–30.

Brockman, D. K. & Whitten, P. L. (1996). Reproduction in free-ranging Propithecus verreauxi: estrus and the relationship between multiple partner matings and fertilization. Am. J. Phys. Anthropol. 100, 57–69.

Brockman, D. K., Whitten, P. L., Richard, A. F. & Schneider, A. (1998). Reproduction in freeranging male Propithecus verreauxi: the hormonal correlates of mating and aggression. Am. J. Phys. Anthropol. 105, 137–51.

Campbell, C. J., Shideler, S. E., Todd, H. E. & Lasley, B. L. (2001). Fecal analysis of ovarian cycles in female black-handed spider monkeys (Ateles geoffroyi). Am. J. Primatol. 54, 79–89.

Carosi, M., Heistermann, M. & Visalberghi, E. (1999). Display of proceptive behaviours in relation to urinary and fecal progestin levels over the ovarian cycle in female tufted capuchin monkeys (Cebus apella). Horm. Behav. 36, 252–65.

Cavigelli, S. A. (1999). Behavioural patterns associated with faecal cortisol levels in free-ranging female ring-tailed lemurs, Lemur catta. Anim. Behav. 57, 935–44.

Chapman, C. A., Saj, T. L. & Snaith, T. V. (2007). Temporal dynamics of nutrition, parasitism, and stress in colobus monkeys: implications for population regulation and conservation. Am. J. Phys. Anthropol. 134, 240–50.

Curtis, D. J., Zaramody, A., Green, D. I. & Pickard, A. R. (2000). Non-invasive monitoring of reproductive status in wild mongoose lemurs (Eulemur mongoz). Reprod. Fertil. Dev. 12, 21–9.

Daspre, A.; Heistermann, M.; Hodges, K., Lee, P. & Rosetta, L. (2009). Signals of female reproductive quality and fertility in colony-living baboons (Papio h. anubis) in relation to ensuring paternal investment. Am. J. Primatol. 71, 529–38.

Emery, M. A. & Whitten, P. L. (2003). Size of sexual swellings reflects ovarian function in chimpanzees (Pan troglodytes). Behav. Ecol. Sociobiol. 54, 340–51.

Emery Thompson, M., Wilson, M. L., Gobbo, G., Muller, M. N. & Pusey, A. E. (2008). Hyperprogesteronemia in response to Vitex fischeri consumption in wild chimpanzees (Pan troglodytes schweinfurthii). Am. J. Primatol. 70, 1064–71.

Fichtel, C., Kraus, C., Ganswindt, A. & Heistermann, M. (2007). Influence of reproductive season and rank on fecal glucocorticoid levels in free-ranging male Verreaux's sifakas (Propithecus verreauxi). Horm. Behav. 51, 640–48.

French, J. A., Bales, K., Baker, A. J. & Dietz, J. M. (2003). Endocrine monitoring of wild dominant and subordinate female Leontopithecus rosalia. Int. J. Primatol. 24, 1281–300.

Fujita, S., Mitsunaga, F., Sugiura, H. & Shimizu, K. (2001). Measurement of urinary and fecal steroid metabolites during the ovarian cycle in captive and wild Japanese macaques, Macaca fuscata. Am. J. Anthropol. 53, 167–76.

Girard-Buttoz, C., Heistermann, M., Krummel, S. & Engelhardt, A. (2009). Hormonal correlates of reproductive seasonality and social status in wild male long-tailed macaques (*Macaca fascicularis*). *Physiol. Behav.* 98, 168–75.

Hagey, L. R. & Czekala, N. M. (2003). Comparative urinary androstanes in the great apes. *Gen. Comp. Endocrinol.* 130, 64–9.

Hauser, B., Schulz, D., Boesch, C. & Deschner, T. (2008). Measuring urinary testosterone levels of the great apes – Problems with enzymatic hydrolysis using *Helix pomatia* juice. *Gen. Comp. Endocrinol.* 158, 77–86.

Heistermann, M., Tari, S. & Hodges, J. K. (1993). Measurement of faecal steroids for monitoring ovarian function in New World primates, *Callitrichidae*. *J. Reprod. Fertil.* 99, 243–51.

Heistermann, M., Finke, M. & Hodges, J. K. (1995a). Assessment of female reproductive status in captive-housed Hanuman langurs (*Presbytis entellus*) by measurement of urinary and fecal steroid excretion patterns. *Am. J. Primatol.* 37, 275–84.

Heistermann, M., Möstl, E. & Hodges, J. K. (1995b). Non-invasive endocrine monitoring of female reproductive status: methods and applications to captive breeding and conservation of exotic species. In *Research and Captive Propagation*, ed. U. Gansloßer, J. K. Hodges & W. Kaumanns, pp. 36–48. Erlangen: Filander Verlag GmbH.

Heistermann, M., Möhle, U., Vervaecke, H., van Elsacker, L. & Hodges, J. K. (1996). Application of urinary and fecal steroid measurements for monitoring ovarian function and pregnancy in the bonobo *(Pan paniscus)* and evaluation of perineal swelling patterns in relation to endocrine events. *Biol. Reprod.* 55, 844–53.

Heistermann, M., Uhrigshardt, J., Husung, A., Kaumanns, W. & Hodges, J. K. (2001a). Measuremnet of faecal steroid metabolites in the lion-tailed macaque (*Macaca silenus*): a non-invasive tool for assessing ovarian function. *Prim. Rep.* 59, 27–42.

Heistermann, M., Ziegler, T., van Schaik, C. P. *et al.* (2001b). Loss of oestrus, concealed ovulation and paternity confusion in free-ranging Hanuman langurs. *Proc. R. Soc. Lond.* B 268, 2445–51.

Heistermann, M., Ademmer, C. & Kaumanns, W. (2004). Ovarian cycle and effect of social changes on adrenal and ovarian function in *Pygathrix nemaeus*. *Int. J. Primalol.* 25, 689–708.

Heistermann, M., Palme, R. & Ganswindt, A. (2006). Comparison of different enzymeimmunoassays for assessment of adrenocortical activity in primates based on fecal samples. *Am. J. Primatol.* 68, 257–73.

Higham, J. P., Ross, C., Warren, Y., Heistermann, M. & MacLarnon, A. M. (2007). Reduced reproductive function in wild baboons (*Papio hamadryas anubis*) related to natural consumption of the African black plum (*Vitex doniana*). *Horm. Behav.* 52, 384–90.

Higham, J. P., Heistermann, M., Ross, C., Semple, S. & MacLarnon, A. (2008). The timing of ovulation with respect to sexual swelling detumescence in wild olive baboons. *Primates* 49, 295–9.

Hodges, J. K., Brown, J. & Heistermann, M. (2010). Endocrine monitoring of reproduction and stress. In *Wild Mammals in Captivity: Principles and Techniques for Zoo Management*, ed. D. Kleiman, K. V. Thompson & C. K. K. Baer, pp. 447–68. Chicago, IL: University of Chicago Press.

Jurke, M. H., Czekala, N. M. & Fitch-Snyder, H. (1997). Non-invasive detection and monitoring of estrus, pregnancy and the postpartum period in pygmy loris (*Nycticebus pygmaeus*) using fecal estrogen metabolites. *Am. J. Primatol.* 41, 103–15.

Khan, M. Z., Altman, J., Isani, S. S. & Yu, J. (2002). A matter of time: evaluating the storage of fecal samples for steroid analysis. *Gen. Comp. Endocrinol.* 128, 57–64.

Löttker, P., Huck, M., Heymann, E. W. & Heistermann, M. (2004). Endocrine correlates of reproductive status in breeding and nonbreeding wild female moustached tamarins. *Int. J. Primatol.* 25, 919–37.

Lynch, J. W., Ziegler, T. E. & Strier, K. B. (2002). Individual and seasonal variation in fecal testosterone and cortisol levels of wild male tufted capuchin monkeys, *Cebus apella nigritus. Horm. Behav.* 41, 275–87.

Miyamoto, S., Chen, Y., Kurotori, H. *et al.* (2001). Monitoring the reproductive status of female gorillas (*Gorilla gorilla gorilla*) by measuring the steroid hormones in fecal samples. *Primates* 42, 291–99.

Möhle, U., Heistermann, M., Palme, R. & Hodges, J. K. (2002). Characterization of urinary and fecal metabolites of testosterone and their measurement for assessing gonadal endocrine function in male nonhuman primates. *Gen. Comp. Endocrinol.* 129, 135–45.

Möhle, U., Heistermann, M., Dittami, J., Reinberg, V. & Hodges, J. K. (2005). Patterns of anogenital swelling size and their endocrine correlates during ovulatory cycles and early pregnancy in free-ranging barbary macaques (*Macaca sylvanus*) of Gibraltar. *Am. J. Primatol.* 66, 351–68.

Moorman, E. A., Mendoza, S. P., Shideler, S. E. & Lasley, B. L. (2002). Excretion and measurement of estradiol and progesterone metabolites in the feces and urine of female squirrel monkeys (*Saimiri sciureus*). *Am. J. Primatol.* 57, 79–90.

Ostner, J. & Heistermann, M. (2003). Endocrine characterization of female reproductive status in wild redfronted lemurs (*Eulemur fulvus rufus*). *Gen. Comp. Endocrinol.* 131, 274–83.

Pride, R. E. (2005). Optimal group size and seasonal stress in ring-tailed lemurs (*Lemur catta*). *Behav. Ecol.* 16, 550–60.

Pryce, C. R., Schwarzenberger, F. & Döbeli, M. (1994). Monitoring fecal samples for estrogen excretion across the ovarian cycle in Goeldi's monkey (*Callimico goeldii*). *Zoo Biol.* 13, 219–30.

Robbins, M. M. & Czekala, N. M. (1997). A preliminary investigation of urinary testosterone and cortisol levels in wild male mountain gorillas. *Am. J. Primatol.* 43, 51–64.

Schwarzenberger, F., Palme, R. Bamberg, E. & Möstl, E. (1997). A review of faecal progesterone metabolite analysis for non-invasive monitoring of reproductive function in mammals. *Z. Säugetierk.* 62, Suppl. II, 214–21.

Setchell, J. M., Smith, T., Wickings, E. J. & Knapp, L. A. (2008). Social correlates of testosterone and ornamentation in male mandrills. *Horm. Behav.* 54, 365–72.

Shideler, S. E., Ortuno, A. M., Moran, F. M., Moorman, E. A. & Lasley, B. L. (1993). Simple extraction and enzyme immunoassays for estrogen and progesterone metabolites in the feces of *Macaca fascicularis* during nonconceptive and conceptive ovarian cycles. *Biol. Reprod.* 48, 1290–98.

Shideler, S. E., Ortuno, A. M., Moran, F. M., Moorman, E. A. & Lasley, B. L. (1994). Monitoring female reproductive function by measurement of fecal estrogen and progesterone metabolites in the white-faced saki (*Pithecia pithecia*). *Am. J. Primatol.* 32, 95–105.

Shideler, S. E., Munro, C. J., Johl, H. K., Taylor, H. W. & Lasley, B. L. (1995). Urine and fecal sample collection on filter paper for ovarian hormone evaluations. *Am. J. Primatol.* 37, 305–15.

Sousa, M. B. & Ziegler, T. E. (1998). Diurnal variation in the excretion patterns of fecal steroids in common marmosets (*Callithrix jacchus*). *Am. J. Primatol.* 46, 105–17.

Strier, K. B. & Ziegler, T. E. (1997). Behavioral and endocrine chracteristics of the ovarian cycle in wild muriqui monkeys, *Brachyteles arachnoides*. *Am. J. Primatol.* 32, 31–40.

van Belle, S., Estrada, A., Ziegler, T. E. & Strier, K. B. (2008). Sexual behavior across ovarian cycles in wild black howler monkeys (*Alouatta pigra*): male mate guarding and female mate choice. *Am. J. Primatol.* 70, 1–12.

Wasser, S. K., Monfort, S. L. & Wildt, D. E. (1991). Rapid extraction of faecal steroids for measuring reproductive cyclicity and early pregnancy in free-ranging yellow baboons (*Papio cynocephalus cynocephalus*). *J. Reprod. Fertil.* 92, 415–23.

Whitten, P. L. & Russell, E. (1996). Information content of sexual swellings and fecal steroids in sooty mangabeys (*Cercocebus torquatus atys*). *Am. J. Primatol.* 40, 67–82.

Whitten, P. L., Brockman, D. K. & Stavisky, R. C. (1998a). Recent advances in noninvasive techniques to monitor hormone-behavior interactions. *Yb. Phys. Anthropol.* 41, 1–23.

Whitten, P. L., Stavisky, R. C., Aureli, F. & Russell, E. (1998b). Response of fecal cortisol to stress in captive chimpanzees (*Pan troglodytes*). *Am. J. Primatol.* 44, 57–69.

Ziegler, T. E. & Wittwer, D. J. (2005). Fecal steroid research in the field and laboratory: improved methods for storage, transport, processing, and analysis. *Am. J. Primatol.* 67, 159–74.

Ziegler, T. E., Scheffler, G., Wittwer, D. J., Schultz-Darken, N. & Snowdon, C. T. (1996). Metabolism of reproductive steroids during the ovarian cycle in two species of Callitrichids, *Saguinus oedipus* and *Callithrix jacchus*, and estimation of the ovulatory period from fecal steroids. *Biol. Reprod.* 54, 91–9.

BENOÎT GOOSSENS, NICOLA ANTHONY,
KATHRYN JEFFERY, MIREILLE JOHNSON-BAWE AND
MICHAEL W. BRUFORD

21

Collection, storage and analysis of non-invasive genetic material in primate biology

WHY NON-INVASIVE?

Non-invasive genetic analysis using new, high-precision molecular tools has been an extremely important recent development in primatology, with the promise of pioneering studies in the early–mid 1990s (see, for example, Morin et al., 1994) now being realized at the level of large-scale population studies over broad spatial scales (see, for example, Constable et al., 2001; Anthony et al., 2007b). However, it remains technically demanding, time-consuming, expensive and prone to error. Here, we introduce the applications of non-invasive genetics in primatology, then cover protocols for the most common non-invasive sample types, including faeces, urine and hair, outlining the limitations, pitfalls, and methodologies required. We also describe storage protocols for other possible sources of DNA (deoxyribonucleic acid), including blood and tissue biopsy samples for occasions when animals are captured and handled (Chapters 7 and 8).

APPLICATIONS

Molecular phylogenetic studies continue to add to our knowledge of primate diversity, evolution and hence adaptation (see, for example, Burrell et al., 2009). Phylogenetic analysis can also be used below the species level to study the underlying biogeographical factors that have contributed to the diversity present in primate populations today. This approach has been used to highlight new, evolutionarily distinct populations within well-studied species and to pinpoint potentially important geographical barriers that may delimit genetic

Field and Laboratory Methods in Primatology: A Practical Guide, ed. Joanna M. Setchell and Deborah J. Curtis. Published by Cambridge University Press. © Cambridge University Press 2011.

divergences across the range of species (see, for example, Gonder *et al.*, 1997). The amplification, sequence analysis and evolutionary study of viral genomes from faecal samples in wild primate populations is now routine since the advent of efficient kit methods for extracting RNA (ribonucleic acid) from faeces, which has special relevance for immunodeficiency virus research (see, for example, van Heuverswyn *et al.*, 2006).

At the population level and below, molecular genetic analysis can be applied in a myriad of contexts from mating systems and paternity through to population history over demographic timescales (review in Di Fiore, 2003). Characterizing the demographic relationships among, and measuring gene flow between, populations has the potential to enable prioritization of distinct population units and can inform conservationists regarding their optimal management. Management and maintenance of genetic diversity is an issue of concern for all biologists studying threatened species and of great relevance in primate populations, where habitat fragmentation, separation and population diminution will soon have major implications for long-term viability of many species. Fragmentation necessarily results in reduced population sizes and may in time result in significant genetic drift (changes in allele frequencies due to demographic stochasticity), inbreeding (reproduction between related individuals) and susceptibility to further stochastic demographic contraction (population bottlenecks). These phenomena can act multiplicatively and may lead to rapid loss of genetic diversity, measured at the genetic level through departure of allele frequencies from expectation under random assortment and through loss of heterozygosity and of alleles at genetic loci relative to larger populations (see, for example, Frankham *et al.*, 2002).

Finally, it is now possible to use non-invasive genetics to gain a picture of the social system of a species, including dispersal, territory range and genetic structure without knowledge of individuals or even seeing them, by using faecal and hair samples for molecular sexing and tracking individual genotypes in space and time. 'Genotype matching' approaches of this nature are still in their infancy and have been successfully carried out in only a few non-primates to date, but such studies are under way in several primate populations. Rapid population census (measurement of the fraction of adults breeding, operational sex ratios and family size variances) and conservation assessments of unstudied primate populations could potentially be improved in the future using these methods.

Molecular tools can also be used to manage populations *ex situ*, both in primate colonies for research and in zoological parks. Faecal DNA from plants and insects also has the potential to provide information on dietary make-up, of particular relevance for the identification of arthropods fed on by primates in the canopy (Symondson, 2002; Chapter 15).

Molecular markers

Different markers have different properties and relative abundance in the material that you are likely to collect in the field. By far the two most commonly used approaches in primatology today are mitochondrial DNA (mtDNA) sequencing and microsatellite fragment analysis. MtDNA is a small, circular plasmid genome found in the mitochondrion and is found in many copies (up to hundreds) per cell. It is inherited clonally (as a single, linked genetic system) through the maternal line only and its DNA sequences evolve approximately an order of magnitude more rapidly than chromosomal DNA in the nucleus of the cell. However, there are plenty of regions that are sufficiently conserved to provide comparisons and enable markers to be developed across closely related species. The primary use of mtDNA is in phylogeny and phylogeography, and in the identification of genetically distinct units for conservation (see, for example, Anthony *et al.*, 2007b), but it can also be used to examine genetic distinctiveness within populations, although its behaviour (as a maternally inherited marker) is strongly affected by patterns of female philopatry and dispersal. Nuclear translocations of mitochondrial DNA may, however, hamper studies of primate mitochondrial variation (Thalmann *et al.*, 2004; Jensen-Seaman *et al.*, 2004). Careful analysis of putative nuclear sequences can aid in their identification and removal (Anthony *et al.*, 2007a), but always be aware of the possibility of co-amplification of nuclear paralogues and take the necessary precautions (e.g. screening for frameshifts or stop codons in coding sequences, cloning and sequence analysis of mixed PCR products) to exclude them from downstream analyses of population genetic variation. In cases where nuclear translocations might be present, examine sequence data for potential instances of *in vitro* recombination between nuclear and mitochondrial DNA templates (see, for example, Anthony *et al.*, 2007a).

Microsatellites are found mainly in the chromosomes in the nucleus and are present in thousands of copies scattered throughout the genome. They comprise simple repetitive elements (e.g. repeated sections of two DNA bases, GAGAGAGA) and their variation is derived mainly through changes in the number of DNA repeats at any given location. They are highly variable, with up to 20 alleles per locus being common, and heterozygosity (possession of two different length alleles) at any given locus, or genome location, can commonly be up to 80%. Application of 10–20 of these loci can provide you with both an individual-specific genotype and an estimate of the genetic similarity between individuals. Microsatellites are commonly used in studies of paternity and social structure (Sunnucks, 2000). Microsatellite primers (short DNA sequences; see 'Non-invasive genotyping', below) can produce polymorphic markers in related species (e.g. across apes and Old World monkeys, Coote & Bruford, 1996; Roeder *et al.*, 2009), and as more species are studied, more markers will become available for general use. You can access these through journals such as *Molecular Ecology Resources* and *Conservation Genetics Resources* or in Genbank ('Useful Internet sites'). However, you may need to isolate, characterize and develop markers in your own study species.

Single Nucleotide Polymorphisms (SNPs) are used increasingly in molecular studies of primates and may prove more reliable than microsatellites in degraded DNA samples, although this is currently not established (Smith *et al.*, 2004; Rönn *et al.*, 2006; Andrés *et al.*, 2008). However, SNPs are predominantly species-specific (Morin *et al.*, 2009), and are less variable than microsatellites (they usually only comprise two alleles per polymorphism and have lower levels of heterozygosity), so you would need many more SNP markers than microsatellites (Morin *et al.*, 2009).

Non-invasive genotyping

Non-invasively sampled DNA is present in much lower amounts than in blood and tissue, posing special problems for the geneticist. The polymerase chain reaction (PCR) involves the enzymatic amplification of DNA through the use of short pieces of DNA ('primers'), which identify and anneal to (pair with) a specified complementary sequence either end of the DNA under study. PCR is carried out by denaturing template DNA (making it single stranded) at high temperatures (94 °C), and allowing it to cool to a predefined temperature, at which point the primers anneal to the template DNA. Once the sequences have annealed, a heat-tolerant DNA polymerase (*Taq* polymerase, isolated from *Thermophilus aquaticus*) can extend this small section of double-stranded DNA and make two copies from the one (template) copy. The DNA template and new copy are both then denatured at high temperature and the process begins again. This PCR reaction is repeated 20–40 times, to produce millions of copies of the DNA template, and, crucially, give the researcher enough genetic material to analyse and genotype.

The problem with non-invasively collected material is that sometimes the DNA is in very low copy number, it is often highly fragmented and PCR may be inhibited by co-extracted compounds present in the material. Contamination of the PCR reaction with extraneous DNA is thus a real possibility. Consequently, DNA extraction must be: (1) highly efficient; (2) not unnecessarily destructive to the integrity of the DNA; (3) able to remove inhibitory material during purification; and (4) relatively quick and uncomplicated, avoiding unnecessary steps.

Low template DNA copy number and PCR inhibition lead to two problems in non-invasive genotyping. First, PCR product may be extremely difficult to generate and resultant product fragments may not be sufficient for analysis. Using more PCR cycles (up to 40 is advisable) or a second round of PCR, using the fragments generated in the first to 'seed' the reaction, may help here. However, the more cycles you use, the greater the possibility that the DNA polymerase introduces a copying error, giving a false polymorphism. Replicate PCRs are imperative to confirm the results. False data may occur in DNA sequences (artificial point mutations) or in microsatellite fragments (false allele lengths due to DNA polymerase 'slippage' during the PCR reaction). DNA polymerase slippage is a general phenomenon of microsatellite PCRs and you can usually allow for this by only recording the one (for homozygotes) or two (for heterozygotes) most intensely amplified fragments. False alleles may confuse this procedure, but are usually very weak fragments, and should in any case be replicated. Second, the stochastic non-amplification of one of the two potential alleles at a microsatellite locus ('allelic dropout') can occur due to low template copy number or DNA degradation. The latter may be a special problem for loci exhibiting a wide range of allele lengths, because longer alleles may not be amplifiable if their length exceeds the maximum fragment size present in the degraded template DNA. Repeated amplifications using several independent DNA extractions (see below) are a minimum requirement in such studies.

COLLECTING, STORING AND TRANSPORTING
MATERIAL FROM THE FIELD

Faeces

Use a stick or sterile plastic or wooden spoon (10–15 cm long) to sample a small pea-sized scraping from the surface of a fresh (< 24 h old) faecal deposit and place it in a small cryovial or 15 ml sterile collection tube containing a suitable preservative (see below). Film pots make useful collection tubes, but avoid plastic bags unless the faeces are dry and will be stored in silica gel. Sample the outer layer of faeces with a shiny surface. Faeces degrade and/or disappear quickly in tropical forest, but may persist longer and desiccate in other environments. Test the efficiency of DNA extraction from faecal samples of different ages before attempting a large-scale study, because this varies for different species and populations. Wear disposable plastic or vinyl gloves and never touch faeces with your bare hands, to avoid cross-contamination of the samples and potential disease transmission (Chapter 8). Hepatitis B vaccination is essential for those likely to come into direct contact with primate faeces, particularly ape faeces. Use an indelible pen to label collection tubes with Global Positioning System (GPS) coordinates, identity, sex and age of the individual (where known), date, the name of the collector and a sample code. Repeat the sample code on the lid of your tube and duplicate all the information in a field-book (preferably waterproof).

There is limited consensus on the best method for preserving faeces because optimal methods vary by species, diet and the environment in which samples are collected (see, for example, Waits & Paetkau, 2005; Soto-Calderon *et al.*, 2009). However, some of the best preservation methods for primate faeces include desiccation on silica gel, or storage in a solution of DMSO–EDTA–Tris–salt solution (DETs; 20% dimethylsulphoxide (DMSO), 0.25 M sodium ethylenediamine-tetraacetic acid (EDTA), 100 mM Tris, pH 7.5, and NaCl to saturation), 70%–95% ethanol or the tissue-stabilizing solution RNAlater (Ambion, Inc.). A combination of initial ethanol preservation followed by silica dessication (Roeder *et al.*, 2004) may yield the greatest amplification success, although this approach adds an additional layer of complexity to field studies. Surgical spirit (which has the advantage of being available almost everywhere) can be used as a last resort.

The quantity of storage solution or silica beads should always greatly exceed the quantity of faecal material collected, preferably by a ratio of c. 5:1 storage material mass to pellet volume. Once you have capped the tube, seal it with sealing film to prevent leakage. Ethanol can remove labelling ink. One way around this problem is to include a pencil-written label in the tube, although this may increase the risk of contamination. Check stored samples periodically for leakage and evaporation. Storage times can affect DNA quality, so extract samples as soon as possible after collection. Finally, if the samples will be analysed by a third party, take about 10 plucked hairs from yourself to check for possible contamination.

Urine

Urine has been used occasiónally as a source of DNA for non-invasive genetic studies (see, for example, Hayakawa & Takenaka, 1999; Valière & Taberlet, 2000; Hedmark et al., 2004; Sastre et al., 2009), but little genetic work has been carried out on samples collected from free-ranging individuals in the field (Valière & Taberlet, 2000). In orang-utans (Pongo spp.) and chimpanzees (Pan troglodytes), you can collect urine from plastic sheets placed under sleeping nests. Individuals often urinate from the side of the nest on awakening and you can transfer the fluid from the plastic sheets to storage vials using disposable plastic pipettes. You can also collect urine opportunistically during field observations or from surrounding vegetation following urination. Epithelial cells suspended in the urine are likely to be the primary source of DNA. Fermentation and presumably DNA degradation of these cells may occur rapidly after urination (Hayakawa & Takenaka, 1999) and you should collect samples comprising as large a volume as possible and transfer them into two volumes of 95% ethanol as rapidly as possible after excretion.

Hair

Hair can be problematic as a source of DNA (comprehensive example in Jeffery et al., 2007), but the following guidelines may help. Plucked hairs are by far the best source of hair DNA for both mtDNA and microsatellite analysis. In studies of humans and gorillas (Gorilla spp.) the majority of plucked hairs are in the active growth or breakdown phase (anagen or catagen) and their roots are therefore packed with mitotic cells. Single plucked hairs with root material should present no problem for genetic analysis, given adequate storage conditions (but if possible, collect more than 10 hairs per individual). However, for shed hair, although mtDNA is present in the root and hair shaft, nuclear

DNA is only present in the root material (Linch *et al.*, 2001). The roots of naturally shed hairs have usually undergone programmed cell death before shedding (telogen phase) and much of the nuclear DNA is degraded. Shed hairs will provide a good source of mtDNA, but a much poorer source of nuclear DNA. However, epithelial tissue may be attached to the root and in the case of freshly shed hairs this can provide a source of undegraded nuclear DNA (Linch *et al.*, 1998).

It is usually only possible to pluck hair in captive or anaesthetized animals (Chapter 8). Plucking hairs from free-ranging primates is rarely attempted, although Valderrama *et al.* (1999) described novel methods for doing so with capuchins (*Cebus olivaceus*) and baboons (*Papio hamadryas*), which included shooting duct tape at the target animal and wrapping food baits with duct tape. Wear gloves and pull multiple hairs out from the root. Place hairs in clean paper envelopes, since plastic produces static that makes hair manipulation difficult. Don't touch the envelopes on the inside and staple or fold them shut (*never* lick them).

For nesting species, you can locate vacated nests and recover hairs shed by the occupant. In gorillas, many nests are located on the ground, but tree dwellers pose a more practical problem of accessibility. Although you can amplify mtDNA from single hair shafts from most nests, regardless of nest age, we have found that microsatellite success can be greatly affected by nest age and field efforts may be better directed to collection from fresh nests only (see Chapter 6 for estimating nest ages). Wear gloves and avoid contamination by collecting hairs with tweezers that you sterilize with a flame between each use. Place hairs into paper envelopes as before. Record the following information: date, GPS co-ordinates, sample ID, nest age, height of nest from ground, collector initials, and reference to any faecal samples collected nearby. Note whether the nest is solitary or in a group. Assign nest groups an ID with the total number of nests recorded, and number each nest within the group. The best storage method for hairs is the simplest: desiccated with self-indicating silica granules, either at room temperature or frozen at −80 °C. Although DNA degradation occurs over time when the hair is in a nest, we have not observed this problem with hairs sealed in envelopes, desiccated and stored for up to 8 years.

Blood and tissue

For invasively extracted material such as blood and tissue biopsy, storage is equally crucial to prevent DNA degradation and damage. Where freezing is not an option, you can store blood at room temperature in an equal volume of $1 \times$ TNE buffer (10 mM Tris, 0.1 M NaCl, and 1 mM EDTA, pH 8.0) or, less optimally, in 1 ml of absolute ethanol. DNA remains intact in TNE for several months but you should use a refrigerator as soon as possible. Tissue biopsies are stored optimally in 25% mass/volume DMSO dissolved in 6 M (saturated) NaCl: score the surface of the tissue with a blade and immerse the biopsy fully in the DMSO solution. Tissue stored this way preserves DNA effectively indefinitely. Queens Lysis Buffer is also highly efficient (Bruford *et al.*, 1998).

Other sources of DNA

Nuclear DNA has also been extracted from male ejaculates, female menstrual blood and wounds (Chu *et al.*, 1999), and from buccal cells extracted from discarded food items ('wadges', see, for example, Hashimoto *et al.*, 1996; Takenaka *et al.*, 1993; Sugiyama *et al.*, 1993). You can store blood and ejaculates in 1 ml of STE buffer (0.1 M NaCl, 10 mM Tris, and 1 mM EDTA, pH 8.0) and 1 ml of absolute ethanol, and you can store about 50 g of wadges in 50 ml polypropylene tubes filled with 90% ethanol and 1 mM Na_3EDTA.

Transporting samples

You may need CITES and health/agriculture department authorizations to transport blood, faecal and hair samples from endangered species, as well as other permits. Pack samples in sealed containers, surrounded by sufficient absorbent material to absorb an accidental spillage. Seal the samples and the absorbent material in a leak-proof plastic bag and place them in a cylindrical metal container, a strong cardboard box or a polystyrene box inserted into an outer cardboard box. Swab the inner and outer packaging with disinfectant before travelling. Pack envelopes containing hairs inside plastic bags or large envelopes, then in well-sealed, air-tight plastic boxes containing silica grains.

LABORATORY METHODS

Contamination with human DNA is a serious problem for non-invasive studies of primates. Carry out faecal extractions in a separate room exclusively dedicated to DNA analysis of faeces, and include negative controls (reagents only) in each extraction to monitor for contamination. We also recommend: physical separation of laboratories where pre- and post-PCR experiments are carried out; avoiding handling concentrated DNA extracts in the pre-PCR room; using pipettes dedicated to non-invasive studies in a laminar-flow hood and aerosol-resistant pipette tips (filter tips); and using face-masks and monitoring all reagents continuously for DNA contamination using negative PCR controls.

DNA extraction

Faeces

Cells containing DNA are not uniformly spread through faeces, so you should make 2–3 extracts per sample. Choose an extraction method that eliminates the need for laborious centrifuging to remove detergents, salts and enzymes from extracts and involves fewer steps and sample transfers, to minimize the potential for contamination. Various kits exist for the extraction of DNA from faeces, all of which come with full instructions. We recommend the QIAamp Stool mini kit (QIAGEN, currently US$171 for 50 reactions (i.e. 25 samples) (Goossens *et al.*, 2000). Other possibilities include: silica-based (Boom *et al.*, 1990), diatomaceous earth (Gerloff *et al.*, 1995), magnetic beads (Flagstad *et al.*, 1999) and Chelex® 100 (Walsh *et al.*, 1991) protocols.

Urine

Centrifuge urine samples and wash the resulting sediment with physiological buffer containing 1 mM EDTA. Re-suspend sediments in 250 µl of STE buffer (0.1 M NaCl, 10 mM Tris, and 1 mM EDTA, pH 8.0) and 30 µl of 10% sodium dodecylsulfate (SDS), and digest them with proteinase K (20 µl of a 5 mg/ml stock) for 2 h at 55 °C. Extract lysates once with 200 µl of phenol–chloroform/IAA (chloroform prepared as a chloroform : isoamyl alcohol mixture of 24:1) and precipitate and redissolve DNA in 20 µl of TE (10 mM Tris, 1 mM EDTA, pH 8.0) buffer. You can also isolate DNA from urine by using the QIAamp DNA stool mini kit following the manufacturer's protocol (see, for example, Hedmark *et al.*, 2004).

Hair

The rule here is the simpler the better. You can extract hairs in Chelex® 100 and Proteinase K (Walsh *et al.*, 1991), but Vigilant (1999) obtained superior results from a simpler method that uses *Taq* polymerase PCR buffer as the extraction buffer (see Allen *et al.*, 1998). We find that using

PCR buffer, water and proteinase K in a small extraction volume works very well for shed hairs.

Blood and tissue

Many standard methods exist for extracting DNA from blood and tissue. See Bruford *et al.* (1998) for details of standard salt extraction and phenol extraction methods, however there are now very many convenient kits available to carry out extractions efficiently (although these can be expensive).

DNA amplification

Improvements in *Taq* polymerases, for example those that become active only after a 10 min incubation at 95 °C (e.g. AmpliTaq Gold™ – Perkin Elmer) reduce non-target amplification via a 'hot start' PCR and allow you to use more cycles while maintaining accuracy in the PCR product. A number of extremely effective kits are available for multiplex PCR reactions involving co-amplification of a number of loci simultaneously (an efficient and cost-saving method), such as the Qiagen Multiplex PCR kit. A multiple tubes procedure (Taberlet *et al.*, 1996) is the best method to provide reliable genotyping when using samples with very small and unknown DNA quantities, to avoid problems associated with allelic drop-out, false alleles and sporadic contamination (see above). This involves repeating PCR experiments 3–7 times using aliquots of the same DNA extract and deducing the genotype by analysis of the whole set of results.

We strongly recommend a pilot study to test storage and extraction methods for non-invasively collected samples (5–10 samples), preferably collected in the field from the species and population under study. This will allow you to compare the efficacy of different storage reagents directly by extracting and amplifying mitochondrial and nuclear DNA. You can also test the effect of removal of dietary inhibitors (Wasser *et al.*, 1997), duration of storage prior to extraction (Soto-Calderon *et al.*, 2009), age of faeces collected from the field (Murphy *et al.*, 2007) and different faecal drying methods (Murphy *et al.*, 2000) at this point.

ADDITIONAL INFORMATION

Once you have optimized your genotyping systems, you can carry out large-scale PCRs with confidence. This is important, since the integrity of genetic data is vital for all studies using non-invasively collected DNA and must be subjected to intense scrutiny prior to publication. 'Manual' genotyping systems, which use radioactivity and silver or other DNA staining methods, and the semi-automated approaches favoured by modern labs due to their increased sensitivity of detection, are beyond the scope of this chapter. However, it is always desirable to use multiple independent scoring of alleles and sequences and, for semi-automated analyses, base-calling, and fragment allele length assignments must always be checked manually by several people and sorted using software such as FLEXIBIN (Amos *et al.*, 2007). You can find a recent list of programs used to detect and quantify genotyping errors associated with microsatellite markers in Pompanon *et al.* (2005). Quantitative PCR allows you to minimize the number of replicates required to recover the correct consensus genotype (Morin *et al.*, 2001).

DATA ANALYSIS

Once you have assembled a matrix of samples and confirmed genotypes, you can start your data analysis. The following free software programs are available and links to the relevant websites are listed under 'Useful Internet sites'.

Population genetics

GENEPOP computes exact tests for Hardy–Weinberg equilibrium, population differentiation and genotypic linkage disequilibrium among pairs of loci. It computes estimates of classical population parameters, such as F_{ST} and other correlations, and is PC (DOS) compatible. GENETIX is in French and is PC-compatible. It estimates standard population genetics parameters and produces factorial correspondence plots. FSTAT 2.9.3 estimates and tests gene diversities and differentiation statistics for co-dominant genetic markers. It computes estimators of gene diversities and *F*-statistics and tests them using randomization methods. It is PC-compatible. Other more sophisticated software packages for methods such as Bayesian clustering, detection of demographic change, admixture and population assignment are available (examples in Chikhi & Bruford 2005).

Parentage and relatedness

KINSHIP 1.3.1 performs maximum likelihood tests of pedigree relationships between pairs of individuals in a population using genotype information for single-locus, co-dominant genetic markers (such as microsatellites). Kinship also calculates pairwise relatedness statistics, and runs on Macintosh only. CERVUS 3.0 is useful for large-scale parentage analysis using co-dominant loci. It can analyse allele frequencies, run appropriate simulations and carry out likelihood-based parentage analysis, testing the confidence of each parentage using the results of the simulation. You can also use simulations to estimate the power of a series of loci for parentage analysis, using real or imaginary allele frequencies. It is PC-compatible.

COST AND TIMING

Population genetic studies can be costly. For example, using 15 microsatellite markers for 300 faecal samples collected in the field will cost approximately US$25 000 (US$2000 for collection, US$4000 for extraction, US$16 500 for amplification, US$2500 for DNA typing) at 2010 prices in the UK, and may take up to 24 months once the genotyping systems have been optimized and depending on how difficult the samples prove to be. A paternity analysis for 50 individuals using 8 microsatellite markers would cost approximately US$16 000 and might take up to six months.

REFERENCES

Allen, M., Engstrom, A. S., Meyers, S. *et al.* (1998). Mitochondrial DNA sequencing of shed hairs and saliva on robbery caps: Sensitivity and matching probabilities. *J. Forensic Sci.* 43, 453–64.

Amos, W., Hoffman, J. I., Frodsham, A. *et al.* (2007). Automated binning of microsatellite alleles: problems and solutions. *Mol. Ecol. Notes* 7, 10–14.

Andrés, O., Rönn, A.-C., Bonhomme, M. *et al.* (2008). A microarray system for Y chromosomal and mitochondrial single nucleotide polymorphism analysis in chimpanzee populations. *Mol. Ecol. Res.* 8, 529–39.

Anthony, N. M., Clifford, S. L., Bawe-Johnson, M. *et al.* (2007a). Distinguishing mitochondrial sequences from nuclear integrations and PCR recombinants: guidelines for their diagnosis in complex sequence databases. *Mol. Phylogenet. Evol.* 43, 553–66.

Anthony, N. M., Johnson-Bawe, M., Jeffery, K. *et al.* (2007b). The role of Pleistocene refugia and rivers in shaping gorilla genetic diversity in central Africa. *Proc. Natl. Acad. Sci. USA* 104, 20432–6.

Boom, R., Sol, C. J. A., Salimans, M. M. M. *et al.* (1990). Rapid and simple method for purification of nucleic acids. *J. Clin. Microbiol.* 28, 495–603.

Bruford, M. W., Hanotte, O. & Burke, T. (1998). Single and multilocus DNA fingerprinting. In *Molecular Genetic Analysis of Populations: A Practical Approach*, ed. A. R. Hoelzel (2nd edn), pp. 225–69. Oxford: Oxford University Press.

Burrell, A. S., Jolly, C. J., Tosi, A. J. & Disotell, T. R. (2009). Mitochondrial evidence for the hybrid origin of the kipunji, *Rungwecebus kipunji* (Primates: Papionini). *Mol. Phyl. Evol* 51, 340–8.

Chikhi, L. & Bruford, M. W. (2005). Mammalian population genetics and genomics. In *Mammalian Genomics*, ed. A. Ruvinsky & J. Marshall-Graves, pp. 539–83. Oxford: CABI Publishing.

Chu, J.-H., Wu, H.-Y., Yang, Y.-J., Takenaka, O. & Lin, Y.-S. (1999). Polymorphic microsatellite loci and low invasive DNA sampling in *Macaca cyclopis*. *Primates* 40, 573–80.

Constable, J. L., Ashley, M. V., Goodall, J. & Pusey, A. E. (2001). Noninvasive paternity assignment in Gombe chimpanzees. *Mol. Ecol.* 10, 1279–300.

Coote, T. & Bruford, M. W. (1996). A set of human microsatellites amplify polymorphic markers in Old World apes and monkeys. *J. Hered.* 87, 406–10.

Di Fiore, A. (2003). Molecular genetic approaches to the study of primate behavior, social organization, and reproduction. *Yb. Phys. Anthropol.* 46, 62–99.

Flagstad, Ø., Røed, K., Stacy, J. & Jakobsen, K. S. (1999). Reliable non-invasive genotyping based on excremental PCR of nuclear DNA purified with a magnetic bead protocol. *Mol. Ecol.* 8, 879–83.

Frankham, R., Ballou, J. D. & Briscoe, D. A. (2002). *Introduction to Conservation Genetics*. Cambridge: Cambridge University Press.

Gerloff, U., Schlötterer, C., Rassmann, I. *et al.* (1995). Amplification of hypervariable simple sequence repeats (Microsatellites) from excremental DNA of wild living Bonobos (*Pan paniscus*). *Molec. Ecol.* 4, 515–18.

Gonder, M. K., Oates, J. F., Disotell, T. R. *et al.* (1997). A new west African chimpanzee subspecies? *Nature* 388, 337.

Goossens, B., Chikhi, L., Utami, S. S., de Ruiter, J. R. & Bruford, M. W. (2000). A multi-samples, multi-extracts approach for microsatellite analysis of faecal samples in an arboreal ape. *Cons. Gen.* 1, 157–62.

Hashimoto, C., Furuichi, T. & Takenaka, O. (1996). Matrilineal kin relationship and social behaviour of wild bonobos (*Pan paniscus*): sequencing the D-loop region of mitochondrial DNA. *Primates* 37, 305–18.

Hayakawa, S. & Takenaka, O. (1999). Urine as another potential source for template DNA in polymerase chain reaction (PCR). *Am. J. Primatol.* 48, 299–304.

Hedmark, E., Flagstad, O., Segerström, P. *et al.* (2004). DNA-based individual and sex identification from wolverine (*Gulo gulo*) faeces and urine. *Cons. Gen.* 5, 405–10.

Jeffery, K. J., Abernethy, K. A., Tutin, C. E. G. & Bruford, M. W. (2007). Biological and environmental degradation of gorilla hair and microsatellite amplification success. *Biol. J. Linn. Soc.* 91, 281–94.

Jensen-Seaman, M. I., Sarmiento, E. E., Deinard, A. S. & Kidd, K. K. (2004). Nuclear integrations of mitochondrial DNA in gorillas. *Am. J. Primatol.* 63, 139–47.

Linch, C. A., Smith, S. L. & Prahlow, J. A. (1998). Evaluation of the human hair root for DNA typing subsequent to microscopic comparison. *J. Forensic Sci.* 43, 305–14.

Linch, C. A., Whiting, D. A., & Holland, M. M. (2001). Human hair histogenesis for the mitochondrial DNA forensic scientist. *J. Forensic Sci.* 46, 844–53.

Morin, P. A., Moore, J. J., Chakraborthy, R. *et al.* (1994). Kin selection, social structure, gene flow, and the evolution of chimpanzees. *Science* 265, 1193–201.

Morin, P. A., Chambers, K. E., Boesch, C. & Vigilant, L. (2001). Quantitative polymerase chain reaction analysis of DNA from noninvasive samples for accurate microsatellite genotyping of wild chimpanzees (*Pan troglodytes verus*). *Mol. Ecol.* 10, 1835–44.

Morin, P. A., Martien, K. & Taylor, B. L. (2009). Assessing statistical power of SNPs for population structure and conservation studies. *Mol. Ecol. Res.* 9, 66–73.

Murphy, M. A., Waits, L. P. & Kendall, K. P. (2000). Quantitative evaluation of fecal drying methods for brown bear DNA analysis. *Wildl. Soc. Bull.* 28, 951–7.

Murphy, M. A., Kendall, K. C., Robinsom, A. & Waits, L. P. (2007). The impact of time and field conditions on brown bear (*Ursus arctos*) faecal DNA amplification. *Conserv. Genet.* 8, 1219–24.

Pompanon, F., Bonin, A., Bellemain, E. & Taberlet, P. (2005). Genotyping errors: causes, consequences and solutions. *Nature Rev. Genet.* 6, 847–59.

Roeder, A. D., Archer, F. I., Poinar, H. N. & Morin, P. A. (2004). A novel method for collection and preservation of faeces for genetic studies. *Mol. Ecol. Notes* 4, 761–4.

Roeder, A. D., Bonhomme, M., Heijmans, C. *et al.* (2009). A large panel of microsatellites for genetic studies in the infra-order Catarrhini. *Folia Primatol.* 80, 63–9.

Rönn, A.-C., Andrés, O., Bruford, M. W. *et al.* (2006). Multiple displacement amplification for generating an unlimited source of DNA for genotyping in nonhuman primate species. *Int. J. Primatol.* 27, 1145–69.

Sastre, N., Francino, O., Lampreave, G. *et al.* (2009). Sex identification of wolf (*Canis lupus*) using non-invasive samples. *Conserv. Genet.* 10, 555–8.

Smith, S., Aitken, N., Schwarz, C. & Morin, P. A. (2004). Characterization of 15 SNP markers for chimpanzees (*Pan troglodytes*). *Mol. Ecol. Notes* 4, 348–51.

Soto-Calderon, I. D., Ntie, S., Mickala, P. *et al.* (2009). Effects of storage type and time on DNA amplification success in tropical ungulate faeces. *Mol. Ecol. Res.* 9, 471–9.

Sugiyama, Y., Kawamoto, S., Takenaka, O., Kumizaki, K. & Norikatsu, W. (1993). Paternity discrimination and inter-group relationships of chimpanzees at Bossou. *Primates* 34, 545–52.

Sunnucks, P. (2000). Efficient genetic markers for population biology. *Trends Ecol. Evol.* 15, 199–203.

Symondson, W. O. C. (2002). Molecular identification of prey in predator diets. *Mol. Ecol.* 11, 627–41.

Taberlet, P., Griffin, S., Goossens, B. *et al.* (1996). Reliable genotyping of samples with very low DNA quantities using PCR. *Nucleic Acids Res.* 24, 3189–94.

Takenaka, O., Takashi, H., Kawamoto, S., Arakawa, M. & Takenaka, A. (1993). Polymorphic microsatellite DNA amplification customised for chimpanzee paternity testing. *Primates* 34, 27–35.

Thalmann, O., Hebler, J., Poinar, H. N., Pääbo, S. & Vigilant, L. (2004). Unreliable mtDNA data due to nuclear insertions: a cautionary tale from analysis of humans and other great apes. *Mol. Ecol.* 13, 321–35.

Valderrama, X., Karesh, W. B., Wildman, D. E & Melnick, D. J. (1999). Non-invasive methods for collecting fresh hair tissue. *Mol. Ecol.* 8, 1749–52.

Valière, N. & Taberlet, P. (2000). Urine collected in the field as a source of DNA for species and individual identification. *Mol. Ecol.* 9, 2149–54.

van Heuverswyn, F., Li, Y. Y., Neel, C. *et al.* (2006). Human immunodeficiency viruses – SIV infection in wild gorillas. *Nature* 444, 164.

Vigilant, L. (1999). An evaluation of techniques for the extraction and amplification of DNA from naturally shed hairs. *Biol. Chem*, 380, 1329–31.

Walsh, P. S., Metzger, D. A., & Higuchi, R. (1991). Chelex-100 as a medium for simple extraction of DNA for PCR-based typing from forensic material. *Biotechniques* 10, 506–13.

Waits, L. P. & Paetkau, D. (2005). Non-invasive genetic sampling tools for wildlife biologists: a review of applications and recommendations for accurate data collection. *J. Wildl. Mgt.* 69, 1419–33.

Wasser, S. K., Houston, C. S., Koehler, G. M., Cadd, G. G. & Fain, S. R. (1997). Techniques for application of faecal DNA methods to field studies of Ursids. *Mol. Ecol.* 6, 1091–7.

USEFUL INTERNET SITES

http://www.ncbi.nlm.nih.gov/Genbank – for Genbank
http://genepop.curtin.edu.au/ – for GENEPOP
www.genetix.univ-montp2.fr/genetix/genetix.htm – for GENETIX
www2.unil.ch/popgen/softwares/fstat.htm – for FSTAT 2.9.3
http://www.gsoft.net.us/gsoft.html – for KINSHIP 1.3.1 and other software
http://www.fieldgenetics.com/pages/aboutCervus_Licensing.jsp – for CERVUS 3.0

22

Tips from the bush: an A–Z of suggestions for successful fieldwork

INTRODUCTION

This guide is designed to be read before you travel. It provides a light-hearted, yet serious, list of environmentally friendly suggestions on how to succeed when conducting fieldwork in isolated locations where the facilities may be relatively basic. It is gleaned from conversations with a wide variety of fieldworkers over a number of years, but it may also be useful for research under less arduous conditions. We end with a few wise sayings relating to the environment – to keep you going in times of adversity. We hope to update the list and keep it topical, so please send your suggestions to the first author for future editions.

A

Adaptors. Check what sort of electrical sockets to expect at your destination and take the right adaptors.

Ant-proof socks. Have you ever had the problem of army ants invading your trousers so that you have to get undressed to pull them out of your skin? Just in case you do, tuck in your trouser bottoms and use Gore-Tex® over-socks. Ants are unable to negotiate the smooth material and never make it to your nether regions. These socks also keep your feet dry, since water can only pass out. Not very glamorous, but at least you can feel smug while your companions disrobe in a hurry or suffer from rotting feet.

B

Back up your data, carry it in different bags, and send copies home. Your bag can be stolen, hard-disks become corrupted, and pen drives fail (see also 'Data', 'Xerox copies').

Field and Laboratory Methods in Primatology: A Practical Guide, ed. Joanna M. Setchell and Deborah J. Curtis. Published by Cambridge University Press. © Cambridge University Press 2011.

Bags. Hip bags of various kinds are invaluable, and cloth/shoe bags of different sizes and colours can help to store things and find them quickly at the bottom of a rucksack. Large polythene bags will keep things dry and double as laundry bags (see also 'Zip-lock bags').

Batteries. Rechargeable batteries are improving rapidly, so take lots of these and battery chargers. Buy the latest versions: in comparison with NiCd rechargeable batteries, the more recently developed NiMH batteries will run for longer, have no 'memory effect', and are more environmentally friendly as they contain no cadmium. At the time of writing, good quality, non-rechargable, alkaline batteries are the most reliable. Remember to remove all batteries from your devices before getting on a plane as they can leak and destroy expensive equipment. Batteries that appear to have 'had it' can be 'recharged' a bit by laying them in the sun – sufficient to power your radio once they are no longer useful in a torch. Use rechargeable batteries round the camp and standard batteries for work.

Beeper. A watch that beeps at set intervals (60 seconds, 120 seconds, etc.) is invaluable for collecting behavioural data.

Binoculars. You cannot spend too much on binoculars. Spend time choosing, buy the best you can afford or place this item at the top of your wish list of presents. If you take care of your binoculars they will last you a lifetime. Use lens covers or blow across the lens before use, to avoid getting debris in your eyes.

Books. A paperback book may be just the thing when sitting under your poncho stranded in the rain or on long trips to town. You can always swap with fellow travellers.

Boots. A good stout pair of boots is of paramount importance (ankle-supporting and waterproof if appropriate). Never scrimp on boots, as they get hammered. Many rainforest researchers use cheap rubber boots fitted with high-quality arch-supporting insoles for working in muddy or flooded areas. Plastic, locally made sandals are popular in humid forests, and stout trainers in drier regions.

Boxes. Plastic storage boxes, particularly those of the size used to store shoes, are ideal to take to the field and will be accepted gratefully as gifts when you leave. They protect small and sensitive equipment from the elements and insects, and are perfect for storing food. Like shoe bags, they also provide order in a top-loading rucksack.

C

Cameras. You will miss that once-in-a-lifetime shot if you don't carry a compact zoom, idiot-proof, flash camera on your belt, and practice your quick-draw technique (see 'Mobile phone'). Also take the best quality digital camera you can afford and spare memory cards. Learn simple maintenance of your camera before you go and remember to take spare batteries. Automatic camera traps can show you things that you may never normally see (Chapter 17). However, there are also plenty of other ways to record your trip, so don't just take a camera (see 'Mobile phone', 'Tape-recorder', 'Video', Chapters 16 and 17).

Clothing. Essential field clothes include: three shirts (one on, one drying and one in the wash), three pairs of socks, two pairs of lightweight field trousers (not the heavy cotton type – quick-drying ones are invaluable and rip-proof ones are available). Make sure that your clothes and raincoat are not brightly coloured because animals usually fear bright colours. Undergarments can be kept to a minimum if you get into the habit of washing the pair you have worn each day. Piles of used underwear can attract mice and ants that will demolish them! Take some smarter clothes too, as you will sometimes need to look smart to avoid giving offence when meeting village elders and permit givers, or for social events and special occasions (Chapter 1). Being a fieldworker does not give you the right to be scruffy.

Compass. See 'Orientation'.

Confidence tricksters. When you arrive in a city where visitors are common you are vulnerable to a group of people who gain a living from tourists by relieving them of their belongings. It pays to be wary of those who seem friendly and charming in their desire to 'help' you. A polite but firm response can, for example, halve your taxi fare. Always check exact prices *before* you travel, and negotiate a fair price. You need not avoid going out, but it is wise to do so with the minimum of possessions so that you do not make an attractive target.

Contraceptives. Remember that these may not be available, or may not be of sufficient quality at your destination. The same applies to tampons, but the eco-friendly re-usable moon-cup is convenient and avoids wastage.

D

Data. Perhaps your most treasured possession – always carry your data with you in your hand luggage when travelling (see 'Back up', 'Xerox copies').

Data sheets are useful for all sorts of work. It's amazing how easy it is to forget to note down the obvious if you don't have a prompt (see also 'Lists').

Dehydration. Few visitors realize how much water they need in a hot climate when exercising more than usual. You can dehydrate quickly in the sweaty tropics. Drinking as much water as you can *in advance* is the answer. By the time you get a headache it is too late but it is wise to carry re-hydration salts. Dehydration can kill and, paradoxically, makes you feel less thirsty. Carry enough water, even if it is heavy. While local people may drink directly from streams or rivers, this is best avoided (see 'Water purification').

Dictaphone. Very useful for recording data in the rain, or when you can't write it all down (see also 'Mobile phone'). But don't get carried away – you have to transcribe it all later!

Driver's licence. If you intend to drive (or even if it's just a remote possibility), take your licence with you, and get an International Driving Licence if you are travelling abroad. You may also need to apply for a local driver's licence if staying abroad for any length of time.

Drying your equipment in the tropics often means using silica gel and airtight containers (see 'Silica gel'). Waterproof bags of all kinds are available from camping shops – or use those boxes. Air-conditioned rooms can result in condensation inside your cameras and binoculars when you take them outside. Keep them in sealed containers until they have warmed up. Dry yourself with fast-drying camping towels.

Duct tape is tough canvas tape that is good for temporary repairs to almost anything. Wind some around your water-bottle.

E

Electricity. Be prepared for power cuts and erratic power surges. Make sure you have a long-life battery and a spare if using a computer. It may also pay to invest in a surge protector, since many computers have been ruined without one (see also 'Adaptors').

E-mail. An excellent idea for your sponsors, friends and family is to send reports whenever you can get to the Internet. Such instant feedback on the joys and woes of fieldwork is an exciting way to keep in touch and to involve all those who have helped you. Scan important documents and send them to your own e-mail account, as well as to a few trustworthy friends. This gives you back-ups in case they are lost. You can also email electronic data to yourself (see also 'Back up').

Embassy. If working abroad, your country may have a representative who can help in difficult times and provide useful advice. Make sure you have the address and telephone number of the local Embassy, High Commission or Consulate. It may be useful to report to them on arrival, so that they know who you are and where you are.

Excess baggage is sometimes unavoidable. It's always a good idea to let the airline know in advance, rather than just turning up with it.

Eye-drops. Sounds ridiculous, but if your animals live in trees, and you spend a lot of time looking upwards, you may need them. They can also be useful if you are anaesthetizing your study animals, to protect the corneas from desiccation (Chapter 8).

F

Field guides. When searching for elusive species, local people may point to the nearest likeness without you asking leading questions. For example, the Congolese 'dinosaur', Mokele Mbembe, was identified as a rhino by this means.

First aid. What you take will depend on your particular needs, but a basic medical kit might include a selection of bandages, paracetamol, itch-relief cream, anti-histamines, starters and stoppers, rehydration salts, antibiotics for skin infections and gut problems (medical advice required), fungicides, betadine solution for wounds (treat all wounds, even small ones), and a venin extractor. Travel stores will provide you with useful information; some will sell you all sorts of exciting emergency medical kits, but these are useless unless you have received training in how to use them. Don't worry too much about those frightening health stories – unless they are very recent. Books like Schroeder (1995) and Werner *et al.* (1992) are extremely useful in remote places (see also 'Health').

Flip-flops. Going bare-foot is not a good idea, owing to the risk of parasitic infection, and other potential injuries.

Food fads. It is often safer to eat the same food as local people when eating out, especially if you eat with your hands. Think twice about consuming ice (not just in drinks, but often used to keep food cold, etc.) and salad because of the risk of stomach problems. Reminders of home provide welcome additions to an unfamiliar diet and include spices, herbs, stock cubes or packet foods. Sachets or tins of drinking chocolate can give you instant energy. Dried fruit bars or packets of figs, raisins and bananas provide treats without being too heavy. Prevention is better than cure, so take multivitamins to supplement your diet and eat as well as you can.

Foot rot. Ugh! Remember those Gortex socks! Regular use of a small nail brush works wonders.

Fridge. Portable fridges and freezers are available and are good for cold beer/soda. More seriously, these can be very useful for sample storage (see Chapters 1, 8, 20, 21).

Fungicide. In case you forget the Gortex socks and to get rid of ringworm.

G

Gadgets. We all have our favourite items under this heading. Why not try one new gadget each trip to see whether it handles a rigorous field test? If not, discard it. For example, how about a folding monopod for your camera or microphone, one of those light sticks that you break to provide lighting around camp, a simple toast maker or a folding walking stick?

Gifts. Before you leave home, try to think of easy-to-carry items that will be appreciated in the field. If travelling abroad, things that are typical of home, or that are of practical use such as pencils, watches or kitchen gadgets, are ideal (see Chapter 1). If you are joining people who are already in the field, take magazines, sweets, etc. with you – you will make yourself very popular immediately!

Gloves. If you need to handle animals, consider bringing along a pair of gardening gloves. These are often unavailable locally, and will protect you from nasty bites (together with a towel). Furthermore, a pair of warm gloves may seem unlikely in the tropics, but can be essential for watching nocturnal animals on chilly nights. Finally, sterile latex gloves are invaluable if you are collecting samples for genetic analysis and do not wish to contaminate them, and to keep you safe when handling biological samples (see Chapters 1, 8, 20, 21).

Gossip. If you are an outsider you may become the centre of attention. You may be the only person from your country that people have ever met. You are an ambassador and a celebrity, so you just have to get used to the idea and be cool about it.

GPS. Global Positioning Systems are a valuable modern tool (Chapter 4), but select one carefully if you need it to work under a forest canopy, and remember that you need spare batteries, gadgets can (and often do) fail, and it is difficult to get them repaired in remote places. Your old fashioned compass remains *essential* (see 'Orientation') and back up vital information by hand each day.

H

Hammock. A variety of very useful hammocks are available for use in rough terrain and forests where tents are impracticable. Mosquito nets and rain guards may be built in and provide a safe haven.

Handkerchiefs. The multi-purpose functionality of a handkerchief is unsurpassed. Not only is it useful when you have a cold, and don't wish to pollute the environment with tissues, but it is also good to wipe the sweat and dust off your brow during hikes or long bus journeys (see 'Quick-wipe tissues'). It can be used to wipe your fingers when cutlery and water are unavailable, to dust off unsavoury bus or roadside seats and benches, or to bind a small wound. Folded into a square it provides a cushion behind the lens of your headlamp, or for other purposes where padding is needed.

Health. Find out about the major health hazards (e.g. chloroquine-resistant malaria) in your field area, get the best medical advice you can on how to deal with them, and bring the right medicines. Get the recommended vaccinations (see 'Jabs'), and also a general health and dental check several weeks before you leave. Get a tropical health screening when you return home too – some diseases (e.g. schistosomiasis) may be symptomless, yet very destructive in the long-term. Take a sterile kit with you, with an assortment of needles and syringes, to avoid the potential of infection at small, poorly funded hospitals. Take a dental kit if you have dodgy teeth. If going abroad, see a health professional in your study country to find out local information. Always check the expiry date when buying medicines.

Hip chains are invaluable for hands-free distance measurement, or simply to make sure that you don't get lost in new territory. Just tie the line to a fixed object, set the counter to zero and start walking. Go for photodegradable line, which breaks down with sun-exposure.

Hippos, crocodiles, elephants and other large animals. Find out what the best reaction to potentially dangerous animals is (preferably by asking people, rather than experimenting). Apparently, if you are in a close encounter with a hippo you should stick your fingers in its nostrils. It is forced to come up for air and therefore lets you go. To be honest, if you are this close it may advantageous to pray a little as well!

I

Insects can carry all sorts of diseases (malaria, dengue fever, Japanese B encephalitis, filaria, to name just a few). It may be hot, but light-coloured (biting insects are attracted to dark colours), tropical-wear cotton shirts with long sleeves and long trousers will help prevent insect bites. These are especially necessary in the evenings and at night. For the exposed bits, all you need is to carry a propelling stick of Avon Skin-so-Soft and you will always be safe – and smooth. If you don't believe this, just try it at home on a summer evening before you go. If it doesn't suit you, and you require something with a bit more kick, purchase a repellent containing the compound DEET (diethyl-toluamide). This is rarely available for purchase in tropical countries. DEET sweats off, but impregnated wristbands might work. It also melts plastic. You can 'bug-proof' clothes with DEET spray or permethrin. Head-nets may be needed in extreme conditions and anaesthetic cream works well to soothe bites. Apparently taking vitamin B makes you less tasty to mosquitoes. Daily doses of halibut oil or garlic tablets are also said to work well.

Insurance that covers medical expenses and repatriation is vital. You may already be insured through your institution or university department, so check before buying a policy. Always carry a card with your name, insurance details, medication taken (e.g. malaria prophylaxis), blood group, allergies, and details of emergency contacts and procedures.

J

Jabs. Prevention is better than cure. Find out what vaccinations are recommended. Remember to tell your doctor or travel clinic that you will be working with animals, and ask people who have visited your study area for advice.

Journal. Keeping a journal can keep you sane, and will provide you with a wonderful reminder of events and experiences you will otherwise forget.

K

Kettle. Place a portable kettle in a small bag with a pocket stove (one that uses solid fuel and folds flat), a disposable lighter, plastic containers of tea and coffee, etc., and your cup and water bottle – for a refreshing (and calming) break while on the move.

L

Language training. This is now much easier thanks to the internet and many 'learn in your car' CDs and tapes are also available.

Take a phrase book (if available). Knowing even a handful of words in the local language is invaluable. The fact that you are trying to embrace what is in many ways the essence of a local culture will make people more accepting of you. Better yet, try learning a song in the local language. In many places, if you can sing a song, your hosts will be very pleased, and your fieldwork will progress even more smoothly. ·

Laptop and palmtop computers are being used more and more in the field for direct data entry. Back up data on a memory stick or as a hard copy (see 'Back up' and 'E-mail').

Leeches. Long over-socks made from fine-weave cotton and tied just below the knee allow you to detect leeches before they get into your shoes. Other ways to deter these bloodsuckers include spraying your boots with roach-killing spray or mashing up tobacco and mixing it with water and rubbing it on your boots; the leeches will fall right off. You might also decide it is better to wear sandals or flip-flops and pick the leeches off as they come; try rubbing your feet with citronella oil, or better yet with soap, to deter them from attaching their mouthparts to you. If none of these works, salt or burning with a lighter will do the trick. Make sure the leech has released its grip before you pull it off or blood will flow. De-leech yourself well outside your residence to avoid re-infestation.

Lighters. Take lots. It's amazing how they disappear! Keep some in a self-seal bag or tin and also put one in each bag and pocket for convenience. They also make welcome gifts.

Lists. Keep lists of everything you need to do. This is especially important when visiting new places, when there is so much to remember. Such lists will prove invaluable when planning your next trip. For example, you won't remember your cravings for particular foods unless you write them down at the time you are starving!

M

Malaria kills. If you're working in a malarial area, start taking prophylaxis before you leave, so that if side effects occur, they occur at home, and you can switch to an alternative medication. Continue taking your prophylaxis. Know the symptoms. Self-medication for malaria gains you time to get to a doctor, and is not an alternative to medical help (see 'Health'). Test kits (using mono clonal antibodies) are commercially available for falciparum malaria (and under development for other species), and may be helpful in remote places.

Marking and measuring tapes. Orange 'surveyor's tape' and biodegradable 'flagging tape' can be used to mark out trails and temporarily label trees (Chapters 3, 12). You can write on the tape with permanent marker pens. Drinks cans can be cut into different shapes, marked with numbers and nailed to trees to provide durable tags. Adding a small square of reflective bicycle tape (stuck on with superglue) makes them highly visible at night. Tape measures may be needed for mapping trails or measuring animals (Chapter 9). If you lose your measuring tape, simply use string and calibrate it later.

Mobile phone. Many of the items on our equipment list are already being combined into this one device, which is now invaluable even in remote places. Choosing your phone to suit your fieldwork will make life easier. Solar battery chargers are available which work for mobile phones and even wind-up chargers can be used in an emergency. A phone that can be used with more than one SIM card is useful since local cards save considerable expense. Recycle or redistribute old phones to save coltan, as mining this metallic ore contributes to the destruction of rainforests.

Money. Each country has different ways of changing money and it is best not to make assumptions. Ask about which currencies are accepted and which facilities available before you travel. Keep receipts for claiming. Credit cards and traveller's cheques may be a useful back-up in cities, but do not assume that they will be accepted.

Mosquito net. It is best to have your own (but see 'Hammock' and 'Tent'). Treat your net regularly with permethrin. A silk sleeping sac is also a help in keeping mosquitoes at bay.

N

Notebooks. Of course, for all occasions! 'Rite-in-the-rain' notebooks are the answer to wet climates and, to save losing all your hard won data, write your notes (and letters) in duplicate books (you can buy carbonless ones). Don't forget to store the copies *separately* or send them home! If 'Rite-in-the-rain' books surpass your budget, store your notebook in a plastic bag; write in pencil, which never runs, and writes on damp surfaces.

O

Orientation. Try to obtain maps and guides to your study area well before you leave and *always take a compass*. Look out for additional material on arrival and ask as many opinions as you can. The more you understand in advance, the easier it will be to settle down and get things done, or to arrive at a decision on whether it is even safe to go at all. Maps of your area may be available in the capital city, and guidebooks may list where. What to do if you get lost will depend on your study area. Don't panic. Follow your compass in a straight line until you hit a trail, walk downhill until you reach a stream, then follow it to the river, etc.

Other people. Many field sites have more than one researcher, plus other staff. You must be able to live and work cooperatively in crowded, difficult conditions, with different characters. This can be one of the best parts of the whole experience but, equally, it can also be the worst. Remember too that other researchers may come after you, and how you act will affect their success as well as your own.

P

Pens. Bring lots and make sure they will write well on damp paper. Forestry Suppliers sell a pen used by astronauts, which is sturdy and has a clip so it can be attached to your notebooks with a string. Pens used and sold locally are likely to be cheap and functional. Indelible markers are necessary for labelling sample tubes and bags, and flagging tape, and are less likely to be easily available.

Permits and regulations. Allow plenty of time to organize research permits, permits to export and import biological samples, etc., and respect regulations for the sake of your own research, and that of future researchers.

Photography. Visual aids are invaluable in enabling others to share in your experiences but remember to ask permission and avoid sensitive subjects. Many remote places that you visit may change drastically in years to come and you may be the only person to have recorded what it was like in the past. Set aside time to take key photographs that will be useful and entertaining when you get home, perhaps years into the future, especially of yourself and other people and close-ups of your animals and their habitats (see Chapter 17).

Pictures of your study animals, home and family are invaluable aids to communication when language is a barrier. People will be interested to see pictures of where *you* are from, and to hear about your culture, just as much as you want to learn about theirs. A few passport photos are always useful for permits.

Pockets. You cannot have too many pockets. Some people swear by the organizer waistcoats worn in the tropics in colonial times and sold today as photographer's jackets, or buy army-style field trousers (see 'Bags').

Political situation. Always check before you go (e.g. your own government's recommendations on travel destinations and your local contacts) and ask lots of questions when you get to a new country to learn what to do and what not to do.

Poncho. For those sudden storms (beware of falling trees and branches in forests), and sometimes the only thing that will keep insects off.

Postcards. Why not buy postcards when you first arrive in a city and bring adhesive address labels for your friends and loved ones. By posting your cards immediately, they actually may arrive before you get home if you have come from abroad.

Procrastination. Fieldworker's Procrastination Syndrome is a psychological affliction that can render you useless for weeks, particularly if you have fallen for the romance of fieldwork rather than the reality. Symptoms include a compulsive desire to get everything just right before you start work. Preparation becomes the end rather than the means. Others conspire to support your delays by tempting you to do more enjoyable things or pointing out all the dangers and pitfalls. You end up doing little or nothing. You will have to be single-minded and at least a bit 'driven' to succeed.

Q

Quick-cook meals. A few dehydrated camping meals take up little room, and provide an excellent source of familiarity and comfort. Select some of your favourite meals and save them to celebrate special occasions or in cases of emergency. If meals are too large an option, bringing several of your favourite herbs stored in film canisters is a fantastic way to make local food taste more like something from home.

Quick-wipe cleaning tissues or travel wipes can be purchased in a flip-top dispenser and provide an excellent way to keep clean at a field site or when on the move. They are also very effective for cleaning stains from clothing. Just dab onto the dirt spot, rub or brush when damp and wipe away with a dry face tissue. Magic!

R

Radio. This may help to keep you sane if you are in the middle of the forest and isolated for a while. Short-wave broadcasts from home may help maintain your sanity when feeling run down and lonely.

Raincoat. See 'Poncho'.

Rest. See 'Y'.

Rucksack. If you can't get your stuff into one manageable rucksack and your overnight bag, *leave it at home*. A single 65 litre rucksack has been used successfully for a one-year trip (see 'Clothing').

S

Sarongs are often available locally, and have all sorts of uses.

Secateurs are quieter than a machete/panga and are useful for botanical sampling and for opening non-invasive transects. Slip them into a belt bag or custom-made holster.

Sewing kit. Should be obvious really.

Silica gel. Stored in boxes where electronic devices such as computer, camera, tape recorder, GPS and compass are kept, the silica gel absorbs all the humidity that can harm your equipment. Reusable silica gel is available, so remember to dry your silica in a pan or in the oven every ten days maximum to keep it efficient (see 'Drying').

Sleeping bag and sheet sleeping bag. It's so much easier to wash a sleeping bag liner than a sleeping bag. Having your own sheets can also be useful in cheap hotels/rest houses.

Soap. Camping stores sell biodegradable soaps made without palm oil (the production of which is a major cause of rainforest destruction), including liquid soap that will wash you and your clothes in fresh or salt water. They also sell antibacterial gel, which cleans hands without the need for water, and may be useful.

Snakes. Love them or hate them, you will feel safer if you carry a crêpe bandage large enough to bind a limb (as if for a sprain). This slows the effects of snakebite for long enough to travel for help, saving you from panic, and is safer than other do-it-yourself remedies.

Solar panels can be bought cheaply and hooked up to a car battery to provide a source of power, or charge rechargeable batteries, thereby being a bit more environmentally friendly.

Specimen tubes. Take lots, of varying sizes.

Spectacles. Daily disposable contact lenses circumvent steamy glasses. If you wear spectacles, wear one of those sports elastics to keep them firmly attached to you, and always carry a spare pair with you (imagine losing your only pair in a swamp 5 km from camp).

String, spare straps and bungee cords take up little room but always prove useful as washing lines, guy ropes, etc.

Sun-cream and sun-hat. High-factor sun protection is *vital* for the fair-skinned, particularly near water, sand or snow, or those taking doxycycline as a malaria prophylaxis.

Swiss army knife or leather-man multi-tool (with scissors, saw and most importantly tweezers). Remember not to carry this with you on the aeroplane – pack it in your hold luggage. A simple dissecting kit in a cloth wallet provides a convenient way to carry most of the other tools that you may require – including a small screwdriver for when your glasses fall apart, or to pick thorns out of your boots.

T

Tape recorder. A small tape recorder and good directional microphone will enable you to 'capture' animals that you cannot see. These are as essential as a camera for the study of cryptic (nocturnal and forest dwelling) mammals and birds, but may help you collect valuable data on most species. Calls can often be used for identification and to provide additional information about social interactions and predator detection. Once recorded, they are available for further analysis – perhaps years later (Chapter 16).

Tent. A small tent with a sewn-in groundsheet and mosquito netting is often essential (see 'Hammock').

Ticks. Don't forget the all-essential tick check on returning from the field each day. These tiny creatures can find their way into the most uncomfortable places, and it is best to remove them with your finger-nails or a pair of tweezers as soon as possible. Wear protective clothing, and don't wander around with your shirt hanging out. Be aware of the possibility of tick bite fever on first exposure, but it's not too serious. Pepper ticks must be squeezed between your thumbnails until they click. Use Vaseline or burn off the larger varieties to avoid leaving their mouthparts in you. Get someone to groom those places you cannot reach.

Toilet paper. Doubles as tissues and should go everywhere with you.

Torches/flashlights. Head torches are now available in a brilliant range of sizes and types with additional red filters and halogen bulbs. Not only will you be able to see animals from their eye-shine, and find your way around more easily than with a hand-held torch, but they are invaluable when removing contact lenses and doing chores around camp. A powerful hand-held torch is an important supplement for night work. Always carry a torch when away from camp. Invest in LED (light-emitting diode) torches that last for ages without new batteries for emergency use; you never know when you may get caught with the light fading. These are not suitable for viewing nocturnal animals as they can damage their eyes.

U

Understanding other cultures. Perhaps the most vital key to success is to be aware of the knowledge and customs of local people (see 'Language'). You will make slow progress if you fail to observe rituals of etiquette and politeness that are normal in an area, so listen, learn, and behave accordingly. It can be an offence, or at least insensitive, to show certain parts of your body in some cultures (i.e. belly, thighs, legs, hair (if female)). It is advisable to spend a few days visiting village elders and important officials at the start, to enable them to understand why you are there, and to listen to their advice. If one line of action fails, try another – don't simply give up! You will eventually find someone who can help. There will often be someone who shares your interests and enthusiasm, who will want to meet you, even if they are difficult to find. When trying to influence people away from home, it is helpful to think how you would react if the situation were reversed, and they visiting you (Chapter 1).

Remember to adjust your pace of life and try to be at least as calm and patient as the people you meet. Shopping for supplies can take all day, because you should not ask a stranger a question, pass a friend in the road, or buy something at a stall without following local etiquette and stopping to chat. Local assistants/informers/translators/friends are invaluable for cultural information, and you will learn a huge amount about other people just by sitting down to talk – one of the best things about fieldwork. Note down their full names so that you can acknowledge them later.

V

Video technology is now so good that you may decide to take a small camcorder that also functions as a still camera and tape recorder. For nocturnal observations, the night shot on a video camera can double as a night vision scope (Chapter 17).

Visas and visa extensions. Check on visa requirements as early as possible since these may take time to organize. Consult with people who have experience in the country and region you are visiting, as well as the appropriate Consulates or Embassies, and ensure that you have all the letters of recommendation and other official documents that you need (see 'Xerox copies'). If it is possible to obtain a visa upon arrival it can save a lot of time.

W

Water containers are *vital*, especially a personal water bottle to keep with you. Folding containers (polythene or canvas) are convenient and even a condom can serve as a lightweight water carrier! Always take plenty of sterilizing tablets (either iodine or chlorine) for emergencies when a water filter is not available. Don't forget to bring flavoured drink powder to cover up that nasty tablet taste.

Water purification. Take a filter with you or make a siphon water filter using a bucket hung on branch of tree, a jerry can and Millbank bags for filtering (but in this case remember you will still need to sterilize the filtered water). A small and magical device called a Steripen® uses ultraviolet light to purify your drinking water.

X

Xerox copies. Photocopy all your important documents, such as passport, insurance details, permits, airline tickets and driver's licence. Leave a copy at home with your family or friends, and take another copy with you. This will make your life easier when they get lost or stolen. It may be safer when out in the city to carry these copies and leave the originals in a more secure place, along with your other valuables. You may need to get the photocopies officially authorized (normally at the police station).

Y

Why? An often-asked question, especially when you are tired, hungry and have just lost your study animals in the middle of nowhere.

You have given up your normal life to struggle around on the side of a mountain in the pouring rain and be eaten alive by all sorts of tiny creatures. It is hard to keep a sense of proportion, so it is very important to take breaks and to eat well. Don't work so hard that you collapse. Put your health as the number one priority. You will be less efficient or useless if you become run-down or sick. To maintain a sustainable workload, plan your working week as if you were at home, with time off for relaxation, socializing and sleep.

Z

Zip-lock or self-seal bags. Take tons of freezer bags if you are collecting data in humid areas or during the rainy season. They can also be used to store specimens.

SOME CLOSING THOUGHTS

- Wastefulness has become a product of civilization – but it is not a sign of a civilized person.
- We don't inherit the earth from our forefathers; we borrow it from our children.
- Life is like a tin of sardines – we are all looking for the key.
- Failure is the path of least persistence.
- If you don't enjoy what you have, how could you be happier with more?
- After all is said and done – there is more said than done.
- Enthusiasm breakfasts on obstacles, lunches on objections and dines on competition.
- Many a false step is made by standing still.
- People may doubt what you say but they will always believe what you do.
- Telling others what to do in their own country is like them telling you what to do in yours.

ACKNOWLEDGEMENTS

We thank our research students, those taking the MSc in Primate Conservation at Oxford Brookes University, UK, and the editors and colleagues for comments and additions.

REFERENCES

Schroeder, D. G. (1995). *Staying Healthy in Asia, Africa, and Latin America.* (5th edn.) Emeryville, CA: Volunteers in Asia and Moon Travel Handbooks.

Werner, D., Thuman, C. & Maxwell, J. (2002). *Where There Is No Doctor: A Village Health Care Handbook.* (Revised edn.) Palo Alto, CA: Hesperian Foundation.

Index

Printed in the United States
By Bookmasters